Managing Intermediate Size Cities

The GeoJournal Library

Volume 69

The titles published in this series are listed at the end of this volume.

Managing Intermediate Size Cities

Sustainable Development in a Growth Region of Thailand

edited by

MICHAEL ROMANOS

and

CHRISTOPHER AUFFREY

School of Planning,
University of Cincinnati, U.S.A.

SPRINGER-SCIENCE+BUSINESS MEDIA, B.V.

A C.I.P. Catalogue record for this book is available from the Library of Congress

ISBN 978-90-481-6103-4 ISBN 978-94-017-2170-7 (eBook)
DOI 10.1007/978-94-017-2170-7

Printed on acid-free paper

To our wives,

Carla Chifos and Ann Auffrey

FOREWORD

I am both pleased and honored to introduce this book to readers, and I want to take a few moments to explain why.

Michael Romanos and Christopher Auffrey have produced a volume which will be of immense value to several different types of people. Planners and other specialists concerned with the development of the Southeast Asian region and the issues and opportunities associated with urban growth and sustainable development will find much to interest them in this book. But the book, I believe, has much wider appeal, and that is what I want to touch on briefly here.

The University of Cincinnati, where Michael, Chris, and I work, is attempting to globalize itself – to develop its institutional capacity for international activities, to infuse its curriculum with international themes, and to promote and increase global competence among its graduates. Many American universities are doing this, of course. In the process, we are seeing some very interesting experiments in pedagogy, as faculty look for "learning moments" in new and sometimes exotic places. Michael, Chris, and their colleagues have, it seems to me, developed an outstanding model for learning across national and cultural boundaries. In the chapters which follow, you will read the results of their work. What will be less apparent, however, is the process by which that work was produced.

To my mind, this process is as important as the results themselves. In planning as in education, what you get depends in large part on how you do something. What is particularly interesting and exciting about this volume is the way in which the work was done, through a field project design which has become a model at the University of Cincinnati for global pedagogy. The process involved establishing and developing collaborative relationships with local municipalities overseas, as well as with diverse disciplines on our own and other campuses. The agenda for research and action then arose from the dialogue between university-based specialists, on the one hand, and municipal groups and agencies on the other. The content of what was learned thus arose from the interactions between these groups at a variety of different levels. Such research is essentially inductive in nature, resembling what Schön has termed "reflective practice." This type of approach differs from so much of our more traditional international education, where established canons, concepts, procedures and syllabi predominate, and where the international context becomes little more than a location where we practice what we already know.

At the University of Cincinnati, we are becoming aware that one of our most important tasks as educators and concerned world citizens is to improve the ways in which our students – and the faculty who teach them – learn about the world outside our borders. And this learning, when it occurs, cannot really serve anyone well unless it is somehow applied directly to areas of human concern. The work that forms the basis for this book is an excellent example of how to do just that. The project involved students and faculty from multiple disciplines, universities, and countries, interacting with the citizens and officials of Chiang Mai. The design of the project provided students, in particular, with an excellent introduction to how professionals work together internationally on issues of common concern. Students not only observed this work, they actually participated in and contributed to it.

All university programs teach content. Few university programs link content with context – that is, with the field or situation in which content knowledge is relevant. And, only a very few programs use the field situation to actually generate new knowledge.

The work described here is an excellent example of "learning how to learn", and doing this in an international, intercultural, multidisciplinary context. Michael, Chris, and all of their colleagues and associates deserve high praise for the care, intelligence, and effort which they put into the design and management of this project.

The book which you are about to read will clearly speak to the concerns of planners and policy-makers in a variety of areas, regions and disciplines. But it will also be of interest to anyone concerned with how we collaborate and learn cross-culturally, and how we use what we know.

Dr. Riall W. Nolan, Associate Provost and Director
Institute for Global Studies and Affairs
University of Cincinnati

May 15, 2002

CONTENTS

Part III
Environmental Management Issues in Development

Part IV
The Built Environment's Role in Sustainable Development

PREFACE

This book is the product of work performed in Thailand by the editors and the authors of its chapters over a period of four years, between 1994 and 1998. The idea for it evolved as the University of Cincinnati Sustainable Development Group was embarking on some of its initial sustainable development projects. It matured while Michael Romanos was an Asia Foundation Fellow at the Asian Institute of Technology in Bangkok, Thailand in 1994, and a Fulbright Senior Scholar at the Bandung Institute of Technology in Indonesia in 1995. During that time, the first exploratory project was conducted in Chiang Mai by Romanos and Dr. Suwattana Thadaniti of Chulalongkorn University. The project focused on the city's approaches to municipal development management and the role of public sector management in promoting sustainable development (Romanos and Chifos, 1996). During the project, many useful contacts were established, and it became obvious that a larger project focus would be of use to the municipal administration.

As a result, a collaborative team of faculty from Chulalongkorn University in Bangkok, Gadjah Mada University in Yogyakarta, Indonesia, and the University of Cincinnati developed a proposal for a joint project to study aspects of sustainable development, and to assess the implementation of sustainable development programs and plans in the cities of Southeast Asia. The United States Information Agency's College and University Affiliations Program funded the project in 1996, and several other grants supplemented the initial award in subsequent years, including a major grant from the University of Cincinnati Faculty Development Council. Part of the project's planned activities were to design and run a series of field projects that would give faculty from the three collaborating academic institutions the opportunity to work jointly on applied research and field training activities. Between 1996 and 1997 the outline of one such field project, to be conducted in Chiang Mai as a continuation of the exploratory research initiated by Romanos and Thadaniti earlier, gradually took shape, with participation of all three academic institutions.

By the early spring of 1998, a two-month field project had been designed to take place in Chiang Mai that summer. The project was based on an interdisciplinary team composition, representation of all three academic institutions, collaboration with local academic institutions and NGOs, and the full participation of the Chiang Mai municipality. During several months of preparatory ground work, Dr. Thadaniti mobilized faculty and students of Chulalongkorn's Department of City and Regional Planning and Social Research Institute, and helped organize the logistics of working in Chiang Mai as a group for an extended period of time. Through her efforts, strong connections were established with faculty from Chiang Mai University and officials from the municipality of Chiang Mai. This latter relationship resulted in an invitation to our team from the lord mayor of Chiang Mai to conduct an exploratory field research project as means for moving the city towards sustainable future development. At that time, additional funding and in-kind support was also secured from Gadjah Mada University's Capacity Building Program, Chulalongkorn University's Faculty Support Program, and the Municipality of Chiang Mai.

As a result of this preparation, an interdisciplinary and international team of scholars and graduate students assembled in Thailand during the summer of 1998. It consisted of faculty from the United States and Indonesia, joined by several Thai colleagues from Chulalongkorn. The team members represented a wide range of disciplines, including geography, biology, economics, philosophy, nursing, photojournalism, urban design, business, education, and urban planning. Because of the interdisciplinary and collaborative nature of the team and its cooperative arrangements with the local government agencies, the project is referred to as the Joint Study throughout this book.

The Joint Study had several objectives. Its first objective was to identify and assess the capacity of the Chiang Mai economy, municipal administration, and regional setting, to support a long-term sustainable development program for the city. Recognizing that a number of foreign scholars working in the field for a short period of time could not possibly produce complete solutions and recommendations for Chiang Mai's government and people, we limited ourselves to uncovering and describing the significant issues that would affect the ability of the city to sustain itself in the future. In doing so, we made a special effort to build real and mutually beneficial collaborative linkages with the local government and all the public and private partners on which we depended for information and exchanges of ideas. In fact, our effort all along was not to supply prescriptions for perceived problems, but rather to provide systematic information and decision tools so as to assist the local planners and decision makers in addressing service provision and development issues and in devising their own approaches to such solutions.

A second objective of the project was to provide an educational environment for the faculty and student participants to gain experience with field research. Chiang Mai offered nearly ideal conditions for this. On the one hand, it is a growing city experiencing demographic and economic pressures as well as rapid tourism growth and uncontrolled physical expansion. On the other hand, it is a relatively small urban area, and its citizens and local government are protective of their cultural and environmental resources and keenly aware of the dangers represented by globalization and its resultant growth pressures. As such, the many forces at play in the city effectively constitute a microcosm of the dynamics of growth and modernization present in the intermediate and large urban areas of Southeast Asia, and thus provided an excellent laboratory for study and training. In fact, "educating the educators" became a significant by-product of the field project, as demonstrated by the participants' enthusiasm for that opportunity (Auffrey and Romanos, 2000).

A third objective of the project was to become a catalyst for its participants to learn and become comfortable with the difficulties and shocks associated with living and working in an international environment and with a multi-cultural group of collaborators. It did not become apparent until well into the first month of the project how powerful the cultural experiences and differences of the various members of the group were, and how easy it was to misunderstand or misconstrue intentions, notions, ideas, attitudes, and behavior. The observations of the editors about these group dynamics have been published elsewhere (Auffrey and Romanos, 2000), but we feel that it is useful to summarize this particular effort in Chiang Mai.

The period of our stay in Chiang Mai and the collaboration among scholars from three different countries, with a variety of disciplinary training, and with different ways of thinking about development and sustainability was for all of us an invaluable learning experience. In the time span of about eight weeks, our group met with and interviewed literally hundreds of individuals in positions with the central, provincial and local governments, the industry, the local business, private organizations and formal and informal interest groups, in Bangkok, Chiang Mai, Lampung, and other locations in Northern Thailand.

The field project proved to be an important educational experience for all members of the international, interdisciplinary team, though for a variety of different reasons. And, many aspects of the experience for the Thai and Indonesian colleagues differed substantially from those experienced by the American faculty, though the experience in a number of areas transcended country and disciplinary boundaries.

Working closely with U.S. colleagues was a new cultural experience for many of the Thai and Indonesian members of the group. They found the Americans to be intensely focused on their work, impatient about delays, and concerned about delays in acquiring data and other information. The Americans' interpretations of global events, particularly those taking place in Asia during the project period, were often seen as naïve and simplistic. But working together, the Thais and Indonesians discovered much in common between their two countries, cultures, and approaches to economic development planning.

The field project provided many of the U.S. faculty with their first experience in working with other disciplines in a non-U.S. setting. This exposure to field research in a different place and culture provided useful insights for integrating a global perspective into the educational and research process. Initially, many of the U.S. team members believed that conducting research overseas would be the same as at home. They were quickly dissuaded of this once in Thailand. Most of the local officials and NGOs, for example, while polite, felt no stake in our project and had no major expectations from it. This had repercussions: evidence that certain data existed did not mean they would be made available; questionnaires left with businesses and NGOs did not mean they would be correctly completed; and, requests for meetings, visits, questioning, and information gathering had to be made frequently through a seemingly endless bureaucratic maze. Along with the frustrations that these conditions caused, the experience was educational. By the end of the project, U.S. faculty members had gained a better understanding of what globalization was and how it impacted local development, and a better appreciation of how it could be meaningfully connected to their research and teaching.

All team members learned through direct experience about the relationship of globalization to the Thai economy, local politics, development priorities, growth pressures, and potential for solutions. Likewise, all members came to better appreciate the impact of globalization on democratic institutions, social justice and human rights, as well as on local politics, attitudes, and actions of the NGOs and citizens. An international team of academics, sponsored by the municipal government office, was interpreted as a political statement, and the related visits and interviews carried that baggage. The interdisciplinary composition caused friction initially, as some participants were frequently unresponsive to issues and approaches they did not recognize. Yet, the strength of the interdisciplinary approach was eventually recognized, and a period of intellectual cross-fertilization took hold. It was at this point that team members were able to explain to each other how disciplines related to globalization and sustainable development, and came to a better understanding of the broader nature of these topics.

At the completion of the project, the team produced and submitted a report to the municipality of Chiang Mai, in which the exploratory research and its findings were summarized, several priority issues were identified, and a number of warnings or suggestions were inserted. The municipality adopted this document and has been using it as one of the guidelines for the tracing of its sustainable development future (Sustainable Development Group, 1998).

In the years following the fieldwork in Chiang Mai we worked on the individual chapters that constitute this volume. Our book does not intend to be a comprehensive coverage of the work we did in Chiang Mai – no one book could contain the amount of material generated during the field study – but it selectively presents what we considered some of the most critical development issues, especially as they pertain to a practical, applied model of sustainable development We must note here that the whole notion of sustainability, while it provided the thinking, analytical and operational framework of our group, was not in the forefront of the priorities of either the government of Thailand or the municipal administration of Chiang Mai in the summer of 1998. With the country in the middle of the Asian financial crisis, the order of the day was survival, and the subtleties of sustainable development, with its strong emphasis on the preservation of the environment, the conservation of natural resources, and social justice, were very secondary in official minds. Even so, both government officials and the public recognized that the economic slump would eventually pass, and that thinking about the long-term future of their communities and regions was the logical thing to do. To a certain extent, this recognition is also reflected in the choice of topics for the book chapters, as well as the treatment of the subjects themselves.

As such, this volume is not a treatise on sustainability or sustainable development. Rather, it looks at sustainable development as a desirable normative model, and explores the circumstances, conditions, and restrictions under which it can be applied to the intermediate size cities of Southeast Asia. Indeed, the project itself was an exploratory effort. As such, we do not offer prescriptions here, but rather have intended to provide useful information about the conditions in Chiang Mai, the dangers threatening its development, and the potential solutions for shaping a sustainable future. We hope that, if nothing else, this effort will help the wonderful people of Chiang Mai, a city full of warmth, good cheer, and amazing cultural resources, to plan its future in a sustainable way. Anything less would be a loss, not only for the citizens of the city, but for all of us.

Michael Romanos and Christopher Auffrey
Cincinnati, May 2002

REFERENCES

Auffrey, Christopher, and Michael Romanos. (2002). Educating the Educators: Global Dimensions of Collaborative Fieldwork in an Urban Region of Southeast Asia. *Journal of Planning Education and Research, Vol. 20, No. 3*, 353-64.

Romanos, Michael, and Carla Chifos. (1996). Contributions of the Urban Informal Sector to Environmental Management. Regional *Development Dialogue, Vol. 17, No. 1*, 122-55.

Sustainable Development Group. (1998). *Sustainable Development in the Chiang Mai Metropolitan Region*. Chiang Mai: Municipality of Chiang Mai.

ACKNOWLEDGEMENTS

This book, and the project that preceded it, is the result of the invaluable support and assistance received from numerous individuals and organizations over a period of more than four years. The first opportunity to work on this subject in Southeast Asia was provided by the University of Cincinnati School of Planning, which awarded Michael Romanos a one year academic leave to pursue research in the field of urban development management and its implications for sustainable development. With the generous support of the San Francisco-based Asia Foundation, Romanos was able to spend the first four months of that leave at the Asian Institute of Technology in Bangkok, Thailand, and travel to several other cities in the region to participate in conferences and lecture. The Asian Institute of Technology, his academic home for that period, was a generous and hospitable institution that facilitated his research and involved him in several exciting academic activities. The opportunity to interact with its faculty and peruse the considerable Southeast Asian resources of its library were invaluable and are remembered with gratitude.

Much appreciation is also extended to the Council for International Exchange of Scholars and the American Council of Learned Societies, and particularly to Mr. David Adams who coordinated the Fulbright Scholar Award Program for the region, and whose efforts to accommodate Romanos' schedule and work plans went beyond expectations. The ten-month award of the Fulbright program gave Romanos the time and flexibility to pursue this and a series of other initiatives that culminated in the Chiang Mai field project three years later. Special thanks are due to the Bandung Institute of Technology in Bandung, Indonesia, which offered facilities and support at the time, and particularly to Dr. Kusbiantoro, the director of the Center for Urban and Regional Development Studies (P3WK), who played a multiple role as mentor, colleague, and emergency rescuer during the many months of their association.

The idea of the project matured following the award of a three-year grant (later extended to five years) by the United States Information Agency (USIA), College and University Affiliation Program. That award funded a proposal jointly prepared and submitted by Dr. Suwattana Thadaniti of the faculty of City and Regional Planning and the Social Research Institute at Chulalongkorn University in Bangkok, Thailand, Dr. Leksono Subanu of the Faculty of Architecture and Urban Planning at Gadjah Mada University in Yogyakarta, Indonesia, and Dr. Samuel Sherrill and Michael Romanos of the University of Cincinnati School of Planning in the United States. We are grateful to the funding agency, and particularly to Ms. Joan Zafarano, the program officer in charge of our project, for her continuous support and encouragement, her patience with the many changes we requested over the five years of our affiliation, and her understanding and sound advice when political and fiscal problems in the Southeast Asian region dictated changes to our plans and work schedules.

The collaborations afforded by the USIA, subsequently United States Department of State award, and the multiple exchanges of faculty among the three academic institutions that ensued, provided the ground work for the field project in the north of Thailand. The logistics of this undertaking were considerable, and the field project succeeded only because of the commitment and hard work of several individuals. Foremost among them was Dr. Thadaniti, the director of the Thai component of the project, who used her many connections and friendships, and mobilized many resources at Chulalongkorn and in Chiang Mai to aid our efforts. Also, significant assistance was received from Dr. Subanu to make possible the larger participation of Indonesian faculty in the project. Finally, the University of Cincinnati Faculty Development Council provided a generous multi-year grant to supplement the Department of State funds and make the participation of more University of Cincinnati faculty possible. We are especially thankful to Dr. Joseph Steger, President of the University, for his personal interest in the project and for the moral and financial support extended by his office. We are also indebted to Dr. Riall Nolan, Vice Provost and Director of the University's Institute for Global Studies and Affairs, for his continuous support and his confidence about the value and the success of the project.

At this point, we feel the need to extend a most heartfelt thanks to three members of the University of Cincinnati faculty who, along with the editors, were the coordinators of the Chiang Mai and subsequent field projects. Dr. Howard Stafford of the Department of Geography was a pleasure to work with, and a wonderful colleague on whom to depend for good advice to select the best possible course of action. Dr. Thomas Wagner represented our group exceptionally well in his preparatory visits and contacts with Thailand, and provided a most assuring and calm presence in times of crisis. Dr. Roger Barry used his considerable managerial skills to ensure that things happened when they were supposed to, and that funding was always available when needed. To all three of them we express special gratitude for their role in the project, and for their contributions to this book.

The Chiang Mai field project was only one of a series of activities that took place over the period from 1996 to 2001 under the original and several derivative grants. Throughout this period, we had the support of Dean Jay Chatterjee of the College of Design, Architecture, Art and Planning, and relied on the good offices of three directors of our own home base, the School of Planning. Dr. Charles Ellison generously committed major resources of the School to the project, and went out of his way to see us succeed in this endeavor. Professor Jack Kleymeyer was always willing to make accommodations to the project's demands. And, Dr. David Edelman played an active role and became a participant in our activities, in addition to being supportive of our time needs. Our warm thanks are extended to all three of them.

An edited book is not just the product of the scholarly aspirations of a group of academics, as we found out early on. It is a major production project in its own right, and in this we had to rely on considerable help from several individuals. We are especially grateful to Ms. Carol Darwiche, our friend and very talented editor, for her insightful comments and suggestions, her systematic review of the manuscript, her continuous probing of the authors to ensure quality and consistency, and her gentle but persistent reminders of the deadlines. Ms. Petra van Steenbergen and Ms. Renee de Boo, our kind and supportive editors at Kluwer Academic Publishers, made our collaboration with Kluwer a most pleasant experience, and were there for us each time we had a problem to resolve. Ms. Dyah Kartikawening, our graduate student from Indonesia, capably provided enthusiastic and helpful support in the use of graphics software, and it is to her we owe the quality of the maps contained in the manuscript.

Each book chapter was reviewed by an average of three internal and external referees, in the same fashion used by academic journals. Several individuals gave generously of their time in making comments on one or more of the chapters, and provided valuable insights and suggestions to the contributing authors. We are indebted for this assistance to: Mahyar Arefi, Roger Barry, Carla Chifos, Erik Cohen, John Disinger, George Galster, Bill Goldsmith, Gordon Huang, Gunter Krume, Richard Levine, Bin Li, Johanna Looye, Hosham Mortada, David Ponitz, Brenda Scheer, Tanya Storch, Thomas Wagner, Harold Ward, and Douglas Watts.

While we are grateful to all of those who have provided assistance, when everything is said and done the responsibility for the book and its contents is ours, as much as we may hate to admit it. Putting the book together and seeing it through to completion has been an adventure. We enjoyed it immensely, most of the time, and hope that we may have been able to pass on some of that enjoyment to the reader.

We have two special people to thank for providing the motivation to complete the book and helping us to continue to enjoy life while doing it. Ann Auffrey and Carla Chifos, our wives, believed in us and in the project and encouraged us to work hard on it when we felt discouraged. To the two of them we dedicate this book with our heartfelt thanks.

MICHAEL ROMANOS AND CHRISTOPHER AUFFREY

CHAPTER 1

MICHAEL ROMANOS
CHRISTOPHER AUFFREY

ASSESSING THE POTENTIAL FOR SUSTAINABLE DEVELOPMENT IN THE INTERMEDIATE SIZE CITIES OF SOUTHEAST ASIA: THE EXPERIENCE FROM THAILAND

Abstract This first chapter sets the tone of, and the parameters for, the field project in Chiang Mai and for the content and emphasis of the book chapters. It begins with a discussion of Chiang Mai as an intermediate size city within the urban system of Thailand, which was formally planned in the 1980s to provide the country with a network of growth poles in an attempt to decentralize the fast growing Bangkok metropolitan region. It uses this designation, and the inadequate support it received from the central government over the last twenty years, to explain why Chiang Mai, the designated regional growth center of the Northern region, has been falling further behind in its efforts to provide quality services to its population. It continues with a brief treatment of the issue of sustainability as it is adopted by the project and is used in the book, and then discusses the potential role of intermediate size cities in the sustainable development of their regions. It then uses the framework provided by these discussions to present and explain the structure and organization of the book and its sections, to introduce the chapters, and to explain the approaches taken by their authors in fulfilling the requirements of the book project.

1. INTRODUCTION

1.1. Chiang Mai

In the urban system of Thailand, Chiang Mai occupies a prominent role. Even though by population size it does not get close to Bangkok, at 5.6 million people in 1993 the primate capital and metropolitan region of the country[1], it has been an important urban center since the end of the thirteenth century, when it was the revered capital city of the ancient kingdom of the Lanna (Romanos, Chapter 7, p.3, this volume). Until the end of the 19th century, it was the capital of the independent state of Chengmai, linked in a loose association with the central government of Thailand based on the payment of a tribute to the king every three years. Since its integration into the modern state of Thailand, it has been playing an important role as the economic and social center of the North of the country, as an agricultural market, an industrial center, and a major tourist destination.

The explosive growth of the Bangkok metropolitan region in the 1960s and 1970s induced the central government at the time to attempt a decentralization of

M. Romanos and C. Auffrey (eds.), Managing Intermediate Size Cities, 1—19.

population and economic growth, in an effort to relieve the growth pressures on Bangkok. The result was the Regional Cities Development Programme, launched in the early 1980s with the assistance of the World Bank and the United Nations Development Program (Boonyabancha and Chalitanon, 1996, 88). Following a model similar to the one promoted by the UNDP in other countries of the region, notably Indonesia in the 1960s, the plan proposed the development of nine regional cities (later increased to 24 cities with the addition of another 13 smaller regional centers) during the 5th and 6th National Economic and Social Development Plans, 1981-1991. (For more on the Five-Year Plans see Romanos et. al., Chapter 8, this volume). The principal idea behind this regional scheme was that development centers would be supported in each region, and they would in turn function as growth poles from which economic development would originate and spread into the surrounding areas (National Economic and Social Development Board, 1994). At least one dominant center would be supported in each economic region, and the centers would be linked among themselves and with Bangkok with tangible economic connections. The support from the central government would be in the form of: (a) infrastructure and facilities, (b) location of manufacturing industries, (c) improved land management to facilitate and accommodate physical growth, and (d) improvement of local government managerial capacity (National Economic and Social Development Board, 1985). Chiang Mai was designated as the capital of the North Region (Table 1).

Table 1: Thailand's Regional Cities Development Programme

Designated Region	*Regional Centers*
North Region	Chiang Mai, Pitsanulok
Northeast Region	Udon Thani, Khon Khan, Nakhon Ratchasima, Ubon Ratchathani
East Region	Chonburi
South Region	Songkhla, Phuket

Source: National Economic and Social Development Board, 1985.

As a result of this plan, during the last twenty years the cities identified above have received priority funding to build up their basic infrastructure and facilities, especially water supply, electricity, telephone service, and roads. Other support, in more modest amounts, came for drainage, housing development, slum improvement, and environmental protection. In Chiang Mai, a major government investment has been the establishment and development of the Lampung Industrial Estate (for more on this topic see Stafford, Chapter 4, this volume). Added to the impressive growth of the tourism sector, thanks mostly to private initiative, the economy of the city has had a strong foundation on which to grow. But in the last fifteen years the rate of growth of the population and the local economy have been faster than the rate of growth of public sector investments in the region, and thus the pressures of Chiang Mai's economic boom have been felt in the loss of quality of services (The World Gazetteer, 2002; UNESCAP, 1998; also, see Romanos et. al., Chapter 8, this volume). When this study was undertaken in the summer of 1998, the efforts of the central government of Thailand to support the development of the infrastructure and services in Chiang Mai so that it would become a regional growth center, had backfired. The growth of the city had overrun the speed with which the government was providing public investments for its improvement, and thus the boom growth experienced by the urban region had brought in the adverse effects of shortage of services, overcrowded and congested housing and transportation conditions, environmental degradation, and a declining quality of life. Its growth path was taking the city away from a sustainable development model of evolution. And, its designation as an intermediate size city and regional center programmed to ameliorate some of the woes of Bangkok did not appear to be effective. In fact, while all the regional centers designated by the Regional Programme appear to grow, they did not succeed in slowing down considerably the continuing growth of the Bangkok conurbations (The World Gazetteer, 2002).

1.2. Sustainable Development for Chiang Mai

Our objective in this chapter, and in the rest of the book, is not to explore in depth notions and definitions of sustainability and sustainable development; others have done so, and by now the literature on definitions and philosophical treatises of sustainability and sustainable development is more than extensive (Moffatt et. al., 2001; Moffatt, 1996; Van den Bergh and van der Straaten, 1994; Costanza and Patten, 1995; Clayton and Radcliffe, 1992). Rather, our purpose here is to present the practical aspects of planning and managing a city in a developing region of a developing nation in the context of sustainability, and assess their potential to help that city develop in a sustainable mode in the future. To accomplish this aim, we have adopted a practical and down-to-earth working definition of what in our view constitutes sustainable development, and we have asked the contributing authors to interpret this definition in their own way and according to their own disciplines. As such, there is variation in the interpretations of the principles and demands of sustainable development, ranging from acceptance to conditional encounters to criticisms. This is fine with us. What the chapters of the book want to contribute is their view of how each of the economic sectors studied, the components of the urban system we surveyed, and the foundations of the social structure we analyzed can participate in a coordinated effort to move the metropolitan region of Chiang Mai towards a sustainable development path. This then determined the need for an applied and broadly defined notion of sustainable development. In our view, sustainable development is a development approach that contains four interconnected components: (a) the need to ensure the future survival of the system in its basic form and structure, that is by ensuring economic conditions supporting the system's population while preserving the social and cultural foundations of the system; (b) the desire to bring improvements into the system so as to better the human condition, that is ensure that the entire community, not just individual units of it are better-off in the future; (c) the expectation of social justice governing the system, so that rights of citizenship, opportunities for participation, and protection by the law are equally accessible to all; and, (d) intergenerational responsibility, that is the obligation to put in place mechanisms that will ensure the expansion of the system's future capacity to survive, be able to solve its own problems, and afford its people a high quality of life. The structure of the chapters and the focus of their authors reflects this view.

1.3. Sustainable Development and the Significance of Intermediate Size Cities

At the end of the 1980s, the total population of the eight countries of Southeast Asia was approximately 436 million, of which the urban population was 134 million, or approximately 31% (See Table 2). Of these 134 million people, 51 million, or 38%, lived in large cities of over one million people. The balance of 83 million people, or 62% of the urban population, lived in smaller, intermediate size cities. At the end of the 1990s, the role of intermediate size cities had expanded considerably more. By then, the total population of the eight countries had increased to 503 million, and of these 184 million were urban, representing 36.6 % of the total population, and indicating a clear urbanization trend in the region.

Table 2: Populations of Southeast Asian Countries, 1990

Country	*Total population (millions)*	*Total urban population (millions)*	*Number of large cities*	*Population of large cities (millions)*
Cambodia	9	2	--	--
Indonesia	178	55	8	20
Laos	4	1	--	--
Malaysia (1991)	18	9	2	3
Myanmar (1983)	42	11	1	3
Philippines	63	31	1	10
Thailand	56	11	1	9
Vietnam (1989)	66	14	3	6
Totals	**436**	**134**	**16**	**51**
		31% of total population		**38% of urban population**

Sources: World Bank, 1998; Rand-McNally & Co., 1996

But what is more revealing is the distribution of that urban population (See Table 3). Even though the data for large cities are 2002 estimates, so we cannot compare them directly with those of Table 2, they clearly indicate a significant trend. They show that the total population of the large cities increased to 52.6 million, up from 51 million more than a decade earlier, but representing only 27% of the total urban population of the region, as compared to 38% more than a decade earlier. The balance of this urban population, at least 73% of the total, lives in cities below the 1 million population mark that is in intermediate size cities. Thus, it becomes clear that, while the countries of Southeast Asia continue to become more urbanized, most of the urban population is retained by the intermediate size cities, which grow much faster than the very large cities of the region. This is confirmed by the "stepping stone" theories of rural to urban migration movements, which argue that rural migrants do not go directly to the very large urban centers in search of employment and new life styles, but go to smaller urban centers first. They also help prove the model of local land regional labour retention advocated by the growth pole theory (REFS). With such large population concentrations, the intermediate size cities of the region inevitably are called upon to play a significant role in the economy of their countries[2].

Table 3: Populations of Southeast Asian Countries, 1998

Country	*Total population (millions)*	*Total urban population (millions)*	*Number of large cities*	*Population of large cities 2002 est. (millions)*
Cambodia	11.7	2.6	1--	1.1
Indonesia	204.4	77.6	8	23.6
Laos	5.3	.9	--	--
Malaysia	22.2	12.4	1	1.4
Myanmar	47.6	12.8	2	5
Philippines	72	41	1	10.1
Thailand	61.2	20	1	6.5
Vietnam	78.9	15.8	2	4.8
Totals	**503.3**	**183.1**	**16**	**52.5**
		36.6% of total population		**27% of urban population**

Sources: The World Gazetteer, 2002; UNESCA, 1998.

Yet, the developing economies of Southeast Asia have treated their large cities more favourably than their intermediate size cities (Osborn, 1997; Rigg, 1991; McGee, 1990). There are several explanations for that preferential treatment. One is the high rate of population growth and the resulting urbanization of the largest cities in the region, in the same way this phenomenon has been occurring all over Asia, Africa, and South America. These urban residents and migrants need jobs, housing, and all types of infrastructure and services (Jones, 1991, 13). National governments have to respond first to the high pressure issues, and thus large cities receive more attention than intermediate size ones. A second explanation is that many of these cities, either because they were centers of colonial administrations or because they were capitals or major export hubs of their countries, were favoured in the early development of physical and institutional infrastructure. That infrastructure, in turn, gave the largest cities a competitive advantage, and helped them attract domestic and foreign investments, educational institutions, corporate headquarters, and political power, all of which ensured that they would continue to receive the largest share of the country's largess. A third explanation is that, once a center of economic and political power is established in a large city, that center tends to safeguard its power. This means that both the state apparatus and the corporate and industrial power elites are not disposed favourably towards the power sharing idea between the center and peripheral cities, and their decisions tend to give priority to the needs of the large centers. The centralized nature of the national administrations in most countries of the Southeast Asian region makes that allocation of priorities even more plausible, especially with regard to the capital city, because all major decisions about the allocation of resources are made there (Romanos, 2000).

Still, intermediate size cities have long been recognized by planners and decision makers as especially effective in promoting regional development objectives (Velasquez, 1991; Ruland, 1988; Rondinelli, 1987; Cheema and Rondinelli, 1983). Since the period of large regional development programs following the decolonisation period after the Second World War, they have been chosen as a preferred tool for the implementation of decentralization schemes encouraged by international aid agencies, and were adopted by many industrializing countries as a way to attract direct domestic and foreign investments in the periphery and to induce infrastructure support investments by the private sector, thus relieving the burden of central governments (Rivkin, 1976). Most frequently encountered arguments in favour of supporting the development of intermediate size cities are:

- They promote the goals of decentralization and increased local autonomy and thus induce a higher degree of democratization in the periphery and the country as a whole.
- They are good intermediate stages in the evolution and transition of local economies from a rural/agricultural base to more sophisticated urban and urbanization economies.
- Their carrying capacity is less overloaded so they have more potential for more orderly growth and development if given attention on time; and, thus face a much smaller likelihood of major social and economic problems than large urban centers.
- Because of their smaller size they are better able to provide effective solutions for both employment generation and regional economic development, and thus can become effective growth poles inducing the development of their regions.
- They can more easily accommodate in-migration of rural labor, as well as the location of industry and services that could not be accommodated as well in large cities because of capacity saturation there.
- If successful, they can attract population, business, and industry from the country's major and most congested and problems-ridden metropolitan regions, thus decongesting the large urban areas, relieving their infrastructure and support systems, and alleviating their fiscal pressures.
- Because of their smaller size and their closer relation with the surrounding hinterland, they can provide a more sheltered, less crowded, and more satisfying life style and a higher quality of life.
- Their regional locations, their smaller size, and their closer family and ethnic connections with the local population and customs, make them better able to protect and preserve local culture and character than larger and more cosmopolitan urban areas.

- At least in the earlier stages of their growth, they have fewer pressures on their resources and their environment and, with the right type of planning and foresight, can capitalize on that to develop procedures, mechanisms, and safeguards that would ensure efficient and effective urban development management, and the management of environmental resources and services in particular.

These are substantial advantages, and they should be enough to convince central governments of the wisdom of allocating resources, autonomy, and political and executive power to selective intermediate size cities, so that they could pursue their own approach to economic and social development. But central governments are also aware that that kind of autonomy may induce cities to approach issues of development and resource conservation in ways not necessarily compatible with national growth goals. Similar to the conflict between local and universal optima, what is good for Chiang Mai under this sort of scenario may not necessarily agree with what the central government has decided is good for the country. Therefore, central governments are still unsure about the treatment of their intermediate size cities. On the one hand, they recognize that these provincial centers can become agents of change and improvement in urban conditions and problems, and thus serve as a counter-force to the large urban centers. On the other hand, they are reluctant to give up any of their control-and-command privileges, concerned that that would challenge the existing, rigid but convenient structure of centralized national governance and the power exerted by its central agencies over the affairs of provincial and local jurisdictions. This uncertainty is precisely the reason for the choice of focus of this book. By concentrating our attention on intermediate size cities, we had the opportunity to explore their capacity to lead their own development if they are given the right circumstances, their ability to harness the rights granted them by the central administration in effective ways, and their commitment to the development of just, democratic, and sustainable societies. We also tried to learn if that decentralized model would be easier or more difficult to implement, both for the local and the central actors. And finally, we attempted to understand and communicate to the host city our assessment of their chances to succeed in this endeavor, if conditions changed only slightly and only gradually, because of the desires and the strong commitment to change of the local public and private sectors. These issues and questions occupied us throughout the period of the Joint Study, and our findings, when we were able to have some, are reported in the chapters of this book.

2. THE PROJECT AND ITS PARTICULARS

2.1. Project Background

The project on which the Joint Study was based, and through which it was mostly funded, was titled “Sustainable economic growth in the APEC region through urban environmental management”, and its focus was to explore how improvements in the management of public sector operations, especially at the local government level and particularly in intermediate size cities, could, in addition to rendering better municipal services and more responsive local administrations, lead to the promotion of sustainable development goals and raise of public awareness about sustainability and its principles and requirements (Romanos, 1998). Operationally, the project had several distinct but interrelated components. Part of it concentrated on purely educational issues such as curricular development, training of junior faculty, methods of teaching courses and seminars on sustainable development and on development management, joint teaching, and syllabus design and preparation. Other parts of it organized lectures, seminars, workshops, and other forums of exchange in all three countries (U.S., Thailand, and Indonesia), and used them to spread the discussion of sustainability and public sector management issues, make faculty and students aware of the alternative models of development represented by these concepts, and induce interest in introducing at least some of these notions into the curriculum. Yet, a third set of activities was based on visits to cities and regions in the three countries where some management and implementation activity of note was taking place, and we could learn from it. Such visits to U.S., Thai, and Indonesian cities, accompanied by extensive interviews with key actors there, produced an immense amount of relevant and practical information, invaluable to all three institutions as they were modifying their curricula and were exploring avenues of research on these topical areas. A fourth and culminating experience of the project was the design and conduct of field projects overseas with the participation of as many team members as possible from all three universities.

The field projects were also quite diverse and ambitious in their stated purposes. First and foremost was the objective of studying a community and uncovering its potential to evolve towards a sustainable development future, and examining the capacity and conditions of the public sector, the economy, and several parameters within them, to allow the types of management that would facilitate this effort. Second, we considered our work in the host cities an opportunity to involve the staff of the local government offices and agencies in the day-to-day activities of the Joint Study and make them as much participants in these activities as they had the time and the inclination, on the assumption that these could be an educational experience for them. Because of the different analytical approaches followed by the members of the team and their familiarity with different issues, and because of the extensive use of English during the work, the staff of the Chiang Mai municipality found their association with our team invaluable (Romanos, 1999). Third, these were seen as educational opportunities for the faculty and student members of the team to learn how to conduct field research in collaboration with colleagues from different disciplines, different educational traditions, and different languages (Auffrey and Romanos, 2001). This proved to be much more of a challenge than the previous two objectives. The Joint Study was designed around the notions of sustainable development outlined in the first section of this chapter. That framework went through many permutations, and still today it is not universally accepted by every member of the team. Discussions and arguments went back and forth during the academic quarter before the filed project began, in the spring of 1998, when members of the team unfamiliar with the orientation of the study joined in. It continued in Thailand through the period of the project, and in some ways it has continued through the successive drafts of the chapters of this book. Of interest was the reaction of the municipality of Chiang Mai and the groups and individuals we worked with on location. The NGOs were by and large familiar with the concepts and considered them part of their environmental agenda, so they accepted them with enthusiasm, even as they did not seem quite optimistic about their chances. The municipality and its employees embraced the idea of a sound theoretical basis for the development of their conceptualization of a potential sustainable development program for the city and its regions, but did not scrutinize its details very much. For them, the exploratory research itself was the paramount priority, because they expected out of it a systematic analysis of the problems and challenges facing the municipality, or about to face it, in the future. The conceptual framework through which we would suggest their addressing the solutions to these problems seemed secondary, and likely to be adjusted in the process anyway. As mentioned elsewhere in this chapter, the summer of 1998 was a particularly difficult one for Thailand and its cities, as the expected financial recovery did not take place, the baht continued to lose part of its value against the U.S. dollar, and the government was taking too long to come up with the necessary laws and policies that would begin to attack the causes of the declining economy. Thus, the details of the development model were not as crucial to the host city as was the fact that there was going to be such a model.

More to the point, the framework was designed to analyze and assess the organization of the government apparatus responsible for the planning, financing, implementing, and regulating of development programs, the running of local government operations, the planning of cities, the provision of services, the distribution of funds, and the treatment of the city residents. Issues of decentralization, inaccessibility to data, non-transparent decision making, the passage and implementation of rules and regulations, the existence of authority to act, and the availability of legal and fiscal tools to do so, became issues of major concern to the team.

As it turned out, the exploratory study was able to reveal many aspects of the local dependence upon central agencies and procedures which impeded the ability of local governments to be effective and carry out their mandate responsibilities. Problems for which mayors traditionally were blamed, proved to be squarely the fault of central government constraints. And thus, the information passed onto the municipality, and reported in the chapters herein, was revealing and useful because it defined the real parameters with which the city of Chiang Mai and its institutions were operating, and the barriers that should be overcome if these operations were to be any more efficient and effective. So, even though the short term of our field project and the extreme difficulty of obtaining factual information prevented our team from providing definitive answers to the questions of the municipality, the understanding of these barriers and limitations we offered proved to be much appreciated knowledge.

2.2. The Joint Study

The focus of the "Chiang Mai sustainable development potential" project (referred to as the Joint Study throughout the book) was the study of the economic structure of the Chiang Mai metropolitan region. Our agenda was as follows: 1) identify the key economic sectors and assess their share and contribution to the economic strength of the region; 2) assess the relationship between these economic sectors and the plans and expectations of the city for its future growth and physical expansion; 3) assess the conditions of the city's infrastructure and services, and the efficiency with which they deliver expected service to the urban and suburban communities; 4) discover the social and cultural background within which growth and change are taking place in the region, and the views and beliefs about growth, development, preservation, and the significance of the built and natural environment, including the acceptance of limits to that growth; and, 5) determine the structure and organization of the local government, the legal framework within which it operates, the power and limitations it faces in the execution of its responsibilities, and the efficiency of the operations by which the city manages its human, natural, and capital resources and ensures their just and prudent use.

The field project effort consisted of data gathering, general information sessions, seminars and lectures to the group, ground surveys and visits, structured interviews with a large number of individuals, consultations with municipal employees and local government officials, and brainstorming sessions with the participation of all the members of the group and invited members of the community, local NGOs and/or the local government. We discuss these in more detail below.

2.2.1. Data Gathering

The project members embarked in a systematic and thorough effort to collect hard data describing the conditions in Chiang Mai and the legal, regulatory, and planning environment within which the city operates. Preliminary library research in the U.S. before the departure of the group was not very productive. Most of the information we sought was supposedly available in Thailand but had not been published in international journals and commercially available books. Much of the information gathering effort took place in Bangkok, where the agencies and ministries of the central government are located, and where decisions are made. Despite the great access to these government bodies afforded us by the good connections of Chulalongkorn University, assembling complete data on any aspect of the Chiang Mai urban system proved next to impossible. Even when the data existed, as in the case of maps and proposed plans, rules of confidentiality and reluctance to share or publicize information proved insurmountable stumbling blocks. One of the reasons for this difficulty in obtaining data is the strict confidentiality governing most land use and locational information for national security reasons. Another is that data collecting and organizing can be the sole assignment of some government professionals, who see these data as their own personal accomplishment and develop proprietary feelings about them. That makes them hesitant to part with these data under normal circumstances. Because data are so difficult to obtain, they constitute a source of great power and control, and the people safeguarding them are not willing to give that power up. And finally, data are not simply expected to give an accurate and objective image of reality, but also to measure change and progress of government programs. As such, they are not being publicized because there may be a need for them to be adjusted at some later date. While some such adjustments are to be expected as more final data become available, data manipulation to meet political objectives is an exercise occasionally practiced by many government agencies in Southeast Asia.

2.2.2. General Information Sessions, Seminars and Lectures to the Group

The group had ample opportunity to become oriented on the government structure and operations in Thailand and Chiang Mai through a series of informational sessions, lectures and workshops, and visits to key ministries and other government agencies (Auffrey and Romanos, 2001). These sessions were presented by knowledgeable individuals in high government positions and by technical personnel. The presentations provided excellent interpretations of the planning and management systems and approaches permeating the various layers of government at the national, provincial, and local levels. The group was also fortunate to pay formal visits to the Ministry of the Interior and its Town Planning Department, the National Economic and Social Development Board, the Thailand Housing Authority, the Thailand Industrial Estates Authority, and the Tourism Authority of Thailand, along with several other central agencies. Similar informational and acquaintance visits took place in Chiang Mai during the first week of the field project, with visits to the governor of the province, the Industrial Estate at Lampung, central government branch offices in the province, the Municipal Sanitary Districts, the Chiang Mai Chamber of Commerce, and Chiang Mai University.

2.2.3. Ground Surveys and Visits

Throughout the period of residence in Chiang Mai, the group or its teams visited most areas of the city, its neighborhoods, informal settlements, municipal facilities, economic districts, central and peripheral business districts, and many cultural and archaeological sites, such as monasteries, temples, and areas protected by the Department of Fine Arts. In some of the initial visits, the teams were accompanied by municipal staff, but in many other instances or in follow-up visits we had our own interpreters. During the visits, in addition to taking a physical inventory of the areas, meetings were held with the residents, business people, community organizers, NGOs active in the neighborhoods, and individual residents. These occasions provided valuable insights into neighborhood variations in life styles, as well as an understanding of the issues, the particular problems, the localized politics, the priorities and the activism of the neighborhoods. Mostly, ground visits and structured interviews took place in the morning, to the extent that times were convenient for the subjects. Afternoons were used for record keeping, material cataloguing, interview transcriptions, brainstorming sessions, exchanges of information, library organization, data management, other housekeeping chores, and planning for subsequent meetings and coordination of interviews and ground visits.

2.2.4. Structured Interviews

In addition to the data we collected from government agencies and official publications, the project collected subjective information and personal assessments of issues, conditions, problems, and potential solutions through extensive use of the technique of structured interviews[3]. The structured interviews lasted an average of three hours each, and were of two kinds: general issues and specific focus. The general issues were of interest to the entire group and were aimed at answering general questions and clarifying general issues. About fifty of these were administered to subjects representing a wide spectrum of positions, knowledge, interests, and viewpoints. With representatives of central government agencies in Bangkok, they were conducted mostly at the Director General level; in other central government agencies, the subjects were Directors, Associate Directors, or Technical Directors; in national professional and business organizations – such as the Chambers of Commerce, Industry, and Tourism – the interviews were conducted at the Director and Associate Director levels; with prominent national business and corporate leaders, the corporate presidents or vice-presidents or directors of public relations were interviewed.

In Chiang Mai, we interviewed several departments of the provincial administration; the municipal administration – mayor, associate mayors, and directors of divisions; the supervisors of the local offices of the central government agencies in Chiang Mai at the director level; the Lampung industrial estate at the director level and with the managers or CEOs of manufacturing plants; leaders of NGOs; local professional organizations; local civic leaders; and, researchers and experts at Chiang Mai University. In addition to these general interviews, each topical area team identified and interviewed a number of subjects with specialized knowledge of a topic or with widely expressed views and positions. Again, the subjects were chosen from among government officers, professionals, business people, and university faculty.

2.2.5. Consultations with Municipal Employees and Other Local Government Officials

The group had periodic discussion sessions with representatives of local government agencies, whom we treated as local information sources, and used these sessions for the exchange of ideas, verification of data, checking of other sources of information, and interpretation of information. Members of the agencies also assisted in the arrangement of ground visits and accompanied the group or its teams on these visits. These interactions were two-way, in the sense that they were also used for formal and informal guidance and training of the agency personnel by explaining the organization of the study, techniques of data management, alternative ways to conduct surveys and interviews, rules governing the interviews, and assistance with the English language.

2.2.6. Brainstorming Sessions with the Participation of Community Members, Local NGOs and/or the Local Government Officers

These sessions were widely considered to be the best part of the project. Several of them were held in the group's work space in Chiang Mai during the conduct of the study, and were used for the exchange of information, the debate of issues, the development of arguments and counter-arguments about the nature and appropriateness of sustainability for Chiang Mai, the comparison of interview insights, the comparison of different ways of data organizing and analyzing by different disciplines, and the challenge to personal beliefs and biases. Records of these meetings were maintained and were later used for the preparation of reports and the book chapters.

2.3. Obligations to the Host City

Even though the informal interaction between the study group and the municipality was continuous, we used many opportunities to make more formal contact as well. Each time visitors came from Chulalongkorn University, for example, we used the occasion to organize an semi-formal meeting during which we briefed our Chulalongkorn colleagues on our progress, and at the same time we made brief presentations to the hosts. Even more frequently, we made such presentations to the liaisons between the municipality and the teams so that the liaisons could then brief their supervisors and the rest of their office on our progress. These presentations served a very useful purpose, in that they were in reality consultations with the hosts, during which misconceptions, misunderstandings, or over-eager interpretations of the team members could be checked and adjusted as necessary.

At the end of the program, the group prepared and delivered a comprehensive presentation of our findings and recommendations to the lord mayor, the city council, the chiefs of the municipal departments, many technical staff of the municipal administration, and invited guests. The presentation, which was carried out by a number of team members, generated a lively discussion and enthusiastic reactions from the host. The presentation was used as a basis for a report later delivered to the office of the mayor. That report has been adopted by the city council and is being used ass one of the guidelines assisting in the development of the city and urban region of Chiang Mai since 1998.

3. ORGANIZATION OF THE BOOK

The remainder of this book provides the findings of the exploratory research conducted by our international and interdisciplinary team of researchers from the U.S., Thailand, and Indonesia. The focus of that research was on the development of an intermediate size Asian city and how the use of a sustainable development framework could provide insight to the strengths and weaknesses of the city's current development path and help identify alternative modes of policy and action that would lead to better outcomes for all present and future residents. The structure of the book is intended to reflect this focus by emphasizing four components that we consider essential to a sustainable approach to development. It is important to emphasize here that while each of the book's parts, and in turn, its chapters are intended to provide insight to critical development issues that stand as important in their own right, our intent is to bring all of these together into a coherent set of ideas, and illustrate the need to address the issues of development in a comprehensive manner. As the sustainable development framework emphasizes, attempts to promote economic expansion to the exclusion of the community's social, cultural, and environmental concerns are unlikely to achieve long-term success. What this suggests is that failure to resolve the problem of any one area may overwhelm the successes in other areas. Consequently, we have chosen to present here findings about those social, cultural, economic, and environmental (both natural and built) issues that have been identified as priorities by those in Chiang Mai, and whose resolution is a necessary condition for sustainability.

Part I is titled *Cultural Dimensions as the Underlying Context for Development* and reflects on the importance that the Joint Study team placed on the social and cultural dimensions of our sustainable development framework. While a wide range of topics and ideas might be subsumed under the social and cultural aspects of development, there were two areas that emerged as particularly relevant to the context of our field work in Chiang Mai: the influence of Buddhist religious philosophy on approaches to development, and the Thai concept of and experience with citizen participation.

In Chapter 2, *Buddhism and Sustainable Development*, Matz considers the extent to which Thai Buddhism is compatible with the international conceptualization of sustainable development. Given the pervasive (nearly 95 percent of Thais follow Buddhist traditions) and the important influence of Buddhism on Thai behavior and thought, this is not an issue that can be taken lightly. As Matz points out, "Non-Buddhist planners in Thailand face the prospect that there is a significant difference between their "western" values and the Buddhist values of the Thai people, a difference that purportedly renders problematic any attempt to implement policy in Thailand" (See Matz, Chapter 2 in this volume). He concludes that Buddhism is compatible with the fundamentals of sustainable development but may emphasize different dimensions of the concept than would western planners, though both approaches are essential to sustainable societies.

The second chapter in this part of the book helps further our understanding of Thai approaches to sustainable development by examining the behavior that results from Buddhist and other influences on Thai citizen participation in public decision-making. In Chapter 3, *The Emerging Role of Citizens in Planning and Public Decision-Making in Thailand*, Kahle and Wagner find citizen involvement is shaped as much by culture or social behavior as by the nation's history and political processes. Understanding Thai social behavior includes Buddhist doctrine and the aspects of Thai culture that flow from that doctrine, such as hierarchal structure, patron-client relationships, and the role of the individual in the community. The authors find that this has shaped a uniquely Thai notion of citizen participation whose emerging should support efforts to promote sustainable development.

Part II, *Key Components of the Local Economic Structure,* presents analyses focused on critical sectors of Chiang Mai's local and regional economy. Over the past half century, Chiang Mai has emerged as the center of economic development for the northern region of Thailand. More recently, the pace of development had accelerated until the financial crisis of the mid-1990s. Despite the slowdown, the city continues to possess the administrative, financial, and industrial infrastructure for rapid economic expansion. This part of the book seeks to build on the previous part's ideas of social and cultural context to explore three of the most important sectors of Chiang Mai's economy: manufacturing, handicrafts, and tourism. The rapid growth of the manufacturing sector in the Chiang Mai region over the past two decades, as a result of the globalization of the Thai economy and attempts to de-concentrate growth in Bangkok, has brought international approaches to business and the need for better-trained workers. In Chapter 4, *Manufacturing in Chiang Mai Province: Products, Patterns, and Sustainable Balance*, Stafford examines the spatial distribution of manufacturing throughout the province of Chiang Mai, and uses this to assess its contribution to a viable regional economy. He concludes that the concentration of manufacturing near the city of Chiang Mai is diverse and strong, but the present form serves to draw workers away from outlying districts whose growth prospects are poor. Such a pattern is unsustainable, as the weaknesses will eventually detract from the core. This problem of the spatial distribution of manufacturing is exacerbated by the need for trained workers. In Chapter 5, *Sustainable Development and Sustainable Employment in the Manufacturing Sector of Chiang Mai: An Analysis of the Thai Educational System and the Workforce Supply for the Manufacturing Sector of Northern Thailand*, Barry notes that the Thailand has been slow to make the needed changes in work-force education. His assessment identifies a shortage of workers with the technical skills necessary for an expanding manufacturing sector, and the need for structural changes in its educational system to address these needs. As such, the current educational system creates serious obstacles for the sustained development of manufacturing and other sector of a modernizing economy.

In Part II's two remaining chapters, the research explores two interrelated economic sectors with special significance for Chiang Mai. Handicrafts traditionally have been an important source of jobs and income for many of the city's residents. As an element of a sustainable development framework, the production and distribution of silk, cotton, silver, wood, and lacquer crafts maintain a connection to important social and cultural influences of Chiang Mai's past. Further, the handicrafts industry is closely linked to the area's national and international tourism that has become one of the most important sectors of the local economy in recent decades. In Chapter 6, *The Handicraft Sector in Chiang Mai: Its Role in Sustainable Urban Development*, Chifos and Looye use a sustainable development framework to assess the relationship of the handicraft sector to the Chiang Mai urban area. Their analyses reveals a web of multi-dimensional connections to environmental, social, and cultural inputs and outcomes, which suggest a more holistic view of urban management and economic development initiatives. They conclude that left on its own, the handicraft sector will likely encounter "unsustainable trajectories" that may lead to items with little if any unique cultural significance, and squeeze out any niche for local art or ethnic crafts. As their assessment of multi-dimensional connections suggests, altering the trajectory of this sector in a more sustainable direction has crucial implications for the character of Chiang Mai's tourism sector. In Chapter 7, *The City of a Thousand Smiles: Sustainability and the Pursuit of Tourism in Chiang Mai*, Romanos reports on the character of the fastest growing sector of the national economy, and discusses related issues that impact the sustainable development of the city. These issues cover a wide range, from the effects on the local economy and culture, to the impacts on the environment and public health, and illustrate the utility of a sustainable development framework that is able to incorporate such diverse elements.

Part III, *Environmental Management Issues in Development,* considers threats to the natural environment, the means for addressing these, and the fundamental importance of doing so if the economic, social, and cultural dimensions of the city are to survive and prosper. Environmental protection can no longer be viewed as a luxury of the industrialized world, that developing countries such as Thailand can only begin to address when they reach that vague point in the future where they become wealthy enough to do so. Further, the process is not simply relegated to a central government bureaucracy with limited monitoring capacity and even more limited enforcement ability. Rather, the environmental protection and conservation of natural resources critical to the sustained well-being of the society must emanate from interdisciplinary and multi-sectorial approaches that encompass both the public and private sectors and the full range of environmental threats.

In Chapter 8, *Environmental Management in Thai Cities: Local Challenges, State Responses*, Romanos, Wagner, and Thadaniti examine the Thai institutional framework for urban planning, development, and environmental management within the context of intermediate size cities. It is within this institutional framework that Chiang Mai's most pressing environmental problems and their potential solutions must be viewed. Among the most critical environmental problems, as identified by Chiang Mai residents and foreign observers alike, are water pollution and solid waste disposal. In Chapter 9, *Water Resources for a Sustainable Chiang Mai*, Wang and Liu evaluate water supply and demand, wastewater management, and water quality issues in the context of the institutional and governmental structures of Chiang Mai. Urbanization has significantly degraded water quality in the city, due in part to the lack of comprehensive and sustainable set of proactive policies to guide urban development. In Chapter 10, *A Study of Changing Traditional Urban Solid Waste Management Approaches to Sustainable Practices*, Schultz evaluates the solid waste management practices in the city and considers how Chiang Mai might convert to more environmentally sustainable practices. Traditional waste handling and disposal practices in Chiang Mai have been technically and politically troublesome, perhaps even more so than in most intermediate size cities in Asia. This might be viewed as an opportunity to invest adequate resources to promote effective programs for waste reduction, reuse, and recycling. However, efforts to guide more effective environmental protection and waste reduction require accurate and timely information concerning the spatial dimensions of the problems. One tool that has proven effective in this regard is geographic information systems (GIS). In

Chapter 11, *Building a Geographic Information System*, Liu and Wang assess the feasibility of using a GIS for enhanced environmental management decision making in Chiang Mai. While the investment in technology for addressing the plight of the urban poor may seem incongruous to some, current methods of decision-making may have even higher costs for those who can afford them the least. As their analysis suggests, the tangible returns on investment in a GIS may be substantial.

Part IV, *The Built Environment's Role in Sustainable Development*, assesses the critical issues of the physical and human patterns of urban expansion, and how those patterns manifest themselves and might be directed through appropriate policy guidelines. Here again, the intent is to build on the book's previous content and the reader's understanding of the social and cultural context, economic structure, and issues of the natural environment to discuss the importance of urban development management as an element within a framework for sustainable development. Crucial issues of urban development in Chiang Mai include urban design, informal settlements and the preservation of historic and culturally significant structures. In Chapter 12, *Land Use and Urban Growth in Chiang Mai: Lessons for More Effective Urban Development Management*, Setiawan and Rahmi make the case for effective urban growth management if the balance necessary for sustainable urban development is to be achieved. They argue that close attention must be given to both the physical and socio-economic dimensions of urban growth because they are inextricably linked; ignoring one will lead to problems of the other. In Chapter 13, *Towards a Sustainable Urban Form in Chiang Mai*, Scheer and Scheer focus on the physical form of different areas to assess how changes in urban design may effect the sustainable development of Chiang Mai. They compare the emerging urban form of Chiang Mai's suburban areas with the long established areas near the center of the city, and consider how issues of design affect several factors including social cohesion, physical activity, and local economic interaction. In Chapter 14, *The Informal Settlements of Chiang Mai: Lessons for Other Intermediate Size Cities*, Auffrey brings the issue of the relatively numerous informal settlements surrounding the city's urban core to the discussion of sustainable development. He argues that such settlements epitomize the essential character of sustainable development, bringing together in a direct and obvious fashion the critical social, cultural, environmental, and economic elements within a concentrated place, where change can have an impact in a relatively short time. Many of the issues of Chiang Mai's informal settlements are issues of urban growth management and urban form, though the social, cultural, environmental, and economic dimensions of these add a great deal of complexity to their resolution. Even the issue of historic preservation has been prominent in Chiang Mai's informal settlements, as several of the communities had been built on vacant government property along the ancient outer wall of the city. However, the issue of historic preservation in Chiang Mai is concerned with more than just the ancient walls. As the ancient capital of the Lanna Kingdom, Chiang Mai has unique local architectural forms that are attractive for maintaining the region's unique cultural identity as well as providing an attractive basis for a higher quality "cultural" tourism. In Chapter 15, *Preservation of Cultural and Historic Heritage as a Tool for Sustainable Development*, Rincon and Rahmi respond to the voiced concern of many Chiang Mai residents about the loss of cultural identity. This chapter focuses on how appropriate controls and incentives can work to the advantage of property owners as well as those concerned about preservation of the city's built heritage. Through the use of a case study, the authors conclude that while the city cannot avoid all influences of cultural and economic modernization, preservation of the built environment could be significantly more effective if appropriate institutional frameworks were put in place as part of a sustainable urban development plan.

Sustainable development requires adapting to the specifics of the local context, and so it was it was decided that as part of the process of documenting the Joint Study, it would be useful to compile a visual record that could effectively convey the unique context of the project. As a result, a faculty member with a photojournalism background was selected to accompany the other faculty researchers in Chiang Mai. The results of the efforts by the photojournalist to convey that context are displayed throughout the book. In addition, the book concludes with a contribution entitled *A Sense of Place* that provides both text and photographic perspectives of the context for the Joint Study and the community where it took place.

NOTES

[1] In 1993, this number represented 10.7 % of the country's population and 57% of its urban population (Boonyabancha and Chalitanon, 1996, 87-88). By the year 2000, the total population of the country was over 62 million, its urban population was over 21 million, and the population of Bangkok was about 6.5 million. Thus, by the beginning of the 21st century, Bangkok still represented 10.6% of the total country's population, and a significant, although lower, 31.3 % of its urban population (UNESCAP, 1998).

[2] For our purposes, intermediate cities are defined as those urban communities with a population of more than 100,000 but less than one million residents. In Southeast Asia, intermediate size cities carry a significant share of their countries' urban population, although this share and the number of intermediate cities varies widely from country to country (Romanos, 2000).

[3] Structured interviews are based on a standardized questionnaire containing mostly open questions and giving an approximate length of response time. The questions are phrased in exactly the same way, so that there is consistency in the questioning part of the interview. Subjects are identified not at random but based on the particular knowledge of the subject on issues or experience with conditions and situations. In a way, structured interviews have many of the characteristics of surveys of experts, but introduce a measure of comparativeness by keeping a standardized format. While the responses are tabulated and compared, no statistical analysis of trends is attempted, because it is recognized that structured interviews produce subjective, judgmental responses rather than statistically objective ones.

REFERENCES

Auffrey, Christopher, and Michael Romanos. (2001). Educating the Educators: Global Dimensions of Collaborative Fieldwork in an Urban Region of Southeast Asia. *Journal of Planning Education and Research, Vol. 20, No. 3*, 353-364.

Boonyabancha, Somsook, and Natvipa Chalitanon. (1996). Case Study of Nakhon Ratchasima, Thailand. In United Nations Centre for Human Settlements (Habitat), *The Management of Secondary Cities in Southeast Asia* (pp. 87-107). Nairobi: UNCHS.

Cheema, Shabbir, and Dennis Rondinelli. (1983). *Decentralization and Development.* London: Sage.

Clayton, A. M. H., and N. J. Radcliffe. (1992). *Sustainability: A Systems Approach*. Edinburgh: Institute for Policy Analysis.

Costanza, Robert. and B. C. Patten. (1995). Defining and Predicting Sustainability. *Ecological Economics, No. 15*, 193-196.

Jones, Gavin W. (1991). Urbanization Issues in the Asian-Pacific Region. *Asian-Pacific Economic Literature Vol. 5, No 2, September*, 5-33.

McGee, Terry G. (1990). *The Southeast Asian City*. London: G. Bell and Sons.

Moffatt, Ian. (1996). *Sustainable Development: Principles, Analysis and Policies*. Carnforth: Parthenon Publishing.

____________, Nick Hanley and Mike D. Wilson. (2001). *Measuring and Modelling Sustainable Development*. New York: The Parthenon Publishing Group.

National Economic and Social Development Board. (1985). *Regional Cities Development Program*. Bangkok: NESDB.

____________. (1994). *National Socio-Economic Development Plan*. Bangkok: NESDB.

Osborn, Milton. (1997). *Southeast Asia: An Illustrated Introductory History*. Sydney: Allen and Unwing.

Rand-McNally & Co. (1996). *Cosmopolitan World Atlas*. Chicago: Rand-McNally.

Rigg, Jonathan. (1991). *Southeast Asia: A Region in Transition*. London: Unwin Hyman.

Rivkin, Malcolm D. (1976). *Land Use and the Intermediate-Size City in Developing Countries*. New York: Praeger.

Romanos, Michael. (1998). *Sustainable Economic Growth in the APEC Region. Annual Report, Year 1997*. Cincinnati: University of Cincinnati School of Planning.

____________. (1999). *Sustainable Economic Growth in the APEC Region. Annual Report, Year 1998*. Cincinnati: University of Cincinnati School of Planning.

____________. (2000). Need for Structural Reform in Intermediate Size Cities: Environmental Management in Chiang Mai, Thailand. In Carla Chifos and Ruth Yabes (Eds.), *Southeast Asian Urban Environments: Structured and Spontaneous* (pp. 275-326). Tempe: Arizona State University, Program for Southeast Asian Studies.

Rondinelli, Denis A. (1987). *National Objectives and Strategies for Urban Development in Asia*. Manila: Asian Development Bank, Regional Seminar on Major National Urban Policy Issues. February 3-7.

Van den Bergh, Jeroen C.J.M., and Jan van der Straaten. (1994). *Toward Sustainable Development: Concepts, Methods, and Policy*. Washington, D.C.: Island Press.

United Nations ESCAP (1998). www.un.org/esa/population/pubarchive/escap98/ca8/fca8.htm.

Velasquez, C., Fabio E. (1991). Local Governments in Intermediate Sized Cities in Colombia: Municipal Government for Whom? *Environment and Urbanization, Vol. 3, No 1, April*, 109-120.

World Bank. (1998). *World Development Report 1998: World Development Indicators*. Washington, DC: World Bank.

The World Gazetteer (2002). www.world-gazetteer.com/fr/fr-th.htm.

PART I

CULTURAL DIMENSIONS AS THE UNDERLYING CONTEXT FOR DEVELOPMENT

CHAPTER 2

LOU MATZ

BUDDHISM AND SUSTAINABLE DEVELOPMENT

Abstract The attempt to study and evaluate development in Thailand within the framework of sustainable development raises the philosophical question of the extent to which Buddhism in Thailand might be amenable to the adoption of a sustainable development approach. I contend that a Buddhist approach to development affirms the core elements of sustainable development and so ought to be receptive to its implementation. In the course of my chapter, I briefly explain the basic ideas of Theravada Buddhismthe form of Buddhism in Thailand; isolate what I believe are the core objectives of sustainable development; clarify what prominent Thai monks and scholars believe are the limitations of traditional western liberalism; and, finally, suggest that sustainable development and Buddhism emphasize different dimensions of sustainability and hence can learn much from each other. Proponents of sustainable development have focused primarily on the realm of policy making and formulation of specific indicators to measure scientifically the sustainability of policies and practices, whereas Buddhists in Thailand have focused far more on attaining moral and spiritual awareness and have neglected the importance of public policy making. Both dimensions are essential to truly free and sustainable societies.

1. INTRODUCTION

The attempt to study and evaluate development in Thailand within the framework of sustainable development raises the immediate philosophical question of the extent to which a Buddhist society might be amenable to the adoption of a sustainable development approach. Non-Buddhist planners in Thailand face the prospect that there is a significant difference between their "western" values and the Buddhist values of the Thai people, a difference which purportedly renders problematic any attempt to implement policy in Thailand. This difference between philosophical orientations might be invoked to dismiss the interpretive methods and normative goals of sustainable development since these cannot possibly be valid or beneficial for a society with fundamentally different ethical and religious values. Indeed, the appeal to Buddhist values is usually directed as a criticism of the values of the west, e.g., regarding its philosophical assumptions about the true nature of human well-being and of a just and desirable society.

It turns out, however, that many of the salient criticisms made by Thai Buddhist scholars and activists of traditional western ideals of development are also advanced by proponents of sustainable development. Both are reacting to the social and environmental problems caused by "traditional" western economic and political thought, which is to be distinguished from the paradigm of sustainable development that has western roots but has resulted from a global process. Moreover, proponents

M. Romanos and C. Auffrey (eds.), Managing Intermediate Size Cities, 23—37.

of sustainable development share many of the goals of a Buddhist approach to development, making it practically compatible with basic Buddhist values.

In the first section of this chapter, I explain the fundamental ideas of Theravada Buddhism which is the form of Buddhism in Thailand. In the second section, I characterize what I believe are the essential aims and methods of the notion of sustainable development. In the next section, I describe what the most prominent Thai monks, intellectuals, and activists understand to be "western" values and what they believe the limitations of these values are with respect to sustainability. It becomes apparent that their criticisms are directed at the very same traditional western values that motivate the recent notion of sustainable development.

In the final section, I describe the philosophical relationship between Buddhism and sustainable development and then contend that both western planners and Buddhists have much to learn from each other.[1] Western planners have focused primarily on the policy level and its implementation, including the creation of specific indicators to measure the sustainability of policies and practices. Buddhists in Thailand, however, have historically focused much less on the political realm of policy-making and devising measurement tools of sustainability; rather, they have emphasized moral and spiritual awareness and the meditative methods to attain these ends. Since the defenders of western sustainability recognize the necessity of personal moral or spiritual transformation, they could benefit from a deeper understanding of the Buddhist approach to moral and spiritual change and attempt to find ways to inculcate these ends in the populace through the various institutions of social life. Likewise, planners in Thailand could profit from the political and technical methods that sustainable development planners have created in order to formulate policies and assess their efficacy.

2. THERAVADA BUDDHISM

The following cursory treatment of the central tenets of Buddhism is necessary to understand its ideals of personal and social existence, which are the bases for a genuinely desirable and sustainable society. At least as early as 1250, when the Thai people broke free from the Khmer Empire at Angkor and set up their first independent state at Sukhothai, they have formally embraced the original, orthodox form of Buddhism known as Theravada ('the path of the elders'), which originated in northern India c. 500 BCE and which spread in the sixth century CE via Ceylon (now Sri Lanka) to what is now Thailand, Burma, Cambodia, and Laos. King Rama Kamhaeng's famous 1292 inscription at Sukhothai indicated the strong Buddhist influence on the king, nobility, and people of rank and the existence of a monkhood (Sangha) divided, according to the Sinhalese custom, into forest and town dwellers. The other prominent form of Buddhism, Mahayana, has been followed in the northern regions of Asia, e.g., in China, Tibet, Mongolia, and Japan. Vajrayana is yet another form of Buddhism that is practiced in Tibet and popularized by the current Dalai Lama. While these forms of Buddhism differ in some beliefs, practices, and observances, their core is essentially the same.

The essence of the Buddhist religion is to cultivate practices, and not forms of worship, which respond to the problem of suffering (Dukkha).[2] Its aim is to help humanity achieve peace and happiness in this world (Nibbana, or 'Nirvana' in Sanskrit). Buddhism emphasizes the human capacity and responsibility to liberate itself from the hindrances or defilements --e.g., greed, lust, hatred, ignorance-- which cause suffering to self and others through the practices of morality (Sila), mindfulness or meditation (Samadhi), and wisdom (Panna). It is a non-theistic religion that rejects the worship of celestial beings and the necessity of performing external rituals and ceremonies, although in Buddhist countries there are ceremonies on religious occasions that express respect and gratitude for the Buddha, Dhamma, and Sangha (the Tipitaka, known as the 'three baskets'). Unlike monotheistic religions such as Judaism, Christianity, and Islam, Buddhism is not fundamentally interested in answering metaphysical questions, such as whether the universe was created or whether there is a soul which is independent of the body and which survives physical death; rather, it is a practical method to overcome the sources of suffering and to achieve happiness in this life.

Buddhism is based on the Four Noble Truths that Siddhattha, who became a Buddha (an 'awakened' or 'enlightened' one), expounded in his first sermon. The first truth is that to live is to experience Dukkha, i.e., suffering or dissatisfaction. All things, whether bad or good, cause some kind of suffering, and to recognize this is to detach oneself from them. Buddhists do not deny that there are satisfying experiences and happiness--the long-standing view in the west that Buddhism is a pessimistic religion is simply a caricature--but that suffering is inevitable and that even happy states, whether physical or mental, are impermanent (annica). There is, however, a deeper philosophical meaning to the first noble truth, which gives rise to the central Buddhist doctrine of no-self or no-soul (Annata). The notion of a real, abiding 'self' or 'I' is an imaginary and false belief. The 'self' is really only a combination of various physical and mental aggregates or forces (matter, sensations, perceptions, volitions, and consciousness), and there is nothing permanent and unchanging which underlies the causal interactions among these forces. It is thus an illusion to believe that one has a substantial and permanent 'self.' This Buddhist doctrine of no-soul or no-self has profound personal and social consequences, for when one recognizes the truth of the doctrine of Annata, one realizes that one's own desires and goals are illusory and that to renounce these attachments and defilements is to overcome suffering and achieve peace through a connection to others and the natural world.

The Second Noble Truth is the understanding of the origin of suffering. The principal source of suffering is thirst for, and attachment to, sense-pleasures, wealth, power, and even opinions and beliefs. This thirst to possess and to become more and to accumulate more is natural, yet it is the basic source of suffering. Without practicing certain ethical and meditative techniques to overcome this thirst or willful action, known as 'Kamma' or 'Karma' in Sanskrit, humans will recreate attachments that entrap themselves in the cycle of continuity, namely, of desiring, acting, suffering the effects of action, and desiring anew. This cycle is known as Samsara. It is thus willful action or Kamma that sustains Samsara. The unwholesome pursuit and attachment to sense-pleasures, wealth, power, and beliefs will inexorably bring suffering or dissatisfaction.

The Third Noble Truth is the Absolute Truth or Reality of Buddhism. It is the complete liberation from the continuity of suffering, which is called Nibbana. It is an awareness of the impermanent or transitory nature of the world and of the illusory belief in an abiding, eternal self. Rahula says, "the realization of this Truth, i.e., to see things as they are without illusion or ignorance, is the extinction of craving 'thirst,' and the cessation of dukkha, which is Nirvana"(Rahula, 1959, 40). Since this ultimate state of being is ineffable, it is usually described in negative terms, i.e., in terms of the cessation or liberation from continuity of Samsara. Nonetheless, when one gains the wisdom (Panna) about the reality of things, one lives fully in the present in a peaceful, joyful, and compassionate way.

The Fourth Noble Truth is the practical method which leads to the cessation of suffering, known as the Noble Eightfold Path or the Middle Way, since it charts a way of living between the extremes of great material comfort and asceticism. This path involves right understanding, thought, speech, action, livelihood, effort, mindfulness, and concentration. They are the means to cultivate and to perfect the three fundamental virtues of Buddhism: ethical conduct (Sila), mental discipline (Samadhi), and wisdom (Panna).

The first virtue of ethical conduct is manifested by right speech, action, and livelihood. Right speech fundamentally involves speaking the truth, but it also includes not engaging in backbiting, malicious, abusive, and frivolous speech. The basis of right action is to cultivate compassion (Karuna) for all sentient beings, which leads to loving-kindness (Metta) towards oneself and others. Right action includes prohibitions against intentionally destroying life, stealing, engaging in illegitimate sexual conduct, lying, and using intoxicants that destroy mindfulness;. Observance of these five precepts is necessary to be a lay Buddhist. Right livelihood means that one's profession should be honorable and not bring harm to others. While ethical conduct is an intrinsic good, it also prepares the way for the higher spiritual states of mental discipline and wisdom. The second virtue of Buddhism, mental discipline, involves goals such as right mindfulness and concentration that are principally achieved through various meditative techniques, like Vipassana meditation. The purpose of mental discipline is to control one's thoughts and desires and to create an inner state of calm or equanimity (Upekkha), which is similar to the ancient Greek stoic ideal of ataraxia or unperturbedness of soul. The final virtue of Buddhism, wisdom, involves right thought and understanding. Wisdom is not simply intellectual but involves a deep penetration into the way things are. Sulak Sivaraksa says, "intellectual growth here does not mean that one becomes clever or intelligent. Rather this is a sense of wisdom (panna or prajna) in that one sees things as they really are...It is not difficult intellectually to argue that exploitation of the poor is evil but through spiritual and intellectual development in this way, one might be aware once and for all how not to exploit others in a multitude of devious ways"(Sivaraksa, 1988, p. 61). Wisdom is manifested practically through selflessness and detachment from the desires and defilements which cause suffering to oneself and to others. It leads to that compassion toward all sentient beings which is the foundation of Buddhist ethics.

3. WESTERN CONCEPT OF SUSTAINABLE DEVELOPMENT

The 1987 report by the United Nations World Commission on Environment and Development defines sustainable development as development that "meets the needs of the present without compromising the ability of future generations to meet their own needs." Its overarching purpose is to find ways to live that promote the total well-being of individuals and communities and to sustain this well-being for future generations, especially by using renewable resources at rates for renewal. The notion of sustainable development is ultimately a philosophical response to the social and ecological problems that have been created by traditional western political, economic, and social, and scientific beliefs. It involves a revolution of thinking regarding humanity's relationship to the natural world and social world. As Maser claims, "sustainable development therefore questions the very purpose of society and our participation with our home planet, and demands social/environmental justice, which challenges the very heart of our perceived relationship with Nature and one another, present and future"(Maser, 1997, 95). Because the concept of sustainable development covers virtually all aspects of a society - political, economic, social, psychological - it has come to mean different things to different people. Nonetheless, there are at least five elements which comprise the core notion of sustainable development.

3.1. Ecological Awareness and Action

Perhaps the most fundamental principle of sustainable development is ecological. Ecological understanding and action imply a reconception of humanity's relationship to the natural world, especially a rejection of anthropocentrism, namely, that human interests are paramount over the interests of other sentient beings and the natural environment as a whole. Contemporary biological science demonstrates the inexorable fact that non-living and living entities exist in an interconnected and interdependent way. Humans exist in delicate ecosystems whose alterations must be calculated carefully so that the functionality of the system is not compromised. For example, harvesting coniferous forests in the Pacific Northwest leaves the spotted owl without a home. The owl preys on flying squirrels, which eat truffles that grow underground. The fecal matter of the squirrels--the fungal spores--are washed into the soil by rain and inoculate the roots of the trees. In turn, the existence of the fungal spores depends on the sugars produced by the roots of the trees, and when trees die, their decomposition is essential to the health of the forest. Thus, the fungi, trees, and mammals of the forest are dependent on one another, the result of years of evolution, and the alteration of any part can gravely affect the functionality of the whole since the parts are complementary (Maser, 1997). The traditional belief that the natural realm is distinct from the human social realm and can be manipulated by humans for their own benefit without deleterious effects on the natural world and their own lives is simply fallacious and harmful. Humans must reconceive their relationship to the natural world. It has intrinsic value and is not simply a means to be commodified and transformed for human ends.

Moreover, ecological awareness includes: the recognition of the finite capacity of natural systems to produce renewable resources, such as food crops, forestry products, and water supplies; the limited capacity of natural systems to absorb emissions and pollutants without side effects such as ozone layer depletion and accumulation of greenhouse gases; and, irreplaceable areas of natural beauty (Roseland, 1998, 5).

3.2. Intergenerational Equity

One of the implications of ecological awareness is the consideration of the effects of past and present human conduct for future generations. A central tenet of sustainable development is that the living has a moral obligation to bequeath a livable world to the next generation. Although future generations do not yet exist, they can be harmed by our actions in the present. Those living today, especially in the so-called 'developed' countries that consume so much of the world's resources, owe it to future generations to minimize the consumption of natural resources and to increase the efficiency of resource and energy use. Ecological processes have primacy over human desires when the desires destroy the system for future generations.

3.3. Social Justice

The nature of the laws, customs, and social attitudes that mediate the relationships among individuals in a society are essential to sustainable development. Social justice concerns the moral character of these relationships in all of their dimensions. It includes meeting the basic human needs of all individuals in society, such as food, housing, education, public safety, health care, and employment. It involves creating policies that benefit the community as a whole and not a segment of it, a legal system that is just and impartial, and distributing both the benefits and costs in a society with fairness. Social justice is ultimately the moral commitment to transcend one's selfish interests to promote the well-being of all individuals in one's society and globally, especially those at risk such as the poor, the malnourished, the illiterate, and the politically marginalized.

3.4. Civic Participation

A sustainable society requires active citizen participation. Citizenship educates individuals beyond their own self-interest so that they understand the institutions and problems of political and social existence and feel a connection to the social whole. It binds people together through what the ancient Greeks called 'political friendship', creating a sense of community through shared commitments. Civic participation fosters virtues such as prudence and self-reliance that are essential for self-government. It also creates a check on those in political power, holding them accountable for their actions.

3.5. Personal Moral Excellence

Sustainable development requires a change of spirituality or personal ideals from the selfish, immoderate, materialistic, and consumeristic tendencies that so frequently characterize the West, to living up to ideals of virtue such as compassion, generosity, self-control over physical pleasures, emotions (such as anger and prejudice), material minimalism, and self-reliance. While these virtues cannot be legislated, they ought to be promoted through the family, schools, and the various social institutions of civil society. As Maser observes, "the more we study the problems of our time, the more apparent it becomes that our social/environmental imbalance is a spiritual crisis. This crisis is brought about by our clinging to the expansionist economic world view and its associated value system, which has spawned profoundly lopsided and unhealthy ideologies, technologies, institutions, and lifestyles" (Maser, 1997, 43).

4. A BUDDHIST CRITIQUE OF TRADITIONAL WESTERN LIBERALISM

Buddhadasa Bhikkhu, Sulak Sivaraksa, and Prayudh Payutto have written extensively on the need to recover Buddhist values and practices amid the rapid modernization and westernization of Thailand. Buddhadasa Bhikkhu is regarded as the most important Theravada thinker in Thailand's history. He established the forest monastery Suan Mokkh, the 'Garden of Liberation,' one month before the inception of democracy in Thailand in 1932. Phra Prayudh Payutto is perhaps the most important Thai monk scholar alive today, and Sulak Sivaraksa is a prominent Thai Buddhist social critic and activist. While they recognize the virtues of western democratic liberalism, such as the legal protection of individuals against injustice and the benefits of science and technology, they contend that political liberalism is incomplete and thereby threatens the spiritual happiness of humanity in general and the Thai people in particular. They have articulated a view of development that is based on the ancient teaching of the Dhamma and warn that any attempts to reform society based on traditional political and ethical values are doomed to be incomplete since they do not get at the root of the problems.

Individual liberty is the deepest-held value in western culture and is the basis of political liberalism. Politically, western liberalism is grounded on the ideas of natural law and natural rights, which receive their most influential defense by the seventeenth century English philosopher John Locke. While different traditions have historically contributed to the emphasis on individualism in America,[3] it is arguably John Locke's assumptions about the priority of the individual and the ends of political life, which he describes in his *The Second Treatise of Government*, that best illustrate the value of individual liberty. For Locke, there is a natural moral law which is to govern all human beings and which must be secured by government. This law is that "no one ought to harm another in his life, health, liberty or possessions" (Locke, 1690, 9). This natural law, which is not humanly created but which is discoverable by reason and consonant with it, establishes limits to social and political control over individuals and implies various natural rights, which allow the free use of things independently of the will of others. For Locke, the most important right is to property, which includes the right over one's own body, mind, desires, and labor. The essential function of government is to secure these property rights and to let individuals alone to pursue their own desires and conceptions of happiness as long as they do not harm the liberty of others. It is the individual who has the authority or 'right' to judge what the best kind of life is, within, of course, the limits of natural law.

One of the basic manifestations of this Lockean view of "possessive individualism"(Macpherson, 1962) in the West is seen in its economic practices and sensibilities, which are capitalistic. According to traditional capitalistic rationality, economic life is motivated fundamentally by self-interest, which is its raison d'etre. It then becomes rational or 'natural' to maximize profit for oneself and to accumulate as much money and material possessions as possible, regardless of the effects on others and the environment. The purpose of a business is to make a profit for oneself. Laborers are often simply means for the self-interested ends of employers.[4] And, the natural world is conceived as a resource to be exploited for human ends and whose value arises solely due to the transformative effects of human labor. Since the accumulation of wealth depends on the production of goods or services, it is necessary to promote these, whether they are really needed or not, and to promote them in excess. Through clever marketing and mass communication, consumers are encouraged to satisfy their unbridled desires, whether they are good for one or not, and these desires easily become limitless since new needs are constantly being created. In fact, advertisers promote the notion that one's individuality is truly represented by the kinds of things one has. Moreover, luxury goods are also created, not simply to make life more convenient but to capitalize on people's desire to demonstrate their wealth and social status. From a societal standpoint, economic development typically means the production of more and new goods, the expansion of industry and building, increased exports, and the accumulation of personal wealth and possessions. One important indicator of economic health is the quantity of output of goods, determined by the Gross National Product (GNP). Economic development is thus typically measured in terms of the quantity of material goods and high levels of consumption.[5]

According to Buddhadasa, Sulak Sivaraksa, and Prayudh Payutto, the inherent flaw with western political and economic principles is that they are not capable of addressing the real cause of personal and social suffering, which is the selfish clinging to material objects, desires, hatreds, and beliefs. Liberalism protects (through rights) and even promotes (through its economic and social values) self-interested gain regardless of its effects on oneself and on the community. Its conception of liberty is the Lockean view of 'conditional liberty,' i.e., the complete license to desire and to do whatever one likes as long as this does not interfere with the liberty of others, regardless of whether it is personally or socially ennobling (Payutto, 1994). In the West, the political and economic values of liberalism give individuals a right to be ruled by their unbridled desires and to accumulate wealth and to consume without limit. Indeed, capitalism requires the promotion of materialism, immoderation, consumerism, commercialism, and self-interested gain. The possession of great material wealth and the exhibition of unlimited consumption are also typically indications of true success and happiness in America. In this regard, Buddhadasa Bhikkhu's example about the rich person is illustrative of the difference between Buddhist and western values. He says,

> A person of great material wealth (Sanskrit: sresthi) in the Buddhist tradition differs greatly from the capitalist (Thai: nai thun) of today. Outside of Buddhism, sresthi has the same meaning as nai thun--one who keeps accumulating material wealth far beyond what he actually needs. In the Buddhist tradition, however, the status of a sresthi was measured by the number of rong than that person had. A rong than was an almshouse, a communal place where those in need could find what they lacked materially. The more rong than one had, the wealthier one was considered to be...Sresthi in the non-Buddhist sense, however, are strictly nai thun. They accumulate endless wealth and reinvest all the profits for themselves, while oppressing their workers. A sresthi in the Buddhist sense, on the other hand, employs workers in a cooperative effort for the welfare of the entire community. (Buddhadasa Bhikkhu 1993, 57)

According to Buddhists, liberalism's conception of freedom cannot ultimately address the problem of selfishness, i.e., of attachments and defilements. Freedom becomes an end in itself rather than a means to develop oneself in noble ways and to improve the welfare of the community. Because of the legal protection of rights and the assumption that happiness or the good life can only be judged by the individual and should not be criticized, liberalism does not have the resources to challenge individuals to master their attachments and defilements which create so much individual and social suffering. Rights allow individuals to pursue their desire independent of governmental and social intrusion, but rights cannot direct individuals to achieve higher states of personal excellence. If individuals in America are challenged to be less materialistic and greedy, to be more moderate, and to master their desires and selfishness, the natural reply is to claim that it is one's right to live as one pleases. These reasons lie behind Buddhadasa's and Payutto's frequent refrain that "the liberal philosophy or ideology of freedom does not have the power to resist the strength of human defilements" (Buddhadasa Bhikkhu, 1993, 81).

For Buddhists, the highest conception of freedom is freedom from one's attachments and defilements. It is not ultimately political or economic, but spiritual, the mastery of one's inner life. Without mindfulness and wisdom, one is subject to these oppressive forces and is unfree, causing suffering to self and to others. With wisdom, the pleasures of the senses and of possessions are enjoyed moderately and are shared with others, and people seek social harmony and promote well-being in others rather than be centered on one's own self-interest. Buddhadasa says,

> Anyone infatuated with the word 'freedom' or 'free democracy' should remember that upholding the personal freedom of individuals who are ruled by kilesa goes against the fundamental meaning of politics which is concerned with the good of the whole. A political system that does not focus on society as a whole is an immoral system. Freedom in the religious or dhammic sense is important to keep in mind here because it means in the most fundamental sense to be free from defilements (kilesa). A free democracy succeeds as a political system only to the extent that the people in it are freed from kilesa... Unfortunately, you will never see freedom referred to in this sense in any political treatise, nor will you hear it mentioned in contemporary political discussions. Such freedom is only discussed in terms of religion. Yet, it is precisely this kind of freedom which is at the basis of a dhammic society and the attainment of Nibbana. (Buddhadasa Bhikkhu, 1993, 85-85)[6]

Unlike western liberalism which is based on the notion of individual rights that protect self-interest, the basis of Buddhist ideals originates in the Dhamma, i.e., in the understanding of the four noble truths, in knowing "what is what"(Buddhadasa Bhikku, 1956). Since all things are impermanent, create suffering, and have no selfhood, it is irrational for human beings to cling to the four attachments, i.e., to sense pleasures and material objects, opinions, rites and rituals, and to the idea of a permanent self. The pursuit of pleasures, material possessions, wealth, social status, and political power is thus a result of ignorance since these attachments cause suffering and are an obstacle to the highest spiritual states of morality, mental discipline, and wisdom. If individuals can free themselves from the attachments that define them as separate from others, then they can gain the intuitive understanding that all lives are inter-related and attempt to live rightly according to nature.The highest law of nature (dhammajati) dictates that one takes only what one needs and that one promotes the well-being of all sentient creatures (Buddhadasa Bhikkhu, 1993, 63). The illusion of a permanent self leads individuals to set themselves apart from others and to cling to the attachments which define themselves but which destroy the natural state of mutual dependence, harmony, and loving-kindness. As Buddhadasa notes, "Modern society lacks compassion, because people have become selfish and are attached to the idea of a self and what belongs to it"(Buddhadasa, 1993, 54). The true understanding of nature (Samma-ditthi) leads to the virtue of self-renunciation, as opposed to the western virtue of self-aggrandizement. The accompanying virtues or ideals of such self-renunciation, the virtues of sila-dhamma, are: self-restraint and self-discipline, simplicity, humility, generosity, patience, calmness, and loving-kindness. As Buddhadasa Bhikkhu says,

> solving social problems is dependent on living in a socially moral way: acting in the best interests of the entire community by living according to Nature's Laws; avoiding the consumption of goods beyond our simple needs; sharing all that is not essential for us to have with others, even if we consider ourselves poor; giving generously of our wealth if we are well-to-do. This is the way we will solve our social problems. (Buddhadasa Bhikkhu, 1993, 72)

5. BUDDHISM AND SUSTAINABLE DEVELOPMENT

Given my characterization of the main tenets of sustainable development, it is apparent that there are deep affinities between it and a Buddhist approach to development. With respect to the aims of development, Buddhism has always affirmed the goals of personal moral excellence and social justice since purifying oneself of attachments and defilements, fostering compassion, and acting toward the good of the community as a whole are the moral principles that are the foundation for enlightenment. Moreover, Buddhism naturally leads to promoting ecological awareness since one of its insights, based on careful and patient observation of thousands of years, is that that all entities are interconnected and interrelated. To set humans apart from nature and from one another is an illusion, and one of the consequences of enlightenment is to recognize that there is no ontologically real self that can be isolated from the whole. Similarly, Buddhists have always respected all living creatures, especially those that are sentient or capable of experiencing suffering and happiness. Hence, anthropocentrism is essentially foreign to Buddhism. Finally, given their belief that there is no real, abiding self and that humans should live lightly on the earth, Buddhism leads to a respect for the welfare of future generations.

It should also be evident that Buddhists and proponents of sustainable development are responding to similar problems that have resulted from the limitations of traditional western values. For example, the western conception of natural rights, especially the right to property, has legally protected the exploitation of natural resources and the commodification of the natural world as long as one's actions do not violate the legal rights of others. Property owners, however, should consider themselves "custodial trustees" over land and resources and not as sovereign lords who have the authority to degrade the productive capacity of the land for future generations regardless of the amount of immediate personal gain (Maser, 1997, 61). Moreover, capitalism fosters the philosophical view that happiness is found in the pursuit of personal wealth that increases the possibilities of material gratification. But empirical experience attests that this conception of happiness is one of the sources of personal discontentment and social ills.

Nonetheless, despite shared agreement about the goals of a sustainable society and the deficiencies of traditional western values, Buddhism challenges the emphasis of sustainable development. The primary impetus for the theory of sustainable development has been the growing realization that the present and future welfare of the earth's inhabitants are jeopardized by the effects of industrialization, capitalism, and materialism, such as global warming, ozone-layer depletion, and the inefficient use of energy and resources. The focus of sustainable development remedies for these problems has been on policy-making for local, state, and national government. Indeed, the strength of sustainable development is its technical approach, namely, its application of current scientific understanding to identify problems of development and to recommend solutions and to operationalize specific indicators that can assess the actual sustainability of implemented policies and practices.

Buddhist critics of western development recognize the value of specific planning methods and that Buddhism has traditionally not emphasized the importance of social and political policy in effecting change. As Sivaraksa notes, "The Buddhist contribution to the modern world, especially through meditation, has been much acknowledged by leading Christians like Thomas Merton, who really cared for a just world order. Buddhists, however, are not at the forefront in planning desirable societies in the near future, nor is any Buddhist community a runner for the ideal example"(Sivaraksa, 1988, 62). Nevertheless, Buddhists believe that the lasting foundation of sustainable communities is to create a social and political society that educates and reinforces the cultivation of the Buddhist moral virtues. For Buddhists, the most important change in any social plan for sustainable development must begin within individuals' self-awareness. *From a Buddhist perspective, the true method toward sustainable development begins not with institutions, laws, and political policies but within the individual who must overcome attachments and defilements through an understanding of the four noble truths and the application of the Eightfold path.* Before a society can transform its practices, individuals must first be committed to overcoming their attachments to sense pleasures, materialism, greed, hatred, social and political power; recognize their inter-dependence and their place in the natural world; and, apply the moral principles of Buddhism to their social and political problems. The reduction of materialism and economic competition and the promotion of self-restraint, self-reliance, cooperation, altruism, and overall mindfulness of self and society begin, as it were, at home, within each of us. And those in positions of leadership must especially exhibit mental discipline and wisdom and be aware of the four noble truths. Since the time of great Buddhist emperor Asoka who ruled in India during the third century BCE, the right or authority to rule is to be based on one's righteousness, which is known as the doctrine of Dhammaraja. Theories of sustainable development that do not aim fundamentally at the self-awareness of individuals will not ultimately get at the root of problems such as ecological destruction, pollution, violence, militarism, unemployment, economic exploitation, fiscal irresponsibility, discrimination, social alienation, and general spiritual unrest. Sivaraksa says,

> From the Buddhist point of view, development must aim at the reduction of craving, the avoidance of violence, and the development of the spirit rather than of material things...I agree with Schumacher that small is beautiful in the Buddhist concept of development, but what he did not stress is that cultivation must first come from within...having developed oneself toward happiness and tranquility rather than toward worldly success and material progress, then a Buddhist is in a position to develop his community, starting with his family and his village. He must first be awake before he can awake others.
> (Sivaraksa ,1986, 71, 74-75 & 76).

How, then, does the sustainable development planner assimilate this Buddhist focus on personal moral virtue and enlightenment into practical policies? The difficulty is that Buddhist ideals of virtue, such as self-restraint, material minimalism, altruism, and tranquility, cannot directly be legislated. This explains the uniform refrain among Thai Buddhist critics that the crisis of development is spiritual. And it is understandable why the focus of sustainable development is almost exclusively on political and legal policy and not on individual self-transformation given the distinct separation of politics and religion in western institutions as well as the practical problems of promoting virtuous conduct in citizens. Nonetheless, I believe that these virtues can be realized in the same ways that any society attempts to shape the moral virtues of its citizens--through families, formal educational systems, the social associations of civil society, religious and secular, and through laws and policies that reward virtuous conduct. And a society of virtuous individuals would hold its leaders accountable, demanding that economic practices and political policies are environmentally sustainable and socially just.

Conversely, I believe that planners in Thailand ought to embrace the adoption of sustainable development since it is consistent with Buddhist values. The scientific spirit of Buddhism is perfectly amenable to the scientific justification of sustainable development and the empirical methods that are used to diagnose problems and to measure policy outcomes. The same cannot be said about traditional religious values in the West, particularly some forms of Judeo-Christian-Islamic religious belief, that have been resistant to a scientific understanding of the natural world. And while a Buddhist view of development ultimately has different metaphysical assumptions than sustainable development, e.g., regarding the illusory nature of the human self and Nibbana as the highest state of existence, I do not believe that these theoretical differences are a hindrance to the practical adoption of sustainable development since their goals are essentially the same. Consider, by analogy, the secular humanist and the Catholic liberation theologian who have different metaphysical beliefs but who can be committed to the same practical end of alleviating human suffering.

Ultimately, I believe that planners in Thailand have a unique opportunity to incorporate the aims and methods of sustainable development while attempting to recover their rich Buddhist tradition, a recovery so often urged by Thailand's most prominent Buddhist voices. Such a recovery indeed faces practical obstacles, such as the tendency of monks in Thailand to pursue individual enlightenment rather than social change, the close connection between the upper hierarchy of the monkhood and the government that is conservative in dealing with social and environmental problems, the power of economic interests, and the problem of political corruption (Sponsel & Natadecha-Sponsel, 1995, 39-40). Nevertheless, Buddhism could realize the ideals of sustainable development in Thailand more fully than in the West where traditional political, economic, and religious beliefs are more discordant with the philosophy of sustainable development than Buddhism.[7]

NOTES

[1] In their informative study, Sponsel and Natadecha-Sponsel (1995) argue that Buddhism is a neglected but potentially critical resource in averting the developing environmental and social crises in Thailand that are caused by modernization. In the course of their analysis, they briefly discuss the similarity between Buddhism and "the ecology of Western Science," but they do not examine the fuller and more recent notion of sustainable development, which does not merely focus on humans' relationship to nature. I was made aware of their article after my chapter was essentially completed.

[2] Like any major world religion, certain beliefs have become identified with Buddhism that are neither a part of Buddhist scripture nor of Buddhist tradition, such as belief in spirits, which is prevalent among the laity as well as many monks who come from rural villages.

[3] In their important study *Habits of the Heart*, Bellah et al argue that four traditions have promoted the value of individualism in America, namely: the Biblical tradition, civic republicanism, utilitarianism, and expressivism. They contend, "In seventeenth-century England, a radical philosophical defense of individual rights emerged that owed little to either classical or biblical sources...John Locke is the key figure and one enormously influential in America. The essence of the Lockean position is an almost ontological individualism. The individual is prior to society, which comes into existence only through the voluntary contract of individuals trying to maximize their own self-interest. It is from this position that we have derived the tradition of utilitarian individualism. But because one can only know what is useful to one by consulting one's desires and sentiments, this is also ultimately the source of the expressive individualist tradition as well."

[4] Of course, the unjust treatment of labor was a central grievance of the 19th century European socialists.

[5] Sulak Sivaraksa describes these features of capitalism in 'Buddhism and Development: Is Small Beautiful?' in his *A Buddhist Vision for Renewing Society.*

[6] Buddhadasa's assessment about western political philosophy neglects those philosophers of the western tradition who are critical of a liberal view of society, such as Jean-Jacques Rousseau and Edmund Burke. Consider Burke's comment, "Men are qualified for civil liberty in exact proportion to their disposition to put moral chains upon their own appetites...It is ordained in the eternal constitution of things that men of intemperate minds cannot be free. Their passions forge their fetters."

[7] I would like to thank Michael Romanos, Chris Auffrey, Tanya Storch and Michelle 'Sandcat' DiGuilio for their helpful comments on drafts of this article.

REFERENCES

Bellah, Robert, Richard Madsen, William Sullivan, Ann Swidler, and Steven Tipton. (1985). *Habits of the Heart: Individualism and Commitment in American Life.* New York: Harper & Row Publishers.

Buddhadasa Bhikkhu. (1956). *Handbook for Mankind.* Bangkok: Dhammasapa.

_____. 1993. *Dhammic Socialism*, Second Edition. Bangkok: Thai Inter-Religious Commission for Development.

Locke, John. (1980). *Second Treatise of Government.* Indianapolis: Hackett Publishing, Co.

Macpherson, C.B. (1962). *The Political Theory of Possessive Individualism: Hobbes to Locke.* Oxford: Oxford University Press.

Maser, Chris. (1997). *Sustainable Community Development: Principles and Concepts.* Delray Beach: St. Lucie Press.

Payutto, Prayudh. (1987). *Looking to America to Solve Thailand's Problems.* Bangkok: Sathirakoses-Nagapradipa Foundation.

_____. (1994). *Buddhist Solutions for the Twenty-First Century.* Bangkok: Buddhadhamma Foundation.

_____. (1995). *Buddhadhamma: Natural Laws and Values for Life*. Albany: State University of New York Press.

Rahula, Walpola. (1974). *What the Buddha Taught*, Second Edition. New York: Grove Press.

Roseland, Mark. (1998). *Toward Sustainable Communities: Resources for Citizens and Their Government*. Stony Creek, Canada: New Society Publishers.

Sivaraksa, Sulak. (1986). *A Buddhist Vision For Renewing Society*. Bangkok: Tienwan Publishing House, Part., Ltd.

_____. (1988). *A Socially Engaged Buddhism*. Bangkok: Thai Inter-Religious Commission for Development.

Sponsel, Leslie E. and P. Natadecha-Sponsel. (1995). The Role of Buddhism in Creating a More Sustainable Society in Thailand. In J. Riggs (Ed.), *Counting the Costs: Economic Growth and Environmental Change in Thailand,* (pp. 27-46). Singapore: Institute for Southeast Asian Studies.

CHAPTER 3

KRISTA J. KAHLE
THOMAS E. WAGNER

THE EMERGING ROLE OF CITIZENS IN PLANNING AND PUBLIC DECISION-MAKING IN THAILAND

Abstract In this chapter, the authors examine the general concept of citizen participation and then consider the factors that are shaping the emerging role of citizens in planning and public decision-making in Thailand. Although public involvement of citizens is ordinary in the United States and other Western democracies, citizen participation is less common in Thailand, and may take different forms. Citizen involvement in Thailand is looked at in a broad perspective, inclusive of various influences, such as culture, religion, governmental structure, and national history. The authors contend that citizen participation is beginning to occur in Thailand, often through informal means, such as non-governmental organizations (NGOs). NGOs have found a niche in involving Thai citizens mainly on local issues. As the concept of Thai citizen involvement continues to emerge, it will take the unique characteristics of Thai culture into account and will build on existing methods of participation.

1. INTRODUCTION

The active involvement of local citizens in the development of public policy affecting their daily lives is a relatively recent phenomenon, even in Western democracies. One can trace democratic participation in the United States to the founding of the nation, but citizens' involvement in public decision-making has become a systematic practice outside the election process only in the last 40 years (Darke, 2000, 385-386; Thomas, 1995, 1-8; Langton, 1978, 104). As Western forms of citizen participation[1] have become more widely accepted and more fully institutionalized, international agencies that provide funding and technical support to developing countries have insisted the countries involve citizens in the planning process. Even so, the Western ideal of citizen participation is less well established in many other parts of the world.

In this chapter, we examine the emerging role of citizens in planning and public decision-making in Thailand. First, we discuss the general concept of citizen participation in planning in the United States. Second, we consider Thailand's political structure, history, cultural themes, and other factors that may influence citizen involvement in that country. Third, we assess the possible nature and future direction of Thai citizen involvement. Finally, we briefly discuss implications of citizen participation for planning and sustainable development in Thailand. Our findings are based on case study methodology employed during a visiting scholar

M. Romanos and C. Auffrey (eds.), Managing Intermediate Size Cities, 39—55.

appointment at Chulalongkorn University in fall 1997 and on the Joint Study field project in Chiang Mai, Thailand, in summer 1998. The methodological tools supporting the case study are literature review, interviews, and field observations conducted during the two visits.

1.1. History of Citizen Participation

Planners and social activists in the United States have long stressed the importance of participation by citizens in public affairs. For many years, the citizen's role in public decision-making was viewed simply as involvement in the elective process through participation in political parties, or in some parts of the country through town meetings. After World War II, however, and especially since the 1960s, citizens' participation in planning and decision-making has become routine in many mainstream public agencies. Today almost all such agencies encourage citizens to participate in planning and policy making in some form, ranging from simple surveys to advisory boards (Thomas, 1995, 1-8).

The rise of public participation over the last 40 years has several explanations, including institutional developments such as the decline of political party strength at the local level, increased bureaucratic decision-making, a shift in public values related to the administrative decision-making process, and widespread publicity about governmental problems in the mass media (Darke, 2000, 385-386; Thomas, 1995, 1-8; Langton, 1978, 104). According to Desario and Langton, the expanding role of government as a provider of public services also has led to greater public decision-making at all levels of government. Bureaucrats now routinely consult citizens as often as they consult technical experts, especially in matters of health, education, transportation, and environmental planning. In many cases, agencies are legally required to seek the views of the public on matters of public policy (Darke, 2000, 385; Desarioand Langton, 1987, 1-5).

1.2. Citizen/Public Participation Defined

Citizen participation, much like the planning profession, is a broad field. This subject has been discussed widely, and numerous definitions have been offered. One of the earliest definitions used by planners comes from the classic article by Sherry Arnstein (1969), who defines citizen participation as citizen power. "It is the redistribution of power that enables the have-not citizens, presently excluded from the political and economic processes, to be deliberately included in the future" (Arnstein 1969, 216). She makes the point that citizen participation without redistribution of power is a meaningless exercise for the powerless. If the power holders are the only ones who make the decisions, then the status quo is maintained and true citizen participation has not occurred. Although Arnstein's definition may imply certain political and philosophical viewpoints, it has merit for planners. The model helps to illustrate the different levels of citizen participation, their gradations, and some of the serious problems present at lower levels. Later writers have modified Arnstein's ladders or proposed their own model for meaningful citizen involvement in public decision-making (Darke, 2000, 388-391).

Public participation has many names: citizen involvement, community involvement, and participatory decision-making, among others. Traditionally the concept was identified with citizens' political activities including voting, campaigning, and lobbying by special interest groups such as unions and environmental groups. In the 1960s and 1970s, it was broadened to include involvement in administrative processes. Encouraged in part by the emergence of grassroots organizations and by the community development planning model, which focuses on helping people to help themselves improve their conditions, public participation is now defined in terms of "influence" or "being part of the process." Public participation also implies that some power or authority will support the participants' preferences and demands. Thus, citizen participation entails being involved before decisions are made rather than merely being allowed to comment after the fact or making choices from a predetermined list of options. Most proponents of public participation, both within agencies and among community members and stakeholders, recognize the advantages of allowing citizens to participate in the planning and decision-making process. Among these benefits are improvements in the quality of decisions and plans, building public support for and increasing the likely success of a project, and satisfying legal responsibilities and obligations (Anderson and Yaffe, 1998, 12).

As planning and other public agencies have moved away from a "decide-announce-defend" approach and toward more open planning and policy development, various methods for involving citizens have emerged, including public meetings and workshops, referendums, surveys, the review and comment process, advisory groups, and public outreach through media and other information channels. A wide array of techniques for soliciting informed comments has been developed, such as Delbecqu's nominal group technique, the Delphi technique, charettes, and presentation of alternative plans (Darke, 2000, 392-93; Thomas, 1995, 153; Kweit and Kweit, 1987, 31-32; Langton, 1978, 121; Rosenbaum, 1976, 55).

Planners from the United States, as well as Thai planners and administrators trained in the United States and other Western nations, approach the concept of citizen participation from the perspective discussed above. Although many conventional Western techniques and methods are beginning to be employed in Thailand, the country's unique history, culture, and political system ultimately will shape and define meaningful citizen participation.

2. THE HISTORICAL, POLITICAL, AND SOCIAL CONTEXT OF CITIZEN PARTICIPATION IN THAILAND

Thailand is one of the few nations in Southeast Asia that was never colonized by a Western nation. As a result, the country's political system has evolved based on its own culture and circumstances. For centuries, a small elite who surrounded the monarch held political power and was believed to possess superior intellectual ability and moral excellence. Even today, under a constitutional monarchy, the king and the royal family are held in high regard by citizens and have considerable influence in social and public policy. Thus, to understand the emergence of public participation in Thailand, we must consider the political circumstances related to the development of a constitutional monarchy and the role of the monarchy, the military, the government bureaucracy, parliament, the political parties, the "powerful families", the peasantry, and the Thai business community. Additionally, economic growth and development, as well as democratization have affected the role of the public in planning and public decision-making (Neher, 1992, 415-25).

Modern Thailand can be traced to 1855, when the country was opened to Western trade with the signing of the Bowring Treaty (United Nations, 2001). As trade increased and as the country grew through royal marriage and political maneuvering, wealth accumulated. A small middle class, composed primarily of Chinese immigrants, gained influence (Hewison, 1997, 98). In 1932, a small group of junior military officers and middle-class civilians seized control of the government in a coup and established a constitutional monarchy. The new political system was not a democracy; rather, bureaucrats replaced the elite ruling class, and the new constitution was used to justify their control of the government. Ultimately, the power of the military increased. Political turbulence and a series of coups ensued, which lasted into the early 1990s. During that period, only four elected parliaments completed their terms; all of the others were interrupted by coups. As a result, political parties had no chance to gain strength, and democratic ideals of public participation had little opportunity to thrive (Hewison, 1997; Muscat, 1994; Samudavanija and Paribatra, 1993).

Until the 1970s, the military (known as the Royal Thai Army or RTA) was the dominant institution in Thai society and played a significant role in the country's political and economic affairs. The military focused on maintaining order, safeguarding key national institutions, and overseeing the nation's progress. Despite political turbulence and military repression, the tides of political change during these years laid the foundations of modern Thai government. By 1977, according to Samudavanija and Paribatra, a "mixed" system of government had evolved, "with significant liberal characteristics but without Western-style liberal democracy". The government possessed three important attributes of democracy: security, development, and participation (Samudavanija and Paribatra, 1993, 128).

In May of 1992, Thailand experienced a different type of revolution after which, in the opinion of many observers, the middle-class emerged as the "new center of power" (Hewison, 1996, 137-38). Tired of corruption, bureaucratic favouritism, and incompetence, an expanded middle class demanded greater participation and fuller representation in the government. The old political and military institutions attempted to maintain their hold, but after several citizens were killed during anti-government demonstrations, the King intervened and called for a new government. In 1997, he approved a new, carefully crafted constitution granting greater power to citizens.

Throughout the evolution to the constitutional monarchy in Thailand, particularly over the last 55 years, the King has been an important stabilizing force within the country. Because of the respect and reverence in which the Thai people hold him, King Bhumibol Adulyadej uses the moral authority of his royal position to constrain the excesses of political factions and encourage democratic trends.

3. GOVERNMENT STRUCTURE AND PLANNING IN THAILAND

When considering citizen participation in Thailand, one must not overlook the kingdom's governmental structure and the process for public planning. Although these factors are probably no more or no less important than in other developing countries (or, for that matter, in developed nations), Thailand contains a complex network of government agencies responsible for development and government administration. (See Romanos, et. al., Chapter 8, Section 3, in this book.)

Until 1997, the administration at the provincial and local level was an extension of the central government rather than another level of government. However, since the Provincial Administrative Organization Act in 1997, the provincial government is comprised of an elected council that functions as a legislative branch and an executive board. The 1997 Act allows the provincial chief executive, or governor, to be elected by the council (rather than appointed by the Prime Minister). The governors are responsible for coordinating and planning provincial development and for governing districts (*amphur*) and villages (*muban*).

There are two other local government units: municipalities (*tesaban*) and sanitary districts. The municipalities perform a variety of duties and functions including: construction and maintenance of local roads, waterways, and street lighting; solid waste collection and disposal; sewer and drain systems and wastewater treatment; public health and social services; and tax collection and other administrative functions. A locally elected municipal council is responsible for administrative supervision. The Lord Mayor, elected by the council, and four executives appointed by the Mayor, serve as an Executive Board (United Nations, 2001; Chiang Mai Municipality, undated).

Sanitary districts are designated areas outside municipalities with fairly high population densities. Their responsibilities, which are similar to those of municipalities, include sanitation services, firefighting, road construction and maintenance, and public health. The districts are governed by a Sanitary Board composed of nine members (three elected, three appointed, and three ex-officio) and an appointed district officer and deputy district officer (United Nations, 2001).

Each level of local government is independent of the others but all are subjected to a considerable degree of control by the central government through the Department of Local Administration under the Ministry of the Interior and through the funding process. Generally, local governmental responsibility includes statutory functions such as law and order and the provision of sanitary services; discretionary functions such as providing hospitals, parks, and improvement of slum dwellings; and, functions required by parliamentary acts such as city planning. Many local governmental functions are performed by agencies of the central government whose departments extend their operations into the provinces. With some minor exceptions, local tax revenues are also determined by the central government (United Nations, 2001). The result is that local government in Thailand is subjected to strong administrative and fiscal control by the central government. These restrictions severely limit the ability of local government to deliver basic services, especially in fast growing urban areas. Furthermore, it leaves most planning and development under the control of central government ministries and administrative departments.

3.1. Introduction of the Five-Year Plans

Economic development and planning in Thailand are the responsibility of the National Economic and Social Development Board (NESDB), a division of the Office of the Prime Minister. The Board was established in the 1950s to consider and approve a fiscal program for economic investments and to cooperate with foreign specialists in planning the country's economic development. With assistance from the World Bank, the Board created a fiscal plan as a baseline for development in the country. The first national economic and social plan was released in 1961. The eighth national plan was released in 1996, marking the fiftieth anniversary of King Bhumibol Adulyadej's accession to the throne.

Each plan covers a five-year period and is intended to develop the country in ways that improve the quality of life of its citizens. The development concepts in the national plans have changed over the years, usually reflecting recent perspectives on development and responding to issues and concerns that have arisen since the last plan was formulated. The first three five-year national economic and social development plans emphasized economic growth as the primary objective. The fourth plan emphasized resource protection and rehabilitation, and management of natural resources. The fifth and sixth plans moved toward greater concentration of administration and development in the country's various regions. They addressed the unevenness in growth and the vast differences in development between regions and between urban and rural areas. The seventh plan moved toward creating more sustainable development by trying to establish a balance between growth and income distribution, industrialization and environmental quality, and urban and rural development. This plan was the first to give attention to the role of citizens in the planning process. The plan recognized that earlier plans had followed a "top-down" planning process; thus, efforts were made to include broader participation by government agencies, universities, private organizations, representatives of the business sector, and non-governmental organizations (NGOs).

3.2. The Eighth Plan

The eighth plan (1997-2001) aims "at enabling the country to be better prepared for sustainable development in the future and to become a fully developed country by 2020". It emphasizes the importance of maintaining Thai culture and building stronger families. It proposes that every Thai child have access to at least 12 years of basic education and receives the opportunity for higher education.

The eighth plan also sets forth two new strategies for successful implementation. The first calls for greater citizen involvement by emphasizing "the participation of all sectors in the society in formulating, programming and implementing the plan".

The second seeks "reform of the development administration for effective translation of the plan into action": in other words, greater collaboration between the many national and local government offices responsible for implementing programs (Mekvichai, 1996; National Economic and Social Development Board, 1996; *Thai Development Newsletter*, 1995; Thanyakit, 1998).

In the plan summary, it is stated that Thai people were involved in drafting the plan, a deliberate change from the top-down approach followed in the past. The NESDB proclaims this as the beginning of a new era in planning, which emphasizes the efforts of the whole population (National Economic and Social Development Board, 1997, 3).

4. THE SOCIAL CONTEXT OF THAI CITIZEN PARTICIPATION

Citizen involvement is shaped as much by the citizens' culture or social behavior as by the nation's history and political processes. The ideological framework for understanding Thai social behavior includes Therevada Buddhist doctrine and the aspects of Thai culture that flow from that doctrine, such as hierarchal structure, patron-client relationships, and the role of the individual in the community.

4.1. Therevada Buddhism

Therevada Buddhism, or Brahmanism, affects how Thais participate in planning and public decision-making. Buddhism as a religion, philosophy, and lifestyle plays a major role in Thai social development by cutting across social classes to serve a need for individual peace and serenity. Buddhism is the official state religion. The royal family must be Buddhist. Nearly 95 percent of all Thais follow Buddhist traditions (Kirsch, 1975, 189; Neher and Marlay, 1995, 30).

Several features of the Buddhist dogma have influenced Thai behavior with regard to participation in planning and public decision-making. Buddhists believe that the universe is governed by *dharma,* an immutable natural law. Dharma relates to the social world, natural order, cosmic laws, and the nature of the world. It includes the concept that the way things are is the way they should be. Buddhists place considerable emphasis on *nirvana*; this state is achieved when one has broken the cycle of birth and rebirth and has reached a state of nothingness, with no suffering or attachments (Kirsch, 1975, 179). This emphasis plays a role in citizens' social and political behavior. It tends, among other things, to encourage tolerance, patience, waiting, and, as a result, the hope for a better future.

One important doctrine of Buddhism, which underlies the Buddhist worldview, is the notion of a sacred hierarchy. Just as the natural and social order is governed by dharma, Thai social behavior is affected by the belief that everyone and everything has been placed in the universe just as they should be. Each person is a "balance" of religious merits and demerits accumulated through actions that took place in previous lives. This belief implies a view of people as intrinsically "unequal" in moral qualities or worth (Kirsch, 1975, 180). Thus, striving for merit to elevate oneself above one's current status is a constant goal of Thai Buddhists.

The main way of accumulating merit (for men only) is by becoming a monk. Therevada Buddhists believe that all men should serve in this role at some time during their lives. Becoming a monk and joining the *sangha,* or priesthood of monks, preserves the natural order by perpetuating spiritual leadership and creating merit. Buddhist religious hierarchy emphasizes the monk's superior moral quality; therefore, monks have an extremely important status in society. They are believed to be closer than other people to Buddhist values and goals. Lay people view monks as super ordinate, to be treated with respect and deference (Kirsch, 1975, 184). This respect gives them great credibility and potential for social leadership. In addition to being spiritual leaders of the community, monks act as teachers and have taken the initiative in social movements. (For more on Theravada Buddhism, see Matz, Chapter 2, Section 2.)

4.2. Hierarchy

Buddhism takes a hierarchical view of the world that is reflected in Thai society. Merit, or *bun,* is a way of accounting for a person's high status and carries strong moral connotations. Thai Buddhists believe that a person's status in life depends on the amount of *bun* he or she has accumulated in previous incarnations; therefore, the hierarchy of positions is part of the universal natural order. Rank and status are a part of the moral order: hierarchy is maintained through the distribution of awards in the form of individual rank and merit (Tantayanusorn, 1998; Rabibhadana, 1975, 104). For this reason, Thai people often feel uncomfortable in situations where status is not defined. In most Western cultures, ideas of status and rank conflict with the notions of democracy and equality. For Thais and members of other Buddhist cultures, however, this is not the case. Hierarchy is a natural part of life, and is not only accepted, but also expected. "It is notable that notions of hierarchical status pervade Thai society. There are virtually no roles or relationships in Thai society which carry notions of formal equality" (Kirsch, 1975, 189-90). Thais, however, categorize a person on the basis of tasks, not on personality. Every person has the potential to change from a lower to a higher occupation. It is possible to move up or down the scale, according to the amount of merit one accumulates (Hanks, 1975, 198).

4.3. Patron-Client Relationship

The relationship of a superior to an inferior is called a patron-client relationship: the superior gives something from his or her greater resources that will please an inferior, and the latter returns the favor with services that are at the superior's disposal. Clientship is an instrumental friendship in which striving for resources, whether natural or social, plays a vital role (Rabibhadana, 1975, 93). The patron acts as the helper and protector of his or her clients. The relationship is voluntary and may be terminated by either party. It is reciprocal, serving both parties as long as it is convenient. Patrons may have one client or an entourage of clients who are loyal to the patron, not to the group (Hanks, 1975, 199-200).

Although not based directly on Buddhist doctrine, the patron-client relationship that is commonly found in Thai society derives from Buddhist ideology. Buddhism urges those who are higher on the ladder to help those who are lower. In this way, the patron gains merit, which in turn raises him or her even higher in the hierarchy. In Thailand, this relationship historically consisted of two levels: the members of the lower class, who were clients of patrons; and those of the upper class, who were clients of the king, the ultimate patron (Rabibhadana, 1975, 113). The patron-client relationship has been integrated into Thai bureaucratic structures and is therefore important in understanding citizen participation in Thailand.

4.4. The Role of the Individual

Although the traditional patron-client relationship is pervasive in Thai society, a sense of individualism is growing among many members of the middle class. Historically, Thai moral and ethical tradition has been based on a community worldview, not on the perspective of a single person (*Thai Development Newsletter* 1995, 20). The development of individualistic values has been more apparent in the urban areas. Such values have brought both positive and negative changes to Thai society. On the one hand, individuals have taken more self-responsibility and are more interested in self-development. On the other hand, personal advancement may be self-centered and at the expense of the welfare of others such as family or community and the public interest (*Thai Development Newslette,* 1995, 21).

4.5. Other Factors

Other aspects of Thai society, although not unique to Thailand, also influence individual citizens' participation in planning and public decision-making. Foremost is the growth of the middle class over the course of Thailand's economic development during the last two decades. Many members of the middle class have entered government service as professionals who are concerned with applying their knowledge and skills without regard to political interests.

A second factor is the centralized nature of the Thai government. The model of power centralized in Bangkok had its merits in 1892, when established by King Chulalongkorn. Gradually, however, more power has been granted to the provincial and district authorities, although many central government departments still retain real and *de facto* veto power (United Nations, 2001; Parnwell and Arghiros, 1996, 8). The growing tension between central control and local autonomy will ultimately affect the degree of public participation at all levels of government.

The nature of Thailand's political parties is also important in considering citizen participation. In theory, the foundation of a democratic political system is the local political party. However, political parties in Thailand generally do not provide the same opportunity to participate in public decision-making and governmental processes as in the United States and some other Western nations. Thai political parties do not possess much strength, or a party platform. Instead they are centered on individuals and gain power through coalitions with other individual-centered parties. National and regional parties form around individuals or issues, with little interaction between the two levels. There is evidence that parties are attempting to build relationships with citizens and to develop loyal followers so they can exercise greater influence over policy and parliamentary decision-making (Neher, 1992).

Two other factors help to explain the lack of political participation on the part of many local citizens in Thailand. First, there are very few legislative or legal means for people to directly participate in governmental decision-making. In principle, citizens participate in the political system through the election of local administrators and legislators and members of parliament. However, there are no formal means for citizens to participate in day-to-day decision-making or voice opinions about public issues; except, of course, by taking the issue into the streets. Second, the general public has little knowledge and information about governmental structure and operations of local government. Most Thai citizens learn very little about the structure and functioning of government in elementary and secondary schools (United Nations, 2001). However, the 1997 constitutional reforms, which grant greater local autonomy, should continue to bring increased citizen awareness and involvement in local government.

5. THE RISE OF PUBLIC AWARENESS AND ACTIVISM

5.1. Influence of Non-Governmental Organizations (NGOs)

Since the early 1970's, several events took place that led to greater public awareness and activisim in Thailand. First, the student coup d'etat that took place in October 1973 gave rise to numerous non-governmental organizations (NGOs) established by people from various professions.[2] Although most were not registered, these public-interest, non-governmental organizations (PINGOs) demanded and won reforms to bring social and economic development to poorer sections of Thai society. A counter-coup in 1976, led by conservatives concerned about the growth of communism slowed, but did not stop, the growing participation in and influence of NGOs.

Students became more involved in rural issues and NGO activity after the political events early in 1970. During that decade, several international relief organizations entered Thailand to work with the Indo-Chinese refugee camps. Members of these organizations began to realize that the villagers living in parts of rural Thailand were very poor and lived almost as meagerly as the refugees. With the help of students and activists, they started programs for development in the rural areas (Phongpaichit and Baker, 1995, 385).

Two trends emerged within the NGO movement: working with bureaucrats to mitigate the effects of top-down development, and developing ideas of community and strengthening local wisdom. These ideas had a major influence on the activities of NGOs. Several NGOs founded in the 1980s were committed to alternative development and to strengthening local power through increased participation and empowerment. The organizations have been instrumental in raising public consciousness about self-reliance, human and individual rights, and equity. At the same time, monks at several local *wats* (temples) had founded local development projects. Within the *sangha*, certain monks had become active in social development. This movement paralleled the growth of NGOs (Phongpaichit and Baker, 1995, 386-388).

NGOs in Thailand have focused primarily on unequal distribution of income, the effects of development on disadvantaged groups, and the impacts of development on the environment. Networking is an important aspect of successful NGO activity. Although operational boundaries are clearly understood, there are examples of successful collaborations such as the efforts against the construction of new dams planned in the northern and southern parts of the country (Pongsapich and Kataleeradabhan, 1995, 23-26).

The "For Chiang Mai Group" (FCM) is an excellent example of grassroots public involvement. The purpose of this group is to conserve and preserve Chiang Mai's physical characteristics, including historic architecture and scenic areas such as Doi Sutep, the holy mountain that dominates the city's landscape[3]. FCM won a major public policy battle in 1985; in that year, it prevented construction of a cable car to the summit of Doi Sutep, which had been supported by the Tourist Authority of Thailand. The organization argued that the cable car would be disrespectful to the sanctity of the temple and the mountain, would encroach on the reserved watershed, and would be unsightly.

After the cable car affair, FCM focused on preventing high-rise condominiums from blocking views of Doi Suthep and the Ping River, and on preservation of traditional architecture. Although FCM failed to prevent high-rise development along the river, a historic conservation zone now exists around the river and inside the wall of the ancient city. FCM has been quite successful at involving the *sangha*, lay and temple youth groups, and academics from Chiang Mai University (Narumit-Rekagarn, 1998).

5.2. Reformist Buddhism Movement

During the 1960s and 70s, religious consciousness also began to change. The growing complexity of society and politics was reflected in the organization and practice of Buddhism. Social conservatives have used Buddhism to reinforce authoritarian ideology; others have interpreted it in support of dissident views on the state. The old form of Buddhism, accumulating merit and living as a monk for a short period, was no longer adequate for some members of the new, educated urban society. The traditional form of Buddhism denied the possibility of ethical self-improvement or social elevation. It supported traditional social hierarchies and authoritarian regimes, and did not satisfy those who were interested in education and participation (Phongpaichit and Baker, 1995, 376).

As a result of this dissatisfaction, a movement known as Reformist Buddhism began to emerge as early as the 1960s. This movement arose in response to the demand that Buddhist philosophy and practices match the aspirations of the urban, educated, white-collar class. Reformist Buddhism could be used to challenge the social and political organization, because if an individual's stock of merit was affected by actions in this world, then royalty and the elite could not claim an infinite stock of merit. Some aspects of Reformist Buddhism even justified political movements that opposed the state's authority over the individual (Phongpaichit and Baker, 1995, 380). One idea that emerged from the Reformist movement was that a good Buddhist should work for the betterment of society in this world. Thus, during the 1980s, monks themselves began to be involved with political affairs.

5.3. Growth of a Middle-Class

Beginning in the 1970s, the growth of an urban, educated white-collar "middle class" led to some major changes in Thai society. This emerging segment held individualist ideas, formed NGOs, established a somewhat free press and media, supported dissident religious views outside and within the *sangha*, challenged state-led development, and fought to protect natural resources. Citizens began to participate (although indirectly) in planning and public decision-making, primarily through NGOs. Although Thais are free to participate in politics through party membership, as stated above, the parties have no clear ideology and are weak outlets for public involvement (Kerdphol, 1997). In addition, political participation is hampered by the complexity of Thailand's government structure. As a result, the major avenues for citizen participation have been outside the system of electoral politics and traditional social movements.

The role of middle-class citizens in the 1992 political crisis was also a pivotal point in Thailand's journey toward increased public awareness and activism. During the crisis, two military-based political coalitions vied for power, threatening the monarchy. When students and other citizens were killed during street protest, the king called the two factions before him and directed them to resolve the conflict. The king's authority and full support of the country's people brought a quick end to the crisis. Until 1992, the Thai military had attempted to maintain control of the government, the people, and the media. Nearly every prime minister was a current or former general in the military. Although support by the military is still essential in gaining political influence, middle-class citizens have taken a more active and influential role in governmental politics and in planning.

5.4. Other Factors

Another notable factor in the rise of public awareness and activisim was the changing media. Newspapers, and later radio and television, became important sources of information about political issues, including corruption. During the 1992 crisis, the government used sympathetic media sources to disseminate its own version of events. The contrasts between the government-controlled media and the open local and international press caused an outcry for the government to stop controlling the media (Phongpaichit and Baker, 1995, 373).

Finally, the Thai business community began to demonstrate greater involvement in planning and politics in the past few decades. The expansion of the middle class, the rise of capitalism, the increase in the number of educated professionals in government, and a recognition of the link between government policies and the health of the country's economy have made businesses big and small aware of their need to be politically and socially active. This community, especially the Chinese business class, now plays a stronger role in governmental affairs (Neher, 1992, 421).

6. FINDINGS AND IMPLICATIONS FOR SUSTAINABLE DEVELOPMENT AND PLANNING IN THAILAND

Many factors influence participation by citizens in Thailand. Certainly the governmental structure is a major influence. Although there is a new constitution in place, Thailand is still controlled by the central government, with little input from the provincial or municipal levels. In essence, citizens have few means of direct input at the central level, and only minimal ways to influence decisions at the local level. While local citizen involvement and influence may grow under the new constitution, if past patterns hold, progress will be slow. Thailand's political history has been marked by several relatively bloodless coups in which the power changed hands from one political institution to another, but not much power has shifted to the people. Political parties are less strong than in the United States and many western European countries. The country has apparently stabilized politically in the last few years, and there seems to be no threat of another coup in the near future. This period of stability may be an opportunity to increase citizens' involvement in planning and public decision-making. The 1997 constitution grants far more individual and civil rights to citizens than did any previous constitution, and thus may increase citizens' participation.

Slowly, because of outside influences such as the growing number of citizens (especially planners) who have been formally educated in the West, the forms and concepts of citizen participation are growing stronger. The United States and other Western nations have well-defined theories and concepts of citizen participation that are based on their history and culture. These theories and practices, however, may be less effective when used in Thailand because the country's culture and governmental structure differ from those in the West.

As aforementioned, a major factor in Thai citizens' view of participation has been the growth of the middle class. Further, the change in the religious beliefs of this class is influencing not only Thai culture, but also citizens' expectations about their participation in government decision-making. The Buddhist ideology has shifted from the traditional worldview based on a sacred hierarchy to Reformist Buddhism which allows people to challenge notions of merit and hierarchy. Also, as more people are entering professional careers, they are placing less value on the traditional Buddhist rituals. For example, fewer men are entering the *sangha* for a brief period before embarking on a career or marriage.

This does not mean, however, that the role of the monks and the *sangha* will become unimportant in Thai culture. Some monks are becoming involved in public policy issues and development projects. Their influence on the Thai people will continue to shape citizens' involvement in planning and public decision-making. Most Thais are still practicing Buddhists, and Buddhist ideology is an inherent part of Thai culture. The hierarchical nature of the society facilitates the strong position of monks in Thailand.

Besides the middle class and the *sangha*, academics hold high status in Thai society and are influential in shaping public involvement. In Chiang Mai, many academics and some other professionals are former residents of Bangkok who moved to Chiang Mai in search of a better quality of life (Narumit-Rekagarn, 1998). Many of these academics have formed NGOs in Chiang Mai to prevent or mitigate some of the problems found in Bangkok, such as air pollution and solid waste disposal. They also work with existing NGOs to give greater credibility to their activities. For example, the Northern Development Foundation works with professors and students from Chiang Mai University and Kasetsart University (in Bangkok) on data gathering so that the government will view their data as academic (Aquino, 1998; Ayudhaya, 1998; Sodawat, 1998).

Non-governmental organizations have found a niche in citizen participation in Thailand. Through their activities, Thai citizens have been able to influence public policy and planning decisions. The For Chiang Mai group, for example, has played a major role in preservation and conservation issues in Chiang Mai. The Northern Development Foundation has made progress in the rural areas outside Chiang Mai, especially in community forestry issues. Other NGOs in Thailand work with women, in AIDS education and prevention, in environmental education, with hill tribe minorities, and on various other issues, all with some success. These groups also have enjoyed modest success in influencing legislation and policies coming from Bangkok (Aquino, 1998; Ayudhaya, 1998; Narumit-Rekagarn, 1998; Sodawat, 1998). Thus, the NGOs have proved to be a viable alternative form of citizen participation, outside the traditional forms that are used in the United States and Western Europe.

Over time, the role of citizens in Thailand has grown stronger. Thais have a great respect for authority, especially that of the king. To most Thais, the king is always right, and officials who work for the government represent the king; therefore, they should not be challenged. Fifteen years ago, Thai citizens would not have challenged political leaders, but now many are seeking ways to influence and participate in public planning and policy decision-making.

7. CONCLUSIONS

Non-governmental organizations currently play a central role in bringing citizens into the planning and decision-making process in Thailand. Experience in Thailand and throughout developing countries demonstrates that participatory development is better-quality development. Groups of citizens coming together on a voluntary basis to address public issues and to take direct action do make a difference in the final outcome. Guaranteeing successful sustainable development in intermediate size cities requires the continued and constructive participation of citizens. In Thailand, NGOs may be the most effective means to ensure active and meaningful involvement by local citizens.

Thai NGOs tend to focus on local issues that are meaningful to the people, such as preservation of historic buildings in Chiang Mai or the pollution of the Ping River. NGOs, working with institutions of the culture, such as the *sangha* and academics operate within the framework of Thai society to accomplish their goals, rather than introducing new forms of citizen involvement that may not be adaptable or successful. If Westerners wish to understand how to involve citizens most effectively in sustainable development plans and programs, they would be wise to study past successes and use them as examples. Planners should build on existing community organizations such as villages, farmers' networks, and NGOs rather than creating new structures.

The importance of recognizing Thailand's distinctive characteristics cannot be emphasized too strongly. To encourage meaningful participation in public affairs, techniques must be used that are sensitive to Thai culture. It would be a mistake to transplant Western-style theories and practices of citizen participation into Thailand and expect successful involvement. For example, any efforts to involve Thai citizens must take into consideration the importance of Buddhism, hierarchy, patron-client relationship, and the role of NGOs.

Planners and other government officials in Thailand have tried to utilize Western methods of citizen participation, with varying degrees of success. For example, large open meetings are held to gather opinions on plans when they come to Chiang Mai from Bangkok, but very few citizens attend these meetings; and even fewer express opinions. For the various reasons discussed in this chapter, some Thais may not be comfortable when speaking out in a large group where government officials are present (Tantayanusorn, 1998).

To increase public awareness and activisim, television and radio are major media with great potential for reaching all types of citizens, both urban and rural. Television programs that disseminate information may be valuable for educating citizens and involving them in community affairs. Televised meetings with community leaders, who can speak for the general public, may be an effective way to learn what the community thinks about an issue. In the rural areas, an asset-based approach (rather than a needs-based approach) may be a positive step in empowering people to take a stronger role in planning and policy issues.

Thailand's citizens have begun to involve themselves in planning and public decision-making, but there is room for further involvement. The most effective approach would be a hybrid composed of participatory techniques that are used in the West (such as citizen advisory boards, community technical assistance, and interviews) which are adapted to Thailand, along with some alternative means of participation (such as NGO activity).

Finally, one must remember that the techniques for encouraging participation from educated professionals in Bangkok or Chiang Mai may not be appropriate, or as effective, with less well-educated groups in rural or less highly developed areas.

NOTES

[1] The terms *citizen participation, public participation*, and *citizen involvement* are used here interchangeably with the same definition: active and meaningful participation by citizens in the public process of planning and decision-making.

[2] Non-governmental organizations or NGOs is a term used to define voluntary citizen organizations formed to provide assistance to groups and individuals or to advocate a particular point of view about important public issues. Although not always the case, it is implied that NGOs will keep their distance from governments. NGOs often do things that governments cannot, or will not, do. The organizations may be large or small, local and national. They range from small citizen groups focused on a local issue to large international organizations such as Amnesty International and Greenpeace. Many NGOs in developing countries are funded by the United Nations, European Union, or agencies such as the United States Agency for International Development and its counterparts in Great Britain, Canada, Germany, and other developed countries.

[3] Suthep Mountain rising 3,520 feet west of Chiang Mai is a forested backdrop to the city. Wat Phra That Doi Suthep is one of the city's most important and visible landmarks. The temple was constructed around 1383 A.D. and contains holy Buddha relics. Although a major attraction in a city dependent upon the tourist industry, many citizens believed a cable car would destroy the mountain and temple's aesthetic and historic qualities.

REFERENCES

Anderson, Jeremy and Steven Yaffe. (1998). *Balancing Public Trust & Private Interest: Public Participation Habitat Conservation Planning*. Ann Arbor, MI: University of Michigan School of Natural Resources & Environment.

Aquino, Karan. (1998). Northern Development Foundation. Interview. Chiang Mai, Thailand, July 7.

Arnstein, Sherry. (1969). A Ladder of Citizen Participation. *Journal of the American Institute of Planners. 35, no.4*, 216-224.

Assavanonda, Anjira. (1997). State to lose its protection from the law. *Bangkok Post*, electronic edition.

Ayudhaya, Kingkorn Narintarakul Na. (1998). Northern Development Foundation. Interview by authors, Chaing Mai, Thailand, July 9.

Chiang Mai Municipality, Public Relations Section. (Undated). Local Government Chiang Mai Municipality. Chiang Mai, Thailand: Dararat Ltd.

Darke, Roy. (2000). Citizen Participation, Equal Opportunities, Planning Policies and Decisions. In Philip Allmendinger, Alan Prior, and Jeremy Raemaekers (Eds.), *Introduction to Planning Practice*, (pp. 385-412). New York: John Wiley & Sons. Ltd.

DeSario, Jack and Stuart Langton. (1987). *Citizen Participation in Public Decision-making*. Westport, CN: Greenwood Press.

Hanks, Lucien. (1975). The Thai Social Order as Entourage and Circle. In G. William Skinner and A. Thomas Kirsch (Eds.), *Change and Persistence in Thai Society* (pp.172-196). Ithaca, New York: Cornell University Press.

Hewison, Kevin. (1997). *Political Change in Thailand: Democracy and Participation*. London: Routledge.

_______. (1996). Emerging social forces in Thailand: New political and economic roles. In Richard Robison and David S.G. Goodman (Eds.), *The New Rich in Asia* (pp.137-160). London: Routledge.

Hewison, Kevin and Richard Robison. (1997). Thailand: Capitalist Development and the State. In Garry Rodan (Ed.), *The Political Economy of Southeast Asia: An Introduction* (pp. 93-120). Melbourne, Australia: Oxford University Press Australia.

Hinshiranan, Narumon. (1999). E-mail to authors. May 24.

Kerdphol,Saiyud. (1997). Path of developing democracy has been long and tortuous. *Bangkok Post*. March 20, electronic edition.

Kirsch, A. Thomas. (1975). Economy, Polity and Religion. In G. William Skinner and A.Thomas Kirsch (Eds.), *Change and Persistence in Thai Society* (pp. 172-196). Ithaca, NY: Cornell University Press.

Kweit, Robert W. and Mary Grisez Kweit. (1990). *People and Politics in Urban America*. Pacific Grove, California: Brooks/Cole.

Langton, Stuart. (1978). *Citizen Participation in America*. Lexington, MA: Lexington Books, D.C. Heath and Company.

Mekvichai, Banasopit. (1994). *Patterns of Government in Thailand*. A paper presented at the Workshop on Urban Governance and Poverty Alleviation in Southeast Asia: A Comparative Framework, Manila, Phillippines, November.

Morley, James W. (1993). *Driven by Growth: Political Change in the Asia-Pacific Region*. Armonk, NY: M.E. Sharpe.

Muscat, Robert J. (1994). *The Fifth Tiger*. Helsinki, Finland: United Nations University Press.

Narumit-Rekagarn, Sirichai. (1998). For Chiang Mai. Interview. Chiang Mai, Thailand, July 8.

National Economic and Social Development Board. (1996). The Eighth National Economic and Social Development Plan (1997-2001), Bangkok, Thailand: Office of the Prime Minister.

_______. (1997). Summary, The Eighth National Economic and Social Development Board (1997-2001). Bangkok, Thailand: Office of the Prime Minister.

Neher, Clark D. (1992). Political Development and Political Participation in Thailand. In Myron L. Cohen (Ed.), *Asia, Case Studies in the Social Sciences: A Guide for Teaching*, (pp. 415-425). Armonk, N.Y.: M.E. Sharpe.

Neher, Clark D. and Ross Marlay. (1995). *Democracy and Development in Southeast Asia: The Winds of Change*. Boulder, CO: Westview Press.

Parnwell, Michael J.G. and Daniel Arghiros. (1996). Introduction: Uneven development in Thailand. In Michael J.G. Parnwell (Eds.), *Uneven development in Thailand* (pp. 1-27). Aldershot, England: Avebury.

Phongpaichit, Pasuk and Chris Baker. (1995). *Thailand: Economy and Politics*. Oxford, England: University Press.

Pongsapich, Amara and Nitaya Kataleradaphan. (1995). The non-profit sector in the 1990s: Thailand and the Southeast Asian context. *Thai Development Newsletter 29*, 23-26.

_______. (1997). *Thailand NonProfit Sector and Social Development*. Bangkok, Thailand: Chulalongkorn University Printing House.

Rabibhadana, Akin. (1975). Clientship and class structure in the early Bangkok period. In G. William Skinner and A.Thomas Kirsch (Eds.), *Change and Persistence in Thai Society* (pp. 172-196). Ithaca, New York: Cornell University Press.

Rosenbaum, Nelson M. (1976). *Citizen Involvement in Land Use Governance*. Washington, D.C.: The Urban Institute.

Samudavanija, Chai-Anan, and Sukhumbhand Paribatra. (1993). Thailand: Liberalization Without Democracy. In James W. Morley (Ed.), *Driven by Growth: Political Change in the Asia-Pacific Region*, (pp. 119-141). Armonk, NY: M.E. Sharpe.

Skinner, G. William and A. Thomas Kirsch. (1975). *Change and Persistence in Thai Society*. Ithaca, NY: Cornell University.

Sodawat, Jirawan. (1998). YMCA for Northern Development Foundation. Interview by authors, Chiang Mai, Thailand. July 10.

Susanpoolthong, Supawadee. (1997). How communities know their communities needs best. *Bangkok Post*. electronic edition.

Tantayanusorn, Niwat. (1998). Interview. Chiang Mai, Thailand, July 7.

Thai Development Newsletter, (1995). *A summary of people's development plan: Its concept and direction. No. 29*, 30-34.

Thomas, John Clayton. (1995). *Public Participation in Public Decisions: New Skills and Strategies for Public Managers*. San Francisco, CA: Jossey-Bass Publishers.

Turakij, Krungthep. (1998). Society needs to follow the principles. *Bangkok Post*, electronic edition.

United Nations, (2001). Country paper: Thailand. *Local Government in Asia and The Pacific: A Comparative Study*. Published by Economic and Social Commission for Asia and the Pacific (ESCAP). Available at http://www.unescap.org/huset/lgstudy/country/thailand/thai.htm.

Vorathanyakit,Voravit, (1998). Presentation by Urban Development Coordination Division Staff of the National Economic and Social Development Board. June 25. Bangkok, Thailand: National Economic and Social Development Board Office.

PART II

KEY COMPONENTS OF THE LOCAL ECONOMIC STRUCTURE

CHAPTER 4

HOWARD STAFFORD

MANUFACTURING IN CHIANG MAI PROVINCE: PRODUCTS, PATTERNS, AND SUSTAINABLE BALANCE

Abstract The spatial distribution of manufacturing within Chiang Mai Province, an "outlying" region in an "emerging" country, reveals extreme urban/rural distinctions; these are similar to the national pattern. The manufacturing profile of the Province is somewhat consistent with expectations. Traditional wood products, fabrics, food, and agro-industries are strong. Within the Province (and the nearby Northern Region Industrial Estate), there are also many large and modern factories, in traditional sectors, especially food, and in newer sectors such as electronics. The long-run sustainability of manufacturing in the Province and its contribution to a viable regional economy are assessed. It is argued that: (1) sector balance, (2) geographic balance, and (3) relatively high capital investments promote economic stability and prosperity. The core area within 50km of Chiang Mai City is diverse and strong. Beyond the core area, outlying districts within this outlying Province have little manufacturing employment, will not likely experience growth, and will continue to be areas of out migration. The little manufacturing in these districts has low levels of capital investment. The industrial viability of outlying districts is questionable, and hinterland weaknesses detract from the core.

1. INTRODUCTION

Over the past several decades Thailand's economy (and accompanying social structure) has experienced significant change. A major component of change has been the absolute and relative increase of manufacturing. The majority of industrial expansion has been in the Bangkok metropolitan region. Bangkok and the ten provinces within a three-hour drive of Bangkok constitute the first two of the geographic zones in which the Thai Board of Investment (BOI) promotes industry. The other sixty provinces are target areas for BOI efforts to decentralize manufacturing away from the over-crowded Bangkok region and toward areas that are thought to be overly reliant on agriculture and basic agriculturally related manufacturing. These provinces are perceived to be areas of limited employment opportunity and, thus, of out-migration. The government of Thailand, and the BOI in particular, want to discover commercial employment opportunities, including manufacturing, that are viable in the "new" Thailand and available to residents of the sixty "outlying" provinces (Thailand Board of Investment, 1997; Investment Plan For Chiang Mai, 1994). Manufacturing is a key component of government planning.

Chiang Mai Province is a useful venue (although it is not necessarily typical) in which to study the nature of manufacturing and economic development of an

M. Romanos and C. Auffrey (eds.), Managing Intermediate Size Cities, 59—73.

"outlying" region in an "emerging" country. It is located some 500km north of Bangkok, and is peripheral to the major growth center of Thailand. The Province contains Chiang Mai City, an intermediate size urban area, but the largest in the region. Chiang Mai City has long been the historical, cultural and economic center of Northern Thailand. The current structure of manufacturing in the province is examined to compare perceptions with reality. The distribution of manufacturing within Chiang Mai Province is examined to determine if the urban/rural distinctions identified at the national scale have regional scale counterparts. The long-term viability of manufacturing in the Province and its contribution to a viable and sustainable regional economy is assessed.

2. SUSTAINABILITY AND ECONOMIC DIVERSITY

It is here argued that a diversified regional economy is more stable over the long run than a highly specialized production and employment system. The argument is similar to that for the merits of regional ecological diversity. It is recognized that overly specialized biological systems are fragile. Regional ecological diversity is seen as desirable in promoting sustainable development (Pielou, 1975). Likewise, it has long been argued that highly specialized economic systems are vulnerable and fragile. Seventy years ago, Zimmermann characterized modern highly specialized market economies as capable of reaching lofty heights but resting on a narrow and fragile base. “Traditional” economic systems, in contrast, were characterized as not being able to reach great wealth, but they reside in “sodden safety” (Zimmermann, 1933). The perils of overspecialization appear to explain why many resource exploitation areas fall on hard times when the mineral cannot be competitively mined because the easily accessible deposits are depleted, or the trees have been harvested at a faster than replacement rate, or the soil has lost its fertility. Similarities include the decline of trading cities suddenly bypassed, or the hard times in manufacturing cities producing commodities no longer in sufficient demand.

Another argument, of course, is that many of the word’s economic success stories have been of specialized regions. An intermediate position is that regions with interrelated clusters of mutually reinforcing activities are most viable (Green and McNaughton, 2000). What the future holds for Chiang Mai is unknown, but examining what manufacturing now exists and its characteristics helps provide indications of possibilities.

Historically, Chiang Mai Province has been traditionally specialized in agriculture throughout the region, with some commercial trading functions centered in a few urban areas. More recently, tourism has become a major component of the regional economic system. There are also significant amounts of manufacturing, both traditional (primary processing of agricultural goods) and modern (in factories with divisions of labor).

Throughout the world, people migrate from agricultural areas to urban areas in search of better lives. Chiang Mai Province is no exception, and the rural-urban migration will continue regardless of any plans or investment schemes to keep people on farms and in rural areas. Decent urban employment opportunities are necessary for rural-urban migrants as well as existing urban residents. Employment in agriculture processing and in tourism will remain important but they will not sustain the Province. Another promising economic sector is manufacturing. Thus, the economic, social, and physical environmental implications of the continued development of manufacturing are relevant to the long-term stability and quality of life of the province.

What are the prospects for the manufacturing sector? Is manufacturing that is competitive in the 21st century viable in Chiang Mai Province? Can all sub-areas within the region share in manufacturing employment opportunities? Will manufacturing enhance or diminish the natural and social environments, the quality of which is critical to the tourism and agricultural sectors of the regional economy?

3. COSTS AND BENEFITS OF MANUFACTURING

Around the world, regions that have significant amounts of manufacturing are relatively wealthy, and most of the world's manufacturing is in the "advanced" regions of the world. In general, manufacturing promotes regional wealth, and most manufacturers choose to locate in the wealthy high-wage regions. This is because the wealthy regions host the markets, the suppliers, the skilled labor, and the well developed infrastructure and communications. Most of the world's manufactured goods are produced in the "advanced" and highly capitalized regions of the world, and the vast majority of trade in manufactured goods is between these same regions (Stutz and deSouza, 1998). But this is not a complete picture. Some manufacturing is exploitive, of people and of the environment. Regions can either "race to the top" or "race to the bottom" with the development of manufacturing. Regions that rely on cheap unskilled labor and lax environmental controls to attract industry will head downward. Regions that develop modern manufacturing that utilizes skilled workers and modern production techniques (high capital investment per worker) can be winners.

People can be exploited or they can be empowered. Industries that rely heavily on low skilled and low paid workers are not region builders. They do not encourage the development of a skilled, literate, and prosperous labor force. They also typically have low capital investments per worker and are transient. Manufacturers that employ more highly skilled workers and have higher investments per worker can be positive region builders. The population in general, those directly employed in the factories and in supporting activities, can be enriched.

All human activities cause environmental problems, and manufacturing is no exception. Furthermore, there will be pressure from some manufacturers to allow easy and cheap disposal of effluents, and if proper precautions are not taken the result can be permanent degradation of the human and physical environments. An area can become so polluted it is no longer attractive for any activity, least of all the desirable types of manufacturing. Regions that rely on lax environmental controls to lure industry are misguided. Upward bound regions deal fairly and effectively with the externalities ensuing from human activities.

Fortunately, most manufacturers do not choose to locate in dirty areas that do not police their environments. Other location factors are far more important, and there is little evidence of flights of manufacturing from areas with strict environmental controls (Duerksen, 1983; Leonard, 1984). Indeed, on balance, the "clean" areas are preferred because these are the places where the sought after skilled workers prefer to live (Stafford, 1985). In Chiang Mai, there is already a shortage of skilled workers, and new workers are not going to be attracted to unsightly or unhealthy locales. A healthy manufacturing sector can deliver an additional environmental benefit because the infrastructure built to supply clean water and to properly dispose of its effluents can also be used to service other sectors of the local economy. Currently, for example, there is insufficient sewerage in Chiang Mai. A healthy manufacturing sector needing modern sewage facilities and creating the tax revenues to finance them can be beneficial to all of Chiang Mai.

4. CHARACTERISTICS OF SUCCESSFUL MANUFACTURING REGIONS

Successful manufacturing regions, those in which the trade-offs are positive over long periods of time, tend to have some common characteristics.

4.1 Sector Balance

A variety of income generating activities helps avoid some of the problems associated with the high/low swings of overly specialized areas. Within manufacturing, a diverse array of suppliers and markets (which usually is other manufacturers) enhances the positive industrial districts/networks potential (Lee, Liu, Stafford, 2000). Economically balanced areas have good and varied supply chain connections. Economically over-specialized areas are prone to boom and bust economic cycles. The current economy of Chiang Mai Province is not sufficiently robust or balanced to easily handle a significant downturn, nor is its manufacturing sufficiently diversified to generate attractive supply chain opportunities. Balanced growth is safer and more appropriate.

4.2 Critical Size

The nature of manufacturing is to operate in urban areas. It thrives on proximity to local suppliers and markets, and to local support activities such as printers or advertisers. Urban places also allow access to shared infrastructure, such as roads, telecommunications, and adequate sewers. Very small and relatively isolated towns typically cannot afford the necessary infrastructure and they do not have local actors up and down the manufacturing supply chain. A successful region will have at least one city of critical size. It will have a manufacturing sector of sufficient size and diversity to allow efficient connections across the supply chain. And, it will also be of sufficient size to develop a skilled and stable workforce. The critical determinants of manufacturing success are skilled workers, the development of functioning industrial districts, and infrastructure investments.

4.3 Relatively High Capital Investments Per Factory and Per Worker

The justification for modern factories and enabled workers is parallel to that in favor of sophisticated urban areas. Productivity per worker is the key to the economic success of employees, industries, and regions. However, productive workers are supported by both investments and their skills. Today there is a worldwide shortage of skilled workers, and this is already true in Chiang Mai Province. The province will have to promote manufacturing skills among the population if it wishes to move upward rather than downward. It cannot rely on low wage labor.

4.4 Spatial Balance

As noted, manufacturing is predominately an urban activity, and the larger the urban region the more competitive it is. This polarization can create difficulties for the outlying areas. There will be "spill over" effects in the closer-in peripheral places but these effects will not reach directly to the most remote locales. For these remote places, the tangible benefit of manufacturing development in the urban core will be the availability of jobs that are at least closer to home than the present lure of Bangkok. Conversely, the manufacturing core benefits from any strength the periphery can provide. Spatial balance is not absolutely necessary for the development of successful manufacturing in the region. The region is already unbalanced in terms of its economic geography and it probably will become even more so. However, this geographic, rural-urban imbalance must be recognized because of the economic development (e.g., labor recruitment and training) and policy implications (e.g., pressures to divert investment from the more appropriate urban places to very small centers).

5. CURRENT MANUFACTURING IN CHIANG MAI PROVINCE

Agriculture and business are the leading economic sectors in Chiang Mai Province. However, there is also already a considerable amount of manufacturing activity. There are almost 2,000 registered plants (those with seven or more employees or more than 5 horsepower installed). This is a relatively high number for a region of 1.6 million persons. By comparison, the Cincinnati, Ohio, USA metropolitan area, an area long noted for manufacturing and with a population approximately the same as Chiang Mai Province, has about 1,600 manufacturing plants with fifteen or more workers. It is variously estimated that manufacturing employs between 12 and 26 % of the Province's workforce (estimated by the Thailand Board of Investment, 1997; from Chiang Mai, 1998).

Over half of the factories in the Province are engaged in agro-industrial product processing and livestock processing, and a large number of these are small rice processing plants. Almost as many workers are employed in wood and wood products plants, but the average number of workers per plant is higher. These figures, coupled with the estimate that 65% of those employed in the province are farmers (presumably full or part-time; Chiang Mai Directory, 1998), conform to the perception of this outlying province of Thailand as being heavily dependent on agriculture and on basic resource related manufacturing (Table 1).

Table 1. Manufacturing in Chiang Mai Province by Sector: 1992

Type (sector)	*No. Plants*	*Capital M Baht*	*No. Workers*	*Plants %*	*Capital %*	*Workers %*
Agro-industries	991	1,075	5,367	56	22	21
Food & beverage	134	1,698	3,812	08	34	15
Wood & products	195	453	5,344	11	09	21
Construction materials	78	445	1,573	04	09	06
Fabrics	54	217	3,216	03	04	13
Metal works & mechanics	78	125	671	04	03	03
Petro-chem-plastics	23	71	385	01	01	01
Other (trans., electronics, paper & printing, ceramics, gems)	204	870	4,324	12	18	19
TOTAL	1,757	4,954	24,962	100%	100%	100%

Source: Modified from Investment Plan for Chiang Mai Province (Executive Summary), Table 4.2, 14.

The third and fourth largest sectors in terms of employment are food & beverages and fabrics and related products (ignoring the composite "Other" category for the moment). These relative concentrations again are not unexpected. Food and beverage manufacturing, on the one hand, can be expected in any area of sufficient population (a locational orientation towards markets), and on the other hand, can be seen as a logical step up the agriculture-food product production chain. Fabric production is somewhat explained by local demand, and is also characteristic of manufacturing in emerging economic systems. The handicraft traditions of the province, and their related value to tourism, bolster the fabrics sector.

The three columns on the right side of Table 1 show the number of factories, amount of capital and number of workers for each sector as a percentage of the province totals. Three of the sectors are notable either because they have relatively low capital investment per factory or per worker: agro-industrial product processing and livestock, wood and wood products, and fabrics and related products. Conversely, the food & beverage, construction materials, and other sectors have relatively high investments in plants and/or employees. In terms of balance across manufacturing sectors, the relatively low capital investment types (agro-industries, wood, fabrics) employ 13,937 while the relatively highly capitalized sectors employ only 9,709. If the proposition is true that the more highly capitalized industries are more likely to be successful in the 21st century, then Chiang Mai Province is sectorally unbalanced which detracts from its future economic viability.

Although the leading manufacturing sectors conform to expectations about an "emerging" and "outlying" and relatively "rural" province, this broad-brush picture is far from the whole story. As indicated in Table 1, the province has a range of industries, from rather basic rice mill operations to some manufacturers with considerable invested capital. Even within the "traditional" industries, there are some large, sophisticated, highly capitalized, world-class operations. Chiang Mai Frozen Foods is a case in point. It is a joint venture between Thai, Taiwanese and Japanese investors organized to export processed foods, especially to Taiwan and Japan. The Thai Board of Investment provides support, including tax abatement and the importation of equipment without duty. The most important location factor, however, is the high degree of integration into the regional economy. In addition to providing employment for almost 2,000 Thai workers, Chiang Mai Frozen Foods contracts with farm families throughout northern Thailand for high quality raw materials. The processing plant is world-class in equipment, technology, and hygiene, all necessary to compete in its markets. Chiang Mai Frozen Foods is an excellent example of the potential for manufacturing to move up the supply chain, thus increasing both profitability and viability. Likewise, the nearby Northern Region Industrial Estate (NRIE) provides many examples of modern manufacturing operations that seem to be thriving in this region of northern Thailand. If the region is to have balanced and sustained economic growth with a significant manufacturing component, then it is necessary to have modern and well capitalized manufacturing operations in the mix.

6. NORTHERN REGION INDUSTRIAL ESTATE (NRIE)

The Northern Region Industrial Estate is, in effect, a part of the Chiang Mai industrial structure even though it is located in neighboring Lamphun Province. Like much of Chiang Mai Province, Lamphun is mainly oriented toward processing of local agricultural and forest products. Of the seven districts (amphoe) within the province, one is outstanding. The northern-most district alone has 47% of the factories, 64% of the labor force employed in manufacturing, and an astounding 84% of the invested manufacturing capital. There are two elements that distinguish this one district from the rest of Lamphun Province. First, it is close to Chiang Mai City, which is the largest city in the north of Thailand and the manufacturing center of Chiang Mai Province; Lamphun and Chiang Mai are known as twin cities and provide reciprocal benefits. Second, the district is the home of the Northern Region Industrial Estate (NRIE).

The Northern Region Industrial Estate was established "in accordance with the National Economic and Social Development Policy of the Royal Thai Government to decentralize industrial development away from the Bangkok Metropolitan Area to the regional zones; the Industrial Estate Authority of Thailand (IEAT) has built the Northern Region Industrial Estate (NRIE)...Construction of NRIE commenced in

April, 1983, and was completed in March, 1985" (Northern Region Industrial Estate, 1998).

The sixty factories in the NRIE are only 7% of the plants in Lamphun Province, but they are distinctive and special. They are large, with an average employment of over 400. They are often branch plants of foreign multi-national corporations or are joint ventures between Thai and foreign firms. The factories are sophisticated and represent large capital investment. Over half of the factories are in either electronics or machinery parts production. Another 23% are in agriculture products and food, but these are also at the high technology end of their category (e.g., Pioneer Hi-Bred seed processing). The NRIE demonstrates the potential for large modern manufacturing when sufficient and high quality infrastructure is provided. As elsewhere in the world, sophisticated and well-capitalized manufacturers prefer to locate operations in well-serviced industrial parks. However, they typically do not like to be geographically isolated. At the scale of the NRIE site, they are not isolated because other modern manufacturing plants surround them. And, in spite of being in an "outlying" area of Thailand, they are not really so far removed from the world. NRIE is a world-class industrial park, on an excellent highway, and less than 20 km from Chiang Mai City, with its urban services, university, transportation facilities, and international airport.

7. PROPOSITIONS REGARDING THE ECONOMIC VIABILITY OF MANUFACTURING IN THE PROVINCE

Not all sub-areas within Chiang Mai, or Lamphun, Province are as well situated as Chiang Mai Frozen Foods or the NRIE in terms of proximity to urban infrastructure, amenities, and modern transportation systems, or in the favorable mix of manufacturing operations. What are the structures of manufacturing across all districts within the Province? What do variations in conditions suggest regarding the prospects for manufacturing not only in the vicinity of Chiang Mai City or NRIE, but also in other districts? What are the prospects for viable manufacturing in the Province in the 21st century in a country that is committed to being involved in international competition? The four propositions previously presented are that viability is enhanced by: (a) spatial (geographic) balance; (b) modern, capital-intensive plants; (c) critical size, of urban areas and of the manufacturing sector; and (d) sector balance.

Following systems analysis and ecological precedents, it can be argued that spatial balance across the Province is beneficial to long-term economic viability. While it is clear that the economic core of the Province is the City of Chiang Mai, and will remain so, it is argued that the long-term strength of the core is intimately related to the economic health of the rest of the Province, and vice-versa. The greater the amount of manufacturing in the hinterland, the higher the potential for local inputs and local markets (localization economies) and the more likely local services will be available (urbanization economies). Also, the stronger the manufacturing base in the "outlying" districts, the more likely local jobs will be provided; thus, reducing migration to Chiang Mai (or Bangkok), and increasing the jobs skills of those who eventually do migrate from the more rural areas. In this regard, the greater the variety of manufacturing jobs available, and the more sophisticated (modern) the manufacturing plant, the higher the probability of residents obtaining relevant skills. Likewise, it is argued that the more modern the manufacturing plants, the higher the probability of them being economically viable. It is also proposed that the more balanced a district is in terms of manufacturing types, and the larger the industrial sector, the less vulnerable it is to economic downturns. Next, it is necessary to have at least one urban area of sufficient size to offer modern infrastructure, mutually supportive industries, and communications and transportation. These factors are examined across the Province. The spatial variations are first discussed. The issues of balance (diversity) are explored. Finally, spatial balance, diversity, and size measures are combined.

8. SPATIAL VARIATIONS WITHIN CHIANG MAI PROVINCE

The twenty-three districts (amphoe) within Chiang Mai Province provide a suitable spatial framework. For manufacturing in total, most districts have at least a few factories, but the concentration is toward the center, in the vicinity of Chiang Mai City. The spatial concentration toward the east central section is most pronounced when total manufacturing employment per district is plotted (Map 1). On average, manufacturers in the vicinity of Chiang Mai City operate larger plants than manufacturers in the rest of the Province.

The specific manufacturing sector data are for the second and third categories of factories in the Ministry of Industry directory. The second category plants have between 20 and 50 installed horsepower. Whereas the first category plants are very small and subject to few location restrictions, second category operations must have a permit. Third category factories have 50 or more installed horsepower and must be licensed. In effect, the sector data shown are for small to very large plants, but not for the very smallest or least mechanized operations in the Province.

The distributions of these specific types of manufacturing exhibit three basic types of spatial patterns: those industries strongly concentrated in the core region, those somewhat more spread out, and those relatively dispersed throughout the Province. Fabrics and related products manufacturing, and petroleum, chemicals, and plastics and plastic products are strongly oriented toward the east-central section of the Province. Metal work and mechanics, construction materials, service industries, and "other" show the same general gravitation toward the Chiang Mai City locus; but a secondary concentration appears in the north of the Province, especially in the Fang Amphoe. The Chiang Mai City orientation even shows up in the third group of distributions, but it is somewhat less pronounced. Wood and wood products manufacturing spreads out of the east-central core, especially to the north and northeast. Food manufacturing shows even more spread, again especially to the north and northeast. The most spatially dispersed type of manufacturing is agro-industrial processing and livestock. This type of manufacturing is especially important in the northern and northeastern portions of the Province.

The relatively more even spatial distributions for agro-industrial processing, wood, and food is not surprising given the importance of agriculture and rural activities in the history and current economy of the Province. Even for these industries, however, the Chiang Mai City core region is significant. For all other industries, the core region is paramount. Districts in the east-central core region have most of the manufacturing jobs and have the best variety of industrial types. Districts outside of this core region tend to be less diversified, or to have virtually no manufacturing activity (the east-central and extreme southern districts).

9. MANUFACTURING DIVERSITY WITHIN CHIANG MAI PROVINCE

The districts vary widely in their degree of homogeneity across the nine manufacturing sectors. Map 1 shows the total manufacturing employment per district and the relative share of each district's employment in each of the nine sectors. Five of the districts have no category 2 or category 3 manufacturing. Others, such as Mai Ai in the far north of the province, have manufacturing in only two sectors, food and wood products. Mai Ai has little manufacturing magnitude or richness (number of sectors), but it does have evenness because the two sectors there are rather evenly balanced. Chiang Mai district, in contrast, has the greatest amount of manufacturing among the Provincial districts, and all nine sectors are represented. Furthermore, the balance between the sectors is relatively even. Chiang Mai district has magnitude and both types of diversity, richness and evenness.

As argued previously, if long-term system stability is positively related to internal diversity the impressions gained from Map 1 have critical implications. To corroborate the initial impressions regarding this important issue, the Shannon Index of diversity has been calculated for each district. The Shannon Index is derived from information theory and is widely used in ecological studies. The formula is:

$$D = - \Sigma (n_i/N) (\ln n_i/N), \text{ where}$$
$$N = \text{the total number of sectors, and}$$
$$n_i = \text{the number of employees in each sector.}$$

The Shannon Index incorporates the two components of diversity, richness (in this case, number of economic sectors present) and evenness (relative degree of balance or, conversely, specialization in the sectors present) (Odum, E. P., 1983, 414).

Map 2 shows the relative value of the Shannon Index for each district (higher = more diverse). The Shannon Index confirms the visual impressions from Map 1. The districts with the largest number of sectors represented in their local manufacturing mix are also those with the most equitable distribution of employment across the sectors. They are mainly located in the east-central core region. Conversely, most of the districts with disproportionate reliance on a few industrial sectors are at some distance from the core. The only area outside the 50 km core radius with at least modest magnitude and good balance is Fang, some 140 km north of Chiang Mai.

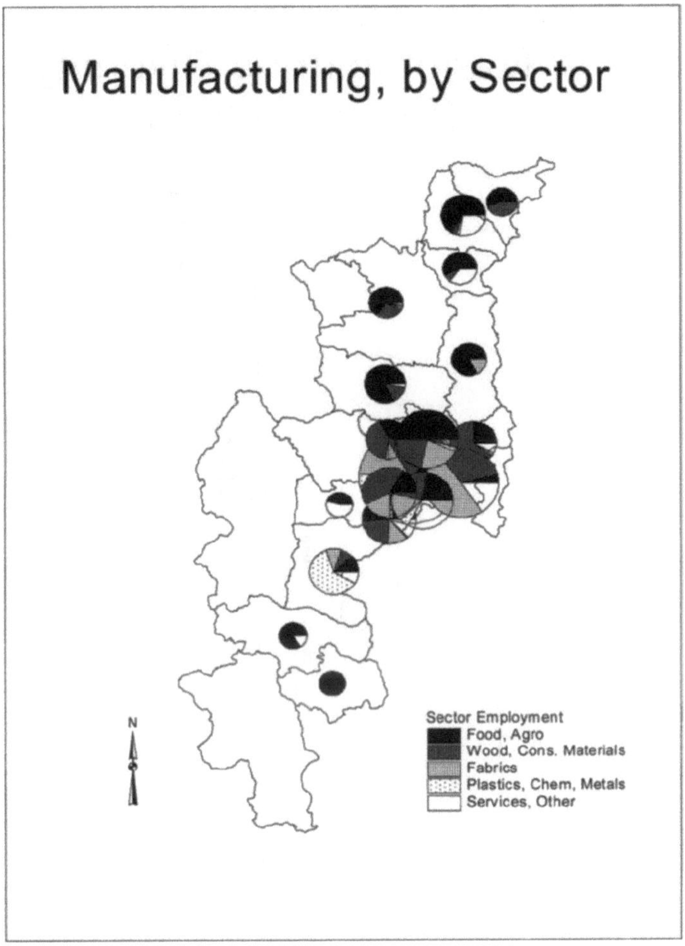

Map 1: Manufacturing employment, district total and by sector.

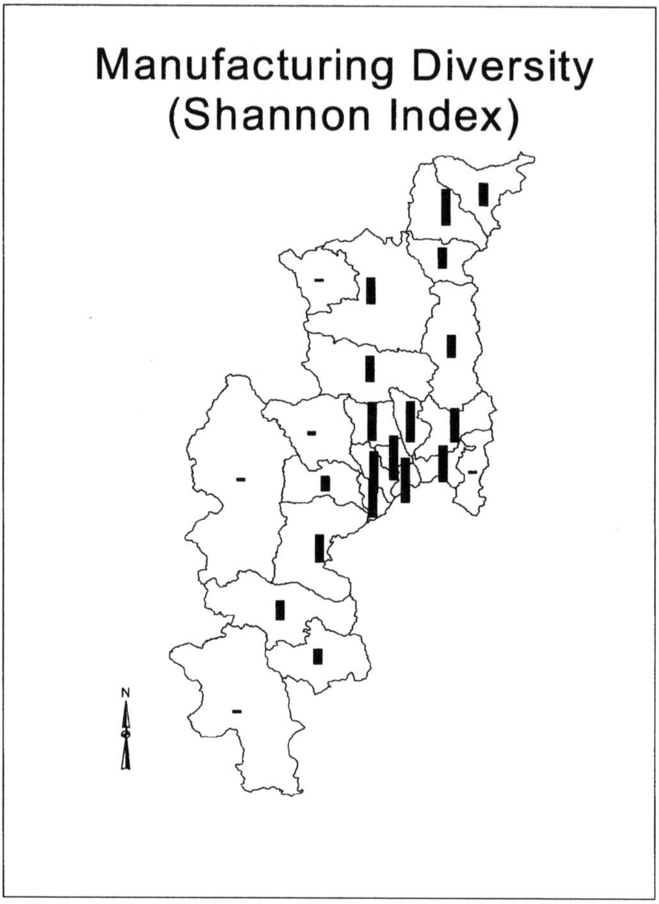

Map 2: Manufacturing diversity within Chiang Mai Province.

10. THE COMPOSITE: SIZE, COMPETITIVENESS, AND BALANCE

To be competitive and stable in a modern economic system, it is proposed that areas must have sufficient size for regional economies of scale (urbanization and localization economies), competitive industries, and balance between the productive sectors. Regarding manufacturing, it is argued that three indices help forecast the future economic success of the districts within Chiang Mai Province. First, magnitude is measured by total employment in manufacturing. Second, overall competitiveness is indicated by capital invested per employee. Third, the Shannon Index of diversity reveals balance. Table 2 lists the value of each district on each measure. Chiang Mai district, for example, ranks first on the number of plants, capital, and number employed in manufacturing criteria, third on the capital/employment ratio, and second on the Shannon diversity index. Conversely, the Mae Chaem and Sameong districts are at the bottom on all criteria. The ranks (inverse, not shown) have been combined for the number employed [M], capital/employment [IE] and Shannon index [D] measures to give a summary measure of the relative robustness of the manufacturing sector in each district of the province. The fourth column is the average rank for each district on these three measures. The average ranks are graphically portrayed on Map 3. Districts with high average ranks are those with substantial current manufacturing, relatively good balance in types of manufacturing, and are the most likely sub-areas for new manufacturing growth. Districts with low average ranks are isolated both geographically and economically. They will continue to be areas of out-migration and will supply unskilled labor to Chiang Mai City, and perhaps to a few other urban places. Manufacturing growth in the most isolated places is unlikely.

Table 2. Chiang Mai Province Manufacturing: Districts Composite

District (amphoe)	*Plants*	*Capital M Baht*	*No. Employ [M]*	*Cap/Emp [IE]*	*Shannon Index [D]*	*Avg. Rank**
Chiang Mai	488	5538	9649	.55	1.75	21.3
Chom Thong	87	254	1429	.18	1.28	13.0
Chiang Dao	36	63	310	.20	1.32	12.0
Doi Saket	113	425	1203	.35	1.40	16.3
Fang	103	431	989	.44	1.44	17.0
Phrao	80	68	363	.19	0.66	10.3
Mae Chaem	0	0	0	.00	0.00	1.0
Mae Taeng	79	163	732	.22	0.68	12.0
Mae Rim	117	833	879	.95	1.49	18.7
Mai Ai	68	37	172	.21	0.72	10.7
Sameong	0	0	0	.00	0.00	1.0
San Kamphaeng	191	1036	4569	.23	1.36	17.3
San Sai	129	1528	2962	.52	1.42	19.3
San Pa Tong	101	613	1901	.32	1.54	18.0
Saraphi	140	2116	2769	.76	1.86	21.7
Hang Dong	109	871	1976	.44	1.83	20.0
Om Koi	0	0	0	.00	0.00	1.0
Hot	29	12	86	.14	0.93	9.3
Doi Tao	11	9	49	.18	0.37	7.3
Wang Haeng	0	0	0	.00	0.00	1.0
Chai Paklan	23	31	329	.09	0.89	9.7
Mae Wang	27	5	54	.09	0.46	6.7
Maeon	0	0	0	.00	0.00	1.0

Average Rank = Rank [(M+IE+D)/3]

Source: List of Factories in Chiang Mai, 1996. Calculations by author.

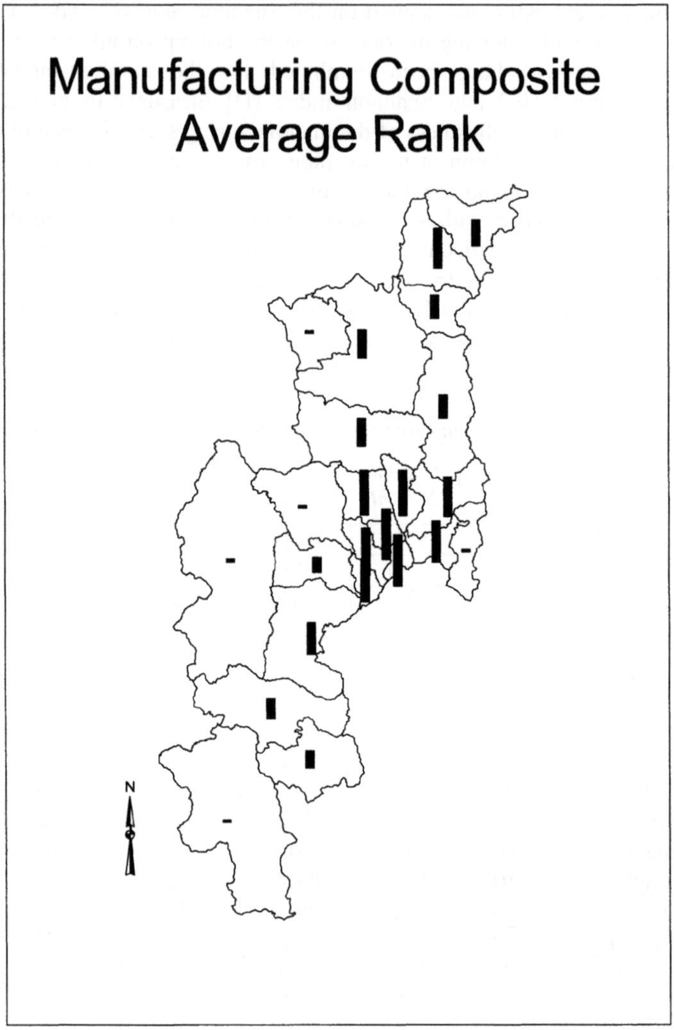

Map 3: Manufacturing composite ranking on size, investment and diversity.

11. SUMMARY

The manufacturing profile of Chiang Mai Province is somewhat consistent with expectations about an outlying region of an economically emerging country. Wood products, fabrics, agro-industries, and food manufacturing are relatively strong.

However, within the Province (and the nearby Northern Region Industrial Estate) there are also many large and modern factories, in both the traditional sectors, especially food, and in newer sectors such as electronics. The core area within 50km of Chiang Mai City is diverse and strong, and can be expected to continue to provide attractive employment opportunities in manufacturing and related industries.

Beyond this core area, however, districts have little manufacturing and/or rely on a few sectors. These outlying districts within this outlying Province are not likely to see local growth of manufacturing employment and will continue to be areas of out migration. These districts typically rely on too few types of manufacturing and have low levels of capital investment per employee. If the assumptions about larger, more modern and capital intensive, and well balanced districts being better able to compete in the international environment of the 21st century are valid, the industrial viability of districts beyond 50km of the core is in question. Furthermore, weaknesses in the hinterland do not help the core region.

Arguments have been advanced that the long-term viability of a regional economy is directly related to its diversity. Regional economic stability is related to regional economic diversity and balance, and stability is desirable for sustained development. Stability helps avoid the physical, environmental, and human costs of boom and bust economic cycles. Manufacturing in the province diversifies the economy and enhances its stability. Environmentally, manufacturing presents trade-offs. Manufacturing activities produce pollution and emissions must be controlled. However, manufacturing also produces revenues that can be used to construct adequate pollution control infrastructures, and these same facilities can be used to treat pollution caused by other economic activities, most notably tourism.

REFERENCES

Chiang Mai Chamber of Commerce. (1998). *Chiang Mai Directory.*

Chiang Mai Provincial Office. (1994). *Investment Plan For Chiang Mai Province* (Executive Summary).

Duerksen, C. J. (1983). *Environmental Regulation of Industrial Plant Siting: How to Make it Work Better.* Washington, D.C.: The Conservation Foundation.

Green, M. B. and McNaughton, R. B. (2000). *Industrial Networks and Proximity.* Aldershot, U.K.: Ashgate Publishing.

Lee, B., Liu, L., Stafford, H. (2000). Industrial Districts: Measuring Local Linkages. In M. B. Green, and R. B McNaughton,. (Eds.), *Industrial Networks and Proximity.* Aldershot, U.K.: Ashgate Publishing.

Leonard, H. J. (1984). *Are Environmental Regulations Driving U.S. Industry Overseas?* Washington, D.C.: The Conservation Foundation.

List of Factories in Chiang Mai. (1996). Chiang Mai Provincial Industry Office, Ministry of Industry.

List of Factories in Lamphun. (1998). Lamphun Provincial Industry Office, Ministry of Industry.

The Industrial Estate Authority of Thailand (brochure). (1998). Northern Region Industrial EstateOdum, E. P. (1983). *Basic ecology.* New York: CBS College Publishing.

Pielou, E. C. (1975). *Ecological Diversity.* New York: John Wiley & Sons.

Stafford, H. (1985). Environmental Protection and Industrial Location. *Annals, Association of American Geographers, 75,* 227-40.

Stutz, F. P. and deSouza, A. R. (1998). *The World Economy: Resources, Location, Trade, and Development.* Upper Saddle River, N.J.: Prentice-Hall.

Thailand Board of Investment. (1997). Untitled pamphlet.

The Federation of Thai Industries, Chiang Mai Chapter. (1996). Directory.

Tourism Authority of Thailand. (1998). *Chiang Mai.*

Zimmermann, E. W. (1933). *World resources and Industries.* New York: Harper & Brothers.

CHAPTER 5

ROGER J. BARRY

SUSTAINABLE DEVELOPMENT AND SUSTAINABLE EMPLOYMENT IN THE MANUFACTURING SECTOR OF CHIANG MAI

An Analysis of the Thai Educational System and the Workforce Supply for the Manufacturing Sector of Northern Thailand

Abstract Chiang Mai and Chiang Mai Province in Northern Thailand have experienced rapid growth, development, and subsequent urbanization, in part as a result of the growth of the manufacturing sector of the local economy. The expansion of the manufacturing sector has been fueled by technological change in agriculture, the rapid expansion of basic literacy, and a national economic policy focused on economic development. Thailand's educational structures have been slow to respond to this change in work-force education needs. Its focus has been on providing "basic education" to make Thailand a literate nation, and a secondary focus has been to strengthen its colleges and universities to supply an educated elite. The concept that educational institutions need to develop the capacity to supply people with marketable work force skills has been slow to develop. Thailand lags behind Malaysia and Singapore in International Standards of Operation (ISO of its local industries. There is a serious shortage of workers with the technical skills necessary to meet the demands of the manufacturing sector. Thailand needs to change its educational policies to support the development of a workforce with the skills needed for an industrializing society that hopes to remains competitive in a world economy.

1. INTRODUCTION

Chiang Mai and Chiang Mai Province in Northern Thailand represent an intermediate-size urban area peripheral to Bangkok that has been experiencing significant growth and development. A major part of this growth, development, and subsequent urbanization of the Chiang Mai area is related to the growth of the manufacturing sector of the local economy (Laopoonpittaya, 2000.) The relationship between the manufacturing sector of the Thai economy, the Thai educational system, and the qualifications of Thai the work force, which are factors in this development and in the urbanization of Chiang Mai, are reviewed in this chapter.

The movement of people from the farm to the city, seeking economic opportunities for themselves and their families is the story of urbanization itself. How successful these people are in finding and sustaining the quality of life they sought is the critical issue. Do they become the slum dwellers of the rapidly urbanizing city, or do they find a job that affords them a decent income and place to live?

M. Romanos and C. Auffrey (eds.), Managing Intermediate Size Cities, 75—88.

Thai central government policies and programs, including incentives to foreign manufacturing companies, have strongly influenced the kind and extent of development that has occurred in the Chiang Mai area. How well prepared people are to take advantage of those developments and to have a sustainable livelihood is also related to government policies and programs.

When the industrialization and the subsequent urbanization of the United States originally occurred, manufacturing companies were organized in such a way as to take advantage of a low-skilled workforce. The early assembly lines could take even illiterate workers and quickly teach them simple tasks that would contribute to the manufacturing process. Since that time, modern manufacturing processes have grown in complexity and sophistication (Sivasomboon, 1998). Globally competitive manufacturing firms with quality assurance processes (like ISO 9000 quality assurance and ISO 14000 environmental standards[1]) and computer-integrated machinery require a literate workforce with higher-level technical work force skills. These are not the skills of a four-year college degree, but work force skills developed usually beyond high school. The primary suppliers of these skills in industrialized countries are technical and community colleges or in-house training programs offered by industry.

Thailand's educational structure has been slow to recognize this change in work-force education requirements as it relates to industrialization. Its main focus to date has been on providing "basic education" and making Thailand a literate nation, (Yuthavong, 1997, 1) while a second focus has been to strengthen its colleges and universities to supply an educated elite. The idea that it also needs to focus on supplying people with marketable work force skills, beyond literacy, has been slow to develop. According to University Affairs Minister Boonchu Trithong in Japan 45 percent of people aged 18-24 attended the university, while in the United States the number is 43 percent. But in Thailand, just 14 percent go on to higher education. If Thailand is to reach newly industrialized status, it is critical that labour skills be improved (Nivatpumin, 1996, 1). This chapter examines the relationship between the growth of the manufacturing sector in Chiang Mai, the subsequent urbanization of Chiang Mai Province, and the current role of Thailand's education system as people in the Chiang Mai region seek to find sustainable employment.

International competition in manufacturing encourages firms to seek three aspects in their search for labor: competitive wage costs, a population with a high degree of literacy, and the potential for training that population for the specialized skills required by the manufacturing processes (Nivatpumin, 1996, 1). Thailand's major international competitor is China (especially after its 2001 admission into the World Trade Organization). The development of a national system that supports work force training skills in Thailand may become a key factor to its success in global competition.

2. TRANSFER OF WORKERS FROM AGRICULTURE TO INDUSTRY

The growth of the manufacturing sector of the Thai economy and the urbanization of the Chiang Mai area of Northern Thailand have been fueled by the triple engines of technological change (especially in the agriculture sector), the rapid expansion of basic literacy over the past thirty years, and a national economic policy focused on economic development. The technological revolution increased agricultural productivity, reduced the need for traditional agriculture workers, and thereby contributed significantly to the urbanization of the country. This trend continues. In 1997, according to the Office of the Board of Investment in the Office of the Prime Minister of Thailand, the rapid structural transformation of the economy away from agriculture will result in a steady transfer of workers from traditional agricultural activities to more modern industry and service sectors. It has been estimated that nearly 60 percent of the labour force is currently employed in agriculture, with manufacturing accounting for 12 percent, and 28 percent in the service sector (Office of the Board of Investment, 1997, 48). The decline of agricultural employment is expected to continue along with the growth of employment in both the manufacturing and service sectors of the economy.

Thailand's economy registered high growth rates since the beginnings of its industrialization in the early 1960's. This growth continued until the crash of 1997. The growth in Thailand was led principally by a rapidly growing manufacturing sector. According to the World Bank, Thailand ranked number one in the world for its average per capita Gross National Product (GNP) growth from 1985 to 1994 (Office of the Board of Investment, 1997, 4).

In 1985, construction was completed for the Northern Regional Industrial Estate in Lamphum (about 20 miles south east of Chiang Mai). Currently there are 24,500 manufacturing related jobs in this Industrial Estate. In interviews with Industrial Estate officials, they estimated that 50% of the current employees came from the local area and that the other 50% moved to the area for new industrial-based jobs. A majority of those who moved came from the northeastern portion of Thailand, which has been a traditionally economically depressed area.

The introduction of universal "basic education" created a plentiful supply of labor with basic literacy skills available to the manufacturing sector. The national plans for education and economic development set a policy framework for rapid expansion of the manufacturing sector (Pongpaibool, 1990, 4). These changes in many respects mirror what happened in industrialized nations around the world in an earlier part of this century. Technological changes in the agriculture sector caused significant increases in agricultural productivity per acre and per agricultural worker, reducing the need for an agricultural workforce and supplying workers for an expanding industrial and urbanizing society.

2.1. Basic Literacy Skills in Thailand

Starting in the late 1960's six five-year plans, called the National Educational Development Plans (NEDPlan) resulted in universal "basic education" in Thailand. In 1937, Thailand had an illiteracy rate of 69%, but by 1960 it had dropped to 29%, and by 1985, to 8.8%. In 1988, the school admission ratio of first graders in proportion to the six-year-old population was reported at 99% (Pongpaibool, 1990, 9). Thailand had become a country of literate people. Expansion of enrollment in secondary and post-secondary education institutions is happening at a slower rate. The significance of this slow expansion of advanced educational opportunities on the Thai economy is discussed later in this chapter.

This previously rural Thai workforce has become the source of industrial workers in the expanding Thai economy. There is a plentiful supply of low-wage labor with basic literacy skills available to the manufacturing sector, because Thai workers do have such basic literacy skills (reading, writing and basic math skills) and are perceived by manufacturing employers to be relatively easy to train in the basic tasks needed for line production in industry.

Interviews and a preliminary workforce survey of Industrial Estate officials and plant managers in the Chiang Mai area found that they were satisfied with workers' basic reading and writing skills, but that some employers were dissatisfied with workers' basic mathematical skills. Those interviewed said that the Thai employees generally had the technical skills to operate machines, read gauges and instruments, and take measurements but that the local workforce lacks people with the technical skills required to operate modern manufacturing processes.

2.2. Economic Development Policy

The industrialization of the Thai economy in the last thirty years has been focused on Bangkok and the surrounding areas. Chiang Mai city and province, located almost 900 kilometers to the North of Bangkok, has seen some growth of the manufacturing sector of its economy but not at the same rate or scope as Bangkok. Attempts to direct growth away from Bangkok and toward the hinterlands have been institutionalized in the Eighth National Plan for Economic Development.

The Eighth Plan (1997-2001): Focusing on Sustainable Development of Natural and Human Resources, adopted in 1996, had the following main objectives:

- Development of human potential in terms of physical health, knowledge, and occupational skills;
- Enhance the development of regions and rural areas by improving quality of life;
- Develop the national economy toward greater stability, prosperity, and balance;
- Provide opportunities for public participation in the process of equitable development;
- Encourage the efficient use, preservation, and rehabilitation of natural resources and the environment; and,

- Reform the system of administrative management to create opportunities for private development organizations, the private sector, and communities to participate in the process of national development (National Economic and Social Development Board, 1996, 6-7).

The early example of this policy approach came with the development of an industrial park near Lamphum (about 20 miles south east from Chiang Mai) that was started in 1983 and completed by 1985. In conjunction with the construction of the infrastructure for an industrial estate (road system, water supply, electricity, telecommunications and central waste water treatment plant), the following incentives and privileges were accorded foreign investors in the industrial estate:

- Permission to own land;
- Permission to bring foreign technicians and experts to work;
- Permission for foreign technicians and experts to stay in the country;
- Permission to take or remit abroad foreign currency;
- Exemption of import duty and value added tax on imported machinery, components, etc., and material imported for factory construction;
- Exemption of import duty and value added tax on raw materials;
- Exemption of export duty and value added tax on exported goods;
- Exemption or refund of taxes and duties for local goods utilized for production (Operational Control Section, 1998, 5).

By 1996, 19,925 million bhat ($805 million, 1996 exchange rate) had been invested with 24,500 persons employed in the industrial estate. When you take into account the multiplier effect of 24,500 manufacturing related jobs, the impact on the local economy was phenomenal (Sawan, 1998).

The growth of manufacturing outside the industrial park and across Chiang Mai Province was also significant, but not as dramatic as found in the government-sponsored industrial park. There are almost 2,000 registered plants (those with seven or more employees or more than 5 horsepower installed), and Chiang Mai Province has about 1,600 manufacturing plants with 15 or more workers. (Pholpipattanaphong, 1998; Teriyapirom, 1998). It is variously estimated that manufacturing employees constitute between 12 and 26% of the Province's workforce (Laopoonpittaya, 2000).

3. HIGHER ORDER LITERACY AND TECHNICAL SKILLS

In interviews, employers did indicate that the Thai employees generally did not have the higher order technical skills necessary to maintain hydraulic, mechanical, or electrical systems. The Office of the Board of Investments confirmed this finding in its 1997 publication, which stated that at the lower skill levels, the low secondary school enrollment ratios are a source of concern, as firms are increasingly demanding secondary school graduates. In addition, vocational graduates, although quite numerous, tend to be concentrated in fields that are not in heavy demand and generally have received little practical training. These problems are lessened to some degree by the easy trainability of Thai workers, something noted by Thai and foreign firms alike (Office of the Board of Investment, 1997, 48).

In terms of work force training, employers indicated that while private consulting firms, local educational institutions, and government agencies conducted some training, the majority of specific job training for industrial line workers is conducted in-house by the company. This included orientation to the industry, specific job training, supervisory training, and team training. Local factories generally train new employees for one or two months, according to the director of the Northern Region Industrial Estate.

A complaint by industrial officials and local plant managers during our interviews was the continuing shortage of Thai workers with advanced technical skills needed by industry. The Office of the Board of Investment in the Office of the Prime Minister of Thailand stated in its 1997 Investment Environment publication that the growth rate and the sophistication of the industrial sector has increased, the supply of workers with higher levels of occupational skills has not kept up with the demand (Office of the Board of Investment, 1997, 48)

In an article in the *Bangkok Post* in June 1997, Raymond Eaton, chairman and chief executive of the Integra Group said past warnings were largely ignored by both the public and private sectors, and consequently factories throughout the country are now forced to face the reality of employing an under-educated, relatively expensive, marginally productive workforce. Currently 70% of Thai workers have only a primary education (*Bangkok Post*, 1997, 1).

Manufacturing is undergoing an evolution of the skills required by its workforce. Basic literacy has been a minimum requirement for most manufacturing firms, but with the introduction of quality assurance management practices, production teams, more sophisticated production techniques and machinery, and implementation of the ISO 9000 quality systems management and ISO 14000 environmental standards for exports to the European Union, the skill level requirements of the worker in the manufacturing plant continues to rise, especially if the plant is to remain competitive in a global market.

3.1. Shortage of Engineering and Other Technologically Advanced Skills

During a National Forum held at the Dusit Thani Hotel in July 1998, which discussed "Can Thailand Be Saved?" expatriate businessmen complained of the shortage of employees with technical skills. David Nardone, President of Herlmaraj Land Plc, which owns and operates several large industrial estates on the Eastern Seaboard, said the shortage of engineers had remained unchanged since he came to Thailand in 1983 and that there is still a shortage of qualified engineers. The government just has not put enough resources into developing more technically qualified people. This shortage, he added, prevented many local firms from upgrading their labour-intensive operations to highly skilled and technologically advanced industries. It is not surprising that the electronics, computer, and heavy technical industries have not done that well (Woo, 1998, A2).

An employer survey conducted in the 1980's by the Office of the National Educational Commission of the Thai labor market found that it was difficult to recruit photoengravers, airplane mechanics, electricians, accountants, textile technicians, mechanical, electrical, or hydraulic technicians, or draftsmen. Employers complained of the quality of the vocational and technical education training programs and recommended more emphasis be placed in the curriculum on actual practical experience and on personal traits such as honesty, industry, pride, and entrepreneurship (Bovornsiri, 1980, 1-18).

3.2. Thailand Lags in ISO Certification of Its Local Industries

Thailand lags behind Malaysia and Singapore in International Standards of Operation (ISO) certification of its local industries. The ISO 14000 is an environmental standard set by the European Union for suppliers of exports. Only 86 Thai companies had ISO 14000 certification as of 1998. This credential is a key requirement for competition in the export market especially in Europe. According to Sukanya Sinsakul, director-general of the Thai Industrial Standards Institute, if local exporters cannot meet this green credential they will lose those markets. The lack of locally trained ISO technicians forces companies to use high cost consulting firms to help them reach the standards (Sivasomboon, 1998, 5).

The National Investment Board dealt with the skill shortage of Thai workers in the Lamphum Northern Industrial Estate by giving specific work force incentives and privileges to investors. Companies that located in the estate were given permission to bring foreign technicians and experts to work on projects, as well as permission for foreign technicians and experts to stay in the country (Operational Control Section, 1998, 5).

4. THAI EDUCATIONAL SYSTEM

To understand the issues of work force skills as they relate to the manufacturing sector and the sustainable development of the Thai economy, we need to understand the Thai educational system. Figure 1 gives an overview of the enrollment structure of Thailand's formal educational system.

4.1. Basic Education

Thailand's early goal of basic education of all citizens through the sixth grade has been met. National efforts in laying groundwork for public education began as far back as 1898 when schools were established to provide preschool, primary, secondary, and vocational education for children in Bangkok and the provinces. Universal "basic education" resulted from the actions and implementations of the six five-year plans called the National Educational Development Plans (NEDPlan). The term "basic education" is confined to education at the primary level, which is six years of formal schooling and literacy education for adults. By the mid-70's, primary level enrollment stabilized reaching above 90 percent. In 1987, 92 percent of Thailand's primary school age population (6 – 11 age group) was enrolled in elementary school (Pongpaibool, 1990, 7-9). Those levels of enrollment continue today. The result of this accomplishment is that Thailand has a literate workforce capable of performing basic low level manufacturing tasks. The high literacy rate makes them trainable for higher order tasks, but they generally do not come equipped with those skills (Pholpipattanaphong, 1998).

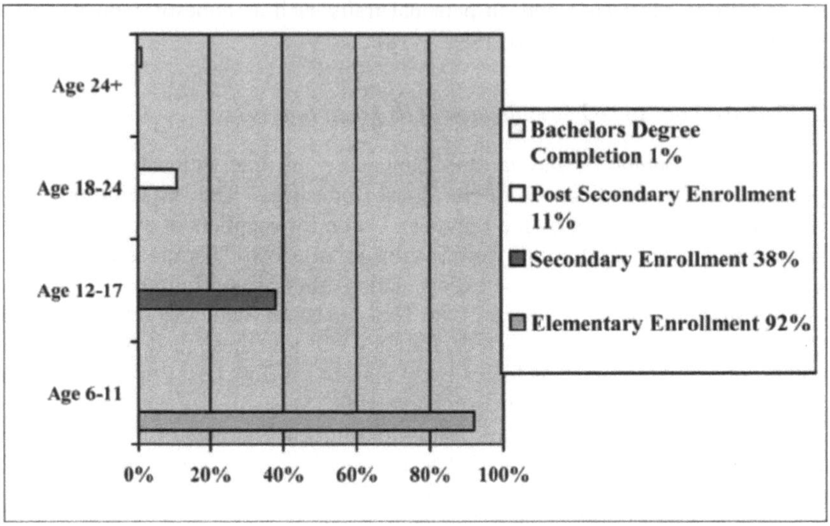

Figure 1. Enrollment in Thai Education System
Source: Pongpaibool, 1990, 7-9.

4.2. Secondary Education

The seventh and eighth NEDPlans called for a new goal of universal "basic education" of all citizens through the ninth grade. Progress has been made but the goal has not been met. Enrollment at the secondary level (12-17 age group) in 1998 was 28% of that age group (Pongpaibool, 1990, 7-9). Currently, 38 percent of the secondary age population is enrolled in secondary school, an increase of 10 percent over the past 10 years (Rangsitpol, 1997, 1).

Politicians feeling the sting of criticism of slow secondary growth enrollment have reacted. Recently, the Deputy Prime Minister and education Minister Sukhavich Rangsitpol proclaimed that the overall educational reform objective is to achieve 12 years of truly universal, compulsory, quality education for all Thai children by the year 2007 (Rangsitpol, 1997, 3).

Enrollment options available for secondary education in Thailand include the traditional college preparatory high school track or a vocational/technical track. These tracks have the same social implications that they have in western societies. The social elite and those who aspire to the middle class status want their students to enroll in college preparatory curriculum. The linkage of education institutions and curriculum development to meet work-force skills needed by business and industry has yet to be developed.

4.3. Secondary School-Technical Education

The Technical Education track in Thailand consists of three levels of study. The "first level" is for three years and the student receives a certificate equivalent to a high school degree. The "second level" of study is two years and is equivalent to an associate degree. The "third level" of study is the professional education certificate after an additional two years of study and is equivalent to a bachelors degree. Our focus is on the vocational/technical track, which should be designed to have the responsibility for supplying the technical workforce for the country.

In Chiang Mai Province in 1990, there were 9,918 secondary students enrolled in secondary vocational/technical programs. About half (51%) of these students were enrolled in private technical schools. The private technical schools are selective in admissions and charge an average tuition of 10,000 bhat ($250) per semester. Private schools specialize in low cost/high return curricula such as business and data processing. More expensive technical training is left to the public sector institutions or not provided (Tiam-Tong, 1998).

There are both seven private and four public vocational/technical schools and one public Institute of Technology in Chiang Mai with a total enrollment in 1990 of 16,109 students. Of the seven private vocational/technical schools, three of the schools are vocational high schools teaching the "first level" students. Two teach "first level and second level" from high school through the associate degree programs, and two of the private technical schools teach all three "levels" of vocational/technical curriculum from secondary school through the bachelors degree level (Chiang Mai Provincial Government, 1990, 16-28).

Three of the public vocational schools operated by the Vocational Education Department and the public Institute of Technology teach all three "levels" from secondary, through associate and bachelors level curriculum. One of the vocational/technical schools, Chiang Mai Multi-Technic School is a non-credit adult vocational/technical education institution (Thiennimitra, 1998).

4.4. Higher Education

The type of degrees offered by state and private colleges in Thailand correspond closely to those offered in the American system. A diploma or associate's degree "second level" is obtained after the second or third year of college. Bachelor's degrees, "third level" is awarded following four years of study in most subjects (Office of the Board of Investment, 1997, 46).

The higher education system in Thailand has expanded in the past decade. A number of public and private secondary institutions have been encouraged to provide college level programs. In 1988, enrollment of the college age group (18 - 24) was 11% of that age group (Chiang Mai Provincial Government, 1990, 16-28).

Access to higher education technical/vocational programs in Thailand however is not similar to the American post-secondary system. Access for both traditional college programs and for vocational and technical programs is highly selective. Access to higher education in Thailand is rationed by test scores, grades, program enrollment caps, and for private schools, high tuition. In an interview with public technical school officials in Chiang Mai, entrance into technical programs is competitive with the college accepting only one out of five applicants because of limited space in programs (Thiennimitr, 1998).

The curriculum at the public Technical Institute has the variety and depth of the technical programs at American community colleges and four-year technical colleges. The faculty is well qualified and has work experience in their technical fields. The facilities are modern and the equipment is up-to-date.

Enrollments in private technical colleges constituted 50 percent of the associate degree enrollment with an average tuition cost of 10,000 bhat ($250) per semester. Public technical associate degree level tuition averages about 1,000 bhat ($25) per semester. Enrollments in 1990 in Chiang Mai Province for bachelor level technical education programs totaled 770 students (Thiennimitr, 1998).

In 1990, there were 2,823 students enrolled in associate degree type programs in both public and private technical colleges, with an additional 1,810 adult students enrolled in the public adult education program. If you were comparing two-year college enrollments with similar levels expected in the United States you would predict enrollments of approximately 45,000 students for a region with the same population base. However, since enrollment is limited by funds provided by the central government, enrollment is competitive and selective. The faculty serves the brightest students. Because of limited resources there is limited enrollment. The technical colleges are severely limited in their ability to provide the workforce needed for the expanding demands of business and industry (Thiennimitr, 1998.).

Looking at higher education enrollment data on university and vocational school graduates by subject for the 1995 academic year, it appears that the public universities' role is focused on education as a process of social mobility. The number of graduates in non-technical programs is similar to any traditional four-year college or university in the United States. Enrollments are strong in education and teacher training, humanities and religious studies, social and behavioral sciences and other traditional liberal arts programs. Engineering and business administration have strong enrollment, the number of graduates in technical programs that supply the workforce for an industrializing economy is relatively low. For example, in public universities for 1995, only eight percent of the graduates had majored in engineering (Office of the Board of Investment, 1997, 27).

Secondary schools, colleges and universities are not yet major players in planning for or supplying the skilled workforce needed by Thai industry. Thailand's colleges and universities, modeled on European and American research universities, have an elitist perspective on their role of educating people for the Thai workforce. Upper and middle class Thai families want their children to be well educated in the upper class European and American tradition. Thai universities focus on traditional academic programs while there is a shortage of graduates in programs needed to supply the industrial sector of the economy. Employment opportunities among college educated Thais in the traditional arts and sciences curriculum are no better in Thailand than in the other industrialized countries. There appears to have been modest growth in post-secondary technical education programs in recent years, but the level of enrollments and the number of graduates appear to have little, if any, relationship with the workforce needs of the Thai economy. The linkage between business and education and Thai educational institutions is weak to non-existent. The concept that Thai educational institutions should train its population for sustainable employment rather than "provide an education" has not become part of the policy of the Thai education establishment (Teriyapirom, 1998).

5. CURRENT THAI LABOR FORCE

The current Thai labor force is made up of a large number of people with basic reading, writing, and math skills (estimated 98%) who have completed the sixth grade and a slowly emerging population with a secondary (high school) education (current enrollments now 38%). The workforce available to the manufacturing sector is literate (sixth grade level), but with a shortage of workers with the technical skills necessary to operate sophisticated manufacturing processes. Workers capable of making measurements of manufactured products and calculating mean, median, mode and standard deviation to ensure product quality are in short supply. Manufacturing firms are setting up their own training programs to develop the skills in the work force or bringing in expatriates from other countries to fill this skill-training gap (Harper, 1998; Nittayajarn, 1998).

This skills-training gap extends to other parts of the Thai economy as well, both from the perspective of employment sustainability for the worker and the sustainability of the Thai economy. All sorts of jobs are requiring higher-level technical skills beyond basic literacy. Ambulance drivers need to become emergency medical technicians, ditch diggers need to become construction machine operators, handicraft makers need to become computer integrated manufacturing technicians, farm laborers need to understand genetic engineered hybrids and non-traditional cropping systems, policemen need a background in law and appropriate law enforcement techniques, and typists need to become computer operators. The link between technical training, work force skills, and employment sustainability become stronger and stronger in an emerging industrialized and post-industrialized society.

The rapid growth in the manufacturing sector as a result of Thailand's economic policies is reflective of the changes that are occurring in other sectors of the Thai economy. For sustainable development to work in an industrializing society, the infrastructure for sustainable employment will need to be developed as well. This does not mean training people for one job, but putting into place the structures that train and re-train people as the structure of the economy and the requirements of jobs change. Currently, Thailand does not have that type of educational infrastructure in place and it does not have the government policies in place that support that sort of educational structure development.

6. CONCLUSIONS

Thailand will need a significant change in direction to produce a workforce that will meet the needs of an urbanizing, industrializing society and to remain competitive in a world economy. These recommendations include:

Elementary Education Emphases
Elementary education will need to place an emphasis on science and math skills. The Thais have been successful in achieving universal literacy. This has also fueled the engine of economic success during the early 1990's. However, now that Thailand is competing in a global economy it must have a workforce that is competitive. A contribution to this success would be a deepened emphasis on ensuring competent basic skills at the elementary level and an emphasis on science and math skills to prepare students to acquire advanced technical skills and to be life-long learners as the need for additional training and re-training occurs.

Secondary Education Expansion
Secondary education needs to continue to expand its enrollment base and should emphasize math and technical skills for a majority of its students and place special emphasis on basic skills necessary for students to become life-long learners. Many of the technical skills required for sustainable employment can be learned in secondary school and the access to these institutions needs to be expanded. The enrollment needs to expand from the 1997 number of 38 percent to twice that number as rapidly as possible. The secondary schools also need to link with employers to ensure that they are matching their curriculum with employment needs. This emphasis on employability skills and the tools to become life-long learners to re-train for new jobs as the structure of the economy changes will be key to sustainable employment in Thailand.

Increase in Resources for Technical Training
Increase resources for technical training at both the secondary and post-secondary level, with a specific emphasis on work force needs of business and industry. Also, creating open access education institutions that permit enrollment on a non-competitive basis both for youth and adults is a critical component of an educational structure that supports sustainable employment. Thailand needs to look at the Singapore model of community college education as a model for potential development.

Public/Private Partnerships
The establishment of school, business, and industry partnerships and the commitment of vocational and technical education to meet the workforce training needs of Thai society. Workforce partnerships are not a tradition in the Thai educational system. The introduction and fostering of this relationship between employers and educators will be a key factor in the development of an educational structure that will support sustainable employment for the Thai people and sustainable development for Thailand.

NOTES

[1] The International Organization for Standardization (ISO) is a worldwide federation of national standards bodies from some 140 countries, one from each country. ISO is a non-governmental organization established in 1947. The mission of ISO is to promote the development of standardization and related activities in the world with a view to facilitating the international exchange of goods and services, and to developing cooperation in the spheres of intellectual, scientific, technological and economic activity.

ISO's work results in international agreements, which are published as International Standards. Standards are documented agreements containing technical specifications or other precise criteria to be used consistently as rules, guidelines, or definitions of characteristics, to ensure that materials, products, processes, and services are fit for their purpose.

The existence of non-harmonized standards for similar technologies in different countries or regions can contribute to so-called "technical barriers to trade". International standardization is well-established for many technologies in such diverse fields as information processing and communications, textiles, packaging, distribution of goods, energy production and utilization, shipbuilding, banking, and financial services.

Development agencies are increasingly recognizing that a standardization infrastructure is a basic condition for the success of economic policies aimed at achieving sustainable development. Creating such an infrastructure in developing countries is essential for improving productivity, market competitiveness, and export capability.

ISO 9000 is concerned with "quality management". This means what the organization does to enhance customer satisfaction by meeting customer and applicable regulatory requirements and continually to improve its performance in this regard. ISO 14000 is primarily concerned with "environmental management". This means what the organization does to minimize harmful effects on the environment caused by its activities, and continually to improve its environmental performance.

REFERENCES

Bovornsiri, Varaphorn and Gerald Fry. (1980). Technical Education and The Thai Labor Market, Office of the National Educational Commission, Bangkok, and the International Institute for Educational Planning, July.

Centre for Engineering Services, Chiang Mai University. (1994). *Investment Plan For Chiang Mai Province (Executive Summary)*, Chiang Mai Provincial Office, Chiang Mai, Thailand.

Chiang Mai Municipality. (No publication date). Local Government Chiang Mai Municipality, Public Relations Section, Chiang Mai Municipality, Thailand.

Chiang Mai Provincial Government. (1997). Chiang Mai Province, Chiang Mai, Thailand.

Chiang Mai Provincial Government. (1990). Data of Technical Schools in Chiang Mai Province, 1990. Chiang Mai, Thailand.

National Economic and Social Development Board. (1996). Summary, The Eighth National Economic and Social Development Plan (1997-2001), Office of the Prime Minister, Bangkok, Thailand.

Nivatpumin, Chiratas, Busaba Sivasomboon and Jirajaree Chaimusik. (1996). "Thailand in crisis for qualified labour force, only 14% of Thais go to college: Boonchu." *Bangkok Post*, September 6.

Office of the Board of Investment. (1997). The Investment Environment in Thailand. Bangkok: Office of the Prime Minister, Royal Thai Government.

Operational Control Section. (1998). Northern Region Industrial Estate, The Industrial Estate Authority of Thailand, Office of the Northern Regional Industrial Estate, The Industrial Estate Authority of Thailand, February.

Pongpaibool, Panom. (1990). *Basic Education for All: A Mission Possible, Thailand's Illustrative Case and EFA Action Plan*, prepared for the World Conference on Education for All, March 5-9.

Rangsitpol, Sukhavich. (1997). "Education for life, Deputy Prime Minister and Education Minister Sukhavich Rangsitpol addressed the Foreign Correspondents Club last night on his government's educational reforms. Below are excerpts." *Bangkok Post*, April 3.

Sivasomboon, Busaba. (1998). "Failure to adopt ISO risks access." *Bangkok Post*, September 28.

Woo, K. I. (1998). "Expatriates seek long-term plans for education." *The Nation, Bangkok*, A5, August 8.

Yuthavong, Yongyuth. (1997). "What Thailand needs to survive, Thailand lags sadly behind its neighbours in education, something it badly needs to improve if it is to continue to thrive in the future." *Bangkok Post*, October 24.

_______. (1998). "Educational minister slammed for 'not achieving much'." *The Nation, Bangkok*, A 2, August 8.

_______. (1997). "Thailand slips behind as warnings ignored." *Bangkok Post*, June 18.

INTERVIEWS

Anumanrajadhon, Tanun. (1998). Professor, Department of Political Science, Chiang Mai University, Chiang Mai, Thailand.

Harper, Cindy. (1998). Manager, Harper Broste Manufacturing, Co., Ltd., Chiangmai, Thailand.

Laopoonpittaya, Winyoo. 2000. Director, Northern Region Investment & Economic Center Office of the Board of Investment, Chiang Mai, Thailand.

Nittayajarn, Jesada. (1998). Parent Seed Manager, Pioneer Hi-Bred (Thailand) Co., Ltd., Northern Regional Industrial Estate, Lampoon, Thailand.

Pholpipattanaphong, Prapas. (1998). Managing Director, Chiangmai Frozen Foods Public Co., Ltd., Chiangmai, (and member of Northern Industrial Federation of Thai Industries Board of Directors), Thailand.

Robert, G. Lamar. (2000). Agriculture and Resource Economist, Senior Advisor, Social Research Institute, Chiang Mai, Thailand.

Sawan, Nakon. (1998). Director, Northern Region Industrial Estate, Lampoon, Thailand.

Staff Briefing. (1998).National Economic and Social Development Board, Bangkok, Thailand.

Teriyapirom, Boonsong. (1998). President, Northern Industrial Federation of Thai Industries, Chiang Mai, Thailand.

Thiennimitr, Prasert. (1998). Assistant Director of Academic Affairs, Rajamangala Institute of Technology, Northern Campus, Chiang Mai, Thailand.

Thiennimitra, Pramoj. (1998). Faculty Member, Printing Division, Rajamangala Institute of Technology, Northern Campus, Chiang Mai, Thailand.

Tiam-Tong, Chongchit Sripun. (1998). Deputy Director, Social Research Institute, Chiang Mai University.

CHAPTER 6

CARLA CHIFOS
JOHANNA W. LOOYE

THE HANDICRAFT SECTOR IN CHIANG MAI: ITS ROLE IN SUSTAINABLE URBAN DEVELOPMENT

Abstract For centuries the city of Chiang Mai has been synonymous with traditional crafts worked from materials as varied as silk, cotton, silver, wood, and lacquer. This former capital of the Lanna Tai Kingdom has been populated with a wide range of craftspeople who could meet the whims and needs of a cultural and religious center backed by the wealth of a powerful kingdom. Now one of the country's primary tourist destinations, it is populated with artisans using the same materials to serve the whims of the tourist market, both locally and globally. This chapter is an attempt in making a first cut at assessing the contribution of handicrafts towards a sustainable future for the urbanizing region of Chiang Mai. We operationalize our conception of sustainable development as it would be applied to an economic sector and its relationship with the urban area in which it operates. We then proceed to tell the story of the handicraft sector in Chiang Mai, first with a brief account of the historical evolution of the sector followed by a description of the sector as we observed it in 1998, with particular attention to operations and relations within the production-distribution/promotion-consumption network and their geographic arrangement. We then analyze Chiang Mai's handicraft sector using our sustainable development framework and conclude with our recommendations for the sector based on our analyses.

1. INTRODUCTION: EXPLORING THE CONTRIBUTION OF HANDICRAFTS TO SUSTAINABLE URBAN DEVELOPMENT IN CHIANG MAI

For centuries the royal city of Chiang Mai has been synonymous with traditional crafts worked from materials as varied as silk, cotton, silver, wood, and lacquer. This former capital of the Lanna Tai Kingdom[1], located in what is now Northern Thailand, has been characterized as "an artisan's town" populated with a wide range of craftspeople who could meet the whims and needs of a cultural and religious center backed by the wealth of a powerful kingdom (Lubeigt, 1994, 120). Nestled in the fertile Ping River valley, which is surrounded by natural beauty of mountains and tropical forests and blessed with an upland climate, and positioned as a key point in traditional regional trading routes, Chiang Mai has maintained an important role in Thailand throughout history. Today, it is considered the second city to Bangkok in size and importance. Even though it is no longer the quaint traditional home of a splendid kingdom, this romantic aura along with its natural amenities continue to draw both domestic and foreign tourists, making Chiang Mai one of the primary tourist destinations in Thailand. The new artisans of Chiang Mai, many

M. Romanos and C. Auffrey (eds.), Managing Intermediate Size Cities, 91—124.

coming from a lineage of the traditional craftspeople of the region, now serve the whims of the tourist market, both locally and globally.

Although it is difficult to estimate the current magnitude of the handicraft sector in the Chiang Mai economy, by all accounts it seems to be important both to the tourism industry and to the manufacturing sector. Tourism brochures claim that "Chiang Mai is, quite simply, Thailand's major center for quality handicrafts" (Souvenir Guide, ND, 14), and Pye asserts that the vast majority of Thailand's craft exports come from either Bangkok or Chiang Mai (Pye, 1988, 68). Finally, the Investment Plan for Chiang Mai Province provides data that permit an estimate of the sector's magnitude. The four industrial classifications that are likely dominated by handicraft industries, (1) wood and wood products, (2) fabrics and related products, (3) gems and jewelry, and (4) ceramics, together account for 16 percent of the number of licensed factories, 17 percent of capital, and 39 percent of workers in the Province (Government of Thailand, 1998, 14). We admit that these figures overestimate the number of businesses in handicraft-type production and hence, the percentage of workers employed in handicraft-type production. However, due to the small-scale nature of the majority of these businesses, a substantial number are not licensed and therefore do not show up in official statistics. Basically, all indications point to handicrafts as a significant sector for the province.

Due to its long history and to its predominant role in the modern tourist trade of Northern Thailand, the structure of the handicraft economy in Chiang Mai is quite intricately articulated. The horizontal and vertical linkages are far reaching in both geographical area and numbers of individuals involved. Social relations within the production and distribution chains vary greatly. The dichotomy between formal and informal activities is blurred in the multiplicity of steps most of the products transcend. Analyzing the structure of this sector at a general level, even though it will miss many of the intricacies, will offer some insights into the predominant characteristics of the current handicraft economy in Chiang Mai, which in turn can be used to make an inquiry into the role of this vital, yet often overlooked economic sector in the urban development of Chiang Mai.

The analysis of urban development in this chapter will be couched within the concept of sustainable development. Much is written on the various definitions of this concept as well as on the continuing debate over the applicability of this vague concept to 'real world' situations. However, despite this, there is no agreed upon definition of a sustainable urban regime.[2] We will not grapple with definitional issues here nor will we contribute to this discussion in a theoretical sense, but will rather take some of the key aspects of sustainable development as commonly defined and seek to identify either the existence of, or potential for, their realization within the context of one economic sector and livelihood strategy in Chiang Mai.

The commonly accepted definition of sustainable development in a developing country situation is the anthropocentric definition from the Bruntland report.[3] This definition emphasizes the importance of meeting the needs of the population (not the demands), implying a system that allows everyone access to the means to acquire and maintain an acceptable bundle of goods, services, and quality of life. This implicates an economic system that can benefit all people at least at a level of minimum requirements, which may be culturally defined. It also emphasizes maintaining this level of benefits across generations, so the support system that feeds this economy and provides an acceptable quality of life is maintained. The basis of this support system is the natural environment. Though this definition sounds simple and even vague, it can be interpreted to suggest an intricately interrelated physical, social, and economic system in which the key to a sustainable future is finding the balance among these components. For this analysis we will look at the interrelationship among these three components -- physical/environmental, economic, and social – which is the commonly accepted tripartite of sustainable development, and we add a fourth dimension – cultural heritage –which we feel is particularly important in our case. The economy, and only the portion related to the handicraft sector, will be the focal point and its relationship to the other elements will be assessed.

It is obvious that crafts have provided both livelihood and cultural identity to the region of Chiang Mai for hundreds of years. However, it is not so clear what promise this sector holds for a rapidly growing and globalizing region. Should handicrafts be supported to become one of the pillars of a sustainable region? In our search for an answer to this question, we strove to understand the structure and role of handicrafts through field work during the summer of 1998 as part of the Joint Study. Our approach during our relatively short time for delving into a new culture and place was to first get a tourist's point of view of handcrafts in the region—What is there to buy and where can you buy it? Once we felt comfortable with that, we delved into the behind-the-scenes story of handicraft production, distribution, and consumption in Chiang Mai. We conducted interviews with government agency representatives, NGO representatives, business organizations (and individual business owners), and individual artisans to gain insight into the dynamics of the sector (See List of Interviews). We were particularly interested in tracking the interactions among the producers, buyers, and consumers. In addition, we sought information about environmental impacts of the raw materials, production processes, and marketing of the handicrafts items. In order to ascertain the historical and cultural role of this sector, we strove to discover as much as possible about the evolution of this sector and the traditions, structures, and locations of production. It was difficult to collect "hard" data on the handicraft sector past and present, given the dearth of literature addressing this topic as well as the limited timeframe for field work and no prior contacts within the sector. However, through our interviews and site visits to the production and distribution points for handicrafts and piecing together whatever written accounts we could find, we were able to construct what we feel is a reasonable story of the handicraft sector in Chiang Mai.

This chapter is an attempt to tell this story and then place our findings into a framework of sustainable development, making a first cut at assessing the contribution of handicrafts towards a sustainable future for the urbanizing region of Chiang Mai. The next section of this chapter operationalizes our conception of sustainable development as it would be applied to an economic sector and its relationship with the urban area in which it operates. We then proceed to tell the story of the handicraft sector in Chiang Mai, first with a brief account of the historical evolution of the sector followed by a description of the sector as we observed it in 1998, with particular attention to operations and relations within the production-distribution/promotion-consumption network and their geographic arrangement. We do this by following the path of discovery that we traveled: we look at what is available, where it is sold, where it is made, and who the actors behind the system are. We focus on three key actors within the handicraft sector in Chiang Mai—individuals, the government, and NGOs—and the form of participation for each, in order to see the social relationships associated with the sector. In section five, we then analyze Chiang Mai's handicraft sector using our sustainable development framework and then conclude with our recommendations for the sector based on our analyses.

2. FRAMEWORK: SUSTAINABILITY AND HANDICRAFTS AS AN ECONOMIC SECTOR AND LIVELIHOOD STRATEGY

Like all the chapters in this book, this chapter seeks to evaluate the current and potential future sustainability of Chiang Mai as an urban region. With this end in mind, we chose to contribute to this evaluation by taking the integrative premise of the concept of sustainable development as the foundation of our approach. By taking an economic sector, that of handicrafts,[4] as a point of departure, our analysis strives to explore the interrelationships among this sector and the broad goals of maintaining environmental integrity and quality, improving social equity and quality of life, and preserving the cultural heritage, while bolstering economic conditions within this multifaceted urban system. These interrelationships may be cause-and-effect linkages or a more complex set of tensions and conflicts among these components of the system.

In developing a framework to carry out this analysis, we apply the key values embedded in the concept of sustainable development to analyze the role of one economic sector and how it affects life and development in an urban area. Despite the on-going debate over the definition of sustainable development and viability of the concept to be a long-term development goal,[5] headway has been made in applying the concept to the planning and development of urban areas and communities and suggestions for operationalization exist (Wheeler, 1996; Beatley and Manning, 1998; Berke and Conroy, 2000). Most of these applications suggest criteria focusing on the interrelationship of the natural and built environment. How the economy and social equity fit in are less well worked out in relation to sustainable places/cities/communities.[6] We are attempting to contribute to the understanding of the interrelationship of the economy with the objectives of sustainable development not by picking up the debates over the integration of environmental aspects with economic analysis as is the most common approach,[7] but by examining an actual sector in a local economy and searching for the linkages to the environmental, cultural, and social aspects of the region.

In searching for the way to carry out our analysis two types of inquiry were consulted: the application of sustainable development to urban areas and to livelihood strategies. The analysis of sustainable development of urban areas has tended to focus on cities as physical entities, elaborating on the interrelationships between physical form and the natural environment.[8] These physical aspects of cities may be as broadly defined as the overall form, densities, and mix of physical development, or as minutely defined as the details of buildings and individual properties. In the less explored area of linking economic activities with sustainable development the trend has been to address the subject dichotomously: with either the firm or the worker/household as the units of analysis. The view of the firm is usually an exploration into the environmental impacts of the firm's behaviors or the ownership of the firm (local vs. global). The analysis of the worker or household has been the focus of studies of the sustainability of livelihoods. The emphases have been on the relationship between poverty and environment through the survival mechanisms of households and communities.[9] Chambers offers some criteria for judging the sustainability of livelihoods, including security against health issues and become poorer, and non-interference with other forms of livelihood (Chambers, 1992, 216-17).

With little guidance from both the urban and livelihood strands of the sustainable development literature, we take a new tack and expand the typical scope of interpretation of a sustainable urban area and a sustainable economy by striving to link these in a multidimensional economic-social-environmental-cultural system. We seek to gain insights into this system at the urban scale, realizing that there are issues at both smaller and larger scales. The interrelationships among these four facets can be examined from the view point of any one of them. In our case, we use the economic facet as our entry point to then explore the others three facets. Figure 1 schematically represents the analysis framework that we propose.

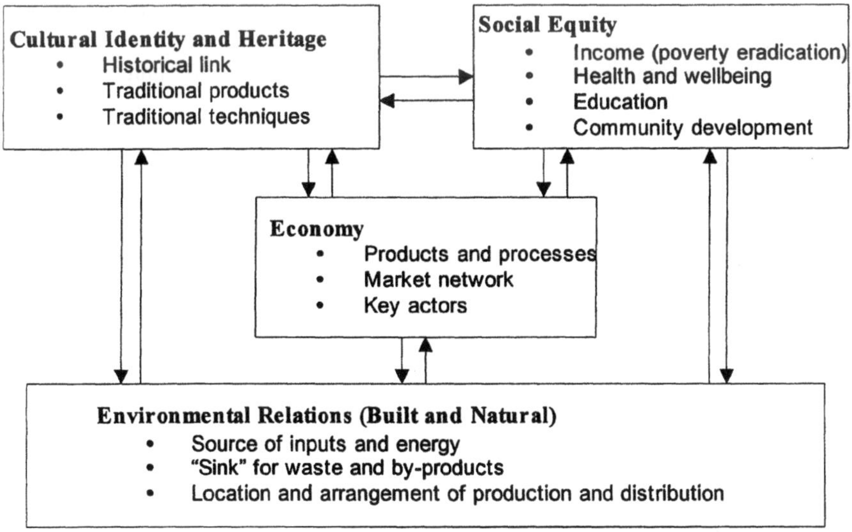

Figure 1: Key elements in the analysis of handicrafts and sustainable urban development

2.1. Economic Structure

The starting point of this analysis is an economic sector, so a firm grounding in the characteristics of this sector anchors the inquiry into the reaches of this sector's tentacles into the other areas of concern for a sustainable urban area. Both qualitative and quantitative descriptors are used to build an understanding of the supply-side and demand-side components of the sector. Within the supply-side component, of key interest are the range of products and the production processes employed to manufacture them. Particular emphasis is placed on any environmental or health concerns associated with the production, the technology employed, and whether it relies on recycled materials or produces by-products that may be re-used or recycled. The demand side component includes investigation into both input markets and sales outlets, whether local or global -- the chain of purchase-sale throughout the sector, with particular emphasis on the trade (or marketing) element of the equation. A third dimension of the handicraft sector is as a labor market, which focuses on which tasks or positions are taken on by which members of the community. Ideally, we would be interested in understanding any policies in place that shape working conditions, wages, and benefits; the extent of labor organization; and the resulting hierarchy of agents of exploitation. However, we approached this analysis from the perspective of "key actors," in part because we lack detailed information about the structure of employment and labor relations in the different (often unregistered) activities, but more importantly because this permits us to focus on how earning an income in the sector (whether as a worker or as an owner) links back to issues of social equity. Finally, we will conclude our discussion of the sector as a whole--as it exists in reality—as part of a supply-demand equation that exists in a particular place, with its historical roots and its future prospects, with its environmental, social, and cultural dimensions.

2.2. Social Equity

A second element of our analysis of the handicraft is the issue of social equity. By focusing on poverty, education, and health and wellbeing of different strata of individuals and communities linked to the handicraft sector, we hope to illuminate ways in which the sector contributes to social improvement and the enhancement of the quality of life of communities where it operates. In addition to exploring the links to the economics of handicraft production, we will also discuss the relationships between social equity and the environment (environmental justice) as well as with culture.

2.3. Environmental Relationships

The third element of our analysis of the handicraft sector is to scrutinize its environmental links and impacts. Here we examine the environment as a source of inputs and energy for the productive process. Our questions include: How is this sector linked to the natural resource base, locally and globally? Is it contributing to the depletion of various resources or is it contributing to the maintenance of various resources? The second focus will be on the environment as a "sink" for waste or by-products of the production process. Our third dimension of examination will be on the location and arrangement of production and distribution. Specifically, we will examine the land use and transportation patterns of the handicraft sector and the environmental implications of these patterns. Finally, as with the discussion of the economic structure, we will step back to look at the environment as it relates to the economy, social equity, and culture, striving to maintain temporal and spatial shifts as a backdrop.

2.4. Cultural Identity and Heritage

The fourth element of our analysis of handicrafts is culture. In order to examine the extent to which the handicraft sector contributes to preservation and enhancement of culture, we will explore handicrafts from the perspective of providing a link to the past.[10] There are two ways in which this is possible: either through the products themselves being traditional or through the production techniques being traditional. These possibilities will be examined particularly as they link to the other three elements of our analysis. Is it economically viable, socially acceptable, and/or environmentally possible to produce traditional items or to employ traditional techniques? In some cases, this may be the case of art versus commercialism; in others it may be that partial or complete mechanization (or substitution of inputs) is simply cheaper, less work, or less environmentally onerous.

2.5. Why Handicrafts?

We believe that this can be an insightful approach, especially with the economic sector we have chosen as the starting point. As the next two sections of this chapter will illuminate, a myriad of characteristics of handicrafts in Chiang Mai make this a rich area for this type of analysis. These characteristics include: (1) the long history of this sector; it has been a significant cultural and economic base for centuries, (2) the obvious influence on spatial structure over this time, both traditional and modern manifestations of craft neighborhoods and villages have emerged, (3) the urban/rural linkages of the sector's production activities, (4) the local/global connections of the products and their points of consumption, (5) the wide range of social relations in every aspect of the sector, (6) the shifting environmental relations as the sector has modernized, (7) its linkages to tourism, and (8) its linkages to a changing cultural identity in multiple cultures. The rich and intricate temporal, spatial, and institutional span of the handicraft sector in Chiang Mai could provide abundant information for analyses for a long time to come.

3. EVOLUTION OF CRAFTS IN NORTHERN THAILAND[11]

Sustainability has been often characterized as a trajectory or dynamic process as opposed to an endpoint or static state (Meadowcroft, 1997; National Research Council, 1999). If we hope to understand the role of the handicraft sector in the development trajectory of Chiang Mai, we need to analyze the sector as a dynamic process. In the case of Chiang Mai, the handicraft sector has been an integral component of the culture, the economy, and social and environmental relations since the 14th century, soon after this place was designated the capital of the Lanna Tai kingdom in 1269.[12] By examining the evolution of the role of the handicraft sector, our interpretations will be enriched with insights into the factors that shaped this sector and contributed to the characteristics we observe today.

3.1. Factors Shaping the Craft Center

Historical and locational factors both contributed to the evolution of Chiang Mai as a significant craft center.[13] By late 14th century Chiang Mai was quite wealthy, mainly due to its midpoint position in the trade route between south China and the Bay of Bengal at the port city of Moulmein[14]. Chiang Mai became a bustling trade center and maintained that role until the late 1800s with many thousands of mules, cattle, elephants, and human porters trekking old caravan routes to the Shan States, Yunnan, lower Burma, and all of northern Siam. Goods, ideas, and skills flowed freely through this crossroads city (Lubiegt, 1994, 119). Traders from the regional kingdoms had to pay taxes which fueled the continuing economic growth of the region, but the traders also brought with them new ideas, new products (such as silk[15] and celadon pottery) and new raw materials spurring the indigenous crafts to evolve, allowing a golden era for religion, culture and art to flourish. Throughout the 15th century these international linkages and exchanges thrived. Europeans scouting for the East Indies Company left records of visits to Chiang Mai in the 1600s, in which they describe a market town with great potential for trade with the Company. Cloth (mainly cotton, but also silk), gems, ivory, gums, tobacco, teak, and sticklac (the resin used for shellac in Europe, but used locally for dying cotton and silk) were available. Cotton was grown locally, and abundant suitable land was spotted to expand this production (LeMay, 1926, 100).[16] Following this high point (the Golden Age of the Lanna Tai Kingdom was 1400-1525) were wars and occupations by the Burmese and by the late 1800's absorption into the Siamese Kingdom, although Chiang Mai remained an independent principality until 1939. However, the couple of centuries of war and uncertainty were not a total dark ages. In fact, Bowie's research points to a thriving market economy during the 1800s, based on textiles.[17] She provides evidence that textile production was not a home subsistence activity, but an articulated economy with specialization at all phases of the process and a large export base for the region. Cotton was grown in the uplands and the spinning and weaving was a lowland activity. In a first hand account from the 1920s, LeMay reports noticing only a few local industries in Chiang Mai. There was production of lacquer ware, in which bamboo boxes were decorated with lacquer (these were predominately betal nut sets), silk weaving, with most of the silk thread being

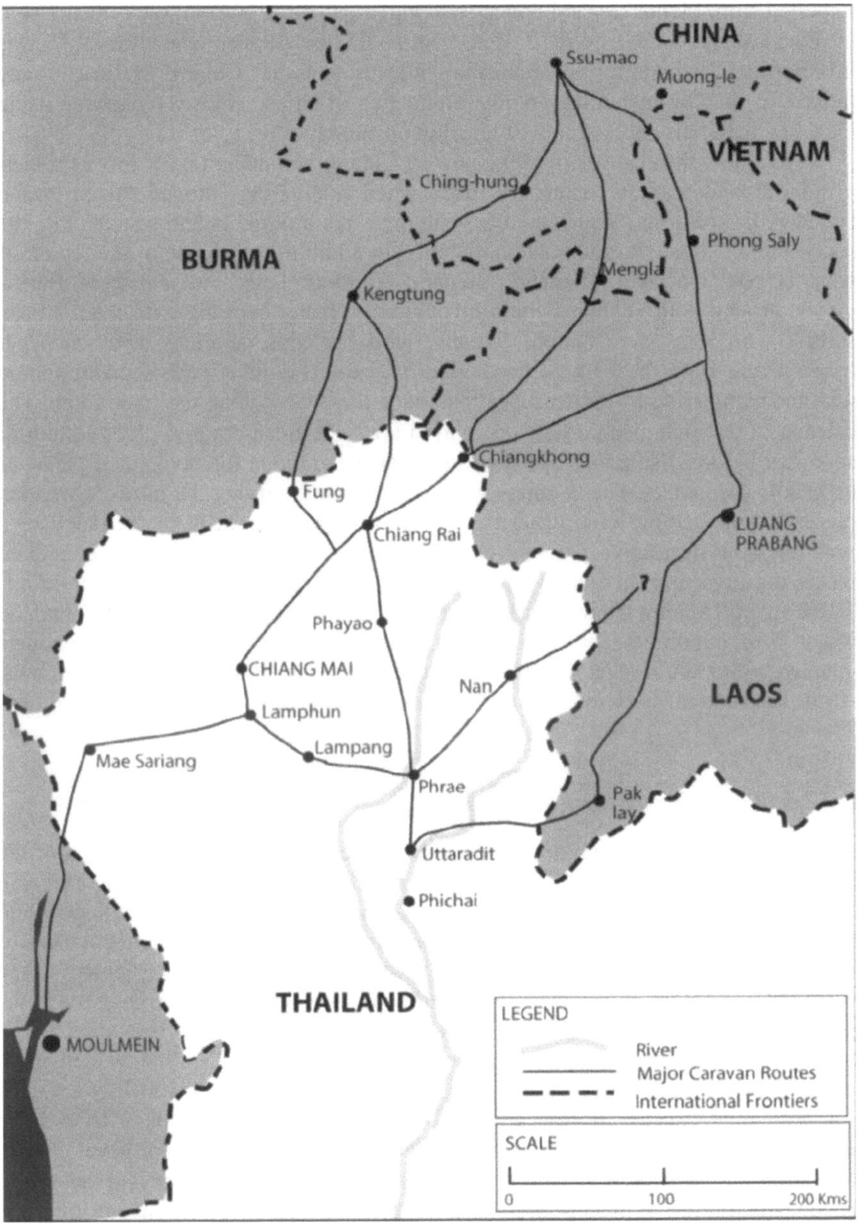

Map 1: Caravan Trade Routes

Source: Forbes, Andrew D.W. (1987). The "Cin-Ho" (Yunnanese Chinese) caravan trade with North Thailand during the late nineteenth and early twentieth centuries. *Journal of Asian History 21, 2,* 1-47. Map on page 45.

imported from Mandalay, and pottery making, a pale grey glazed pottery being sold in Bangkok as Sung (LeMay, 1926, 100). By the 1950s, when the ILO was inventorying Thailand for potential small industry to assist, Chiang Mai was found to have as its only manufacturing the production of an ash colored stoneware using local raw materials and cotton weaving, but no spinning.[18]

Throughout the changes in ruling powers Chiang Mai never totally lost its role as a cultural and market center. At times when populations dipped due to wars, outsiders from Burma, Shan, and other cultures were brought in and inserted into the social fabric[19] many bringing skills in crafts. In addition the people of the region of what is now Northern Thailand, Yunnan, Northern Laos, and Northern Burma moved around both voluntarily and involuntarily, many carrying with them artisan skills[20]. The Shan were famous for silverwork, celedon ceramics, and saa paper (paper made from the bark of the mulberry tree). The silverwork actually started with the hammering of Indian rupee that made there way along the trade route. The celedon of the Shan people was a revival of the light green cracked glaze pottery of centuries before. Saa paper has been used for generations for books and parasols, especially for monks. The Burmese were woodcarvers, carving the intricate wooden pieces of the temple complexes.[21] The Khoen brought skills in lacquerware,[22] providing the decorative bamboo containers used for both home and ceremonial pieces until recently. In addition, the weaving of both cotton and silk was prolific.[23] These are still among the products that comprise the craft skills in Northern Thailand today. This importation of skills and extended exposure to international products coupled with a multi-level demand (everyday utilitarian, aristocratic status symbols, royal ceremonial, religious) nurtured the development of crafts in Northern Thailand.

3.2. Locational Organization of Crafts

Although little seems to be documented about the institutional organizations of the artisans, remnants of a spatial organization of crafts still remains in Chiang Mai, and has continued to provide an organizing principle for "new" villages and neighborhoods. Lubeigt (1994, 120) surmises the origin of the locational choices of artisans as follows: "Seeking to serve the needs arising from the development of the royal town and its pagodas, the artisans set themselves up outside the ramparts to the south of the royal city, between Chiang Mai Gate and Wat Nantaram." The identifiable remnants of artisan neighborhoods are outside the first walled royal capital in the urban overflow area. Land to the east and south of the walls were ideal for settlement because it was protected from flooding and outsiders by a large earthen levee. This wall also allowed a reclamation of swamp land for development. The access to the inner city with its religious and administrative uses and the wealth accompanying this provided immediate outlet for their wares. The names of roads offer a clue to some of these locations, such as Wua Lai Road (or silversmith road). Other areas lacked the obvious names, but have long tradition of a specific craft, such as the lacquer makers congregating in the neighborhoods around Wat Nantaram. There is no clear evidence of other craft neighborhoods within such close proximity to the walled city with such long histories, though they could have existed.

In addition to these urban craft neighborhoods, there were villages in the region which specialized in specific crafts. The village of San Kampaeng, 13 kilometers from the city center, east of Chiang Mai, has been renowned for its silk and cotton weaving possibly since 1800s, if not earlier. Only one other local village, Hot, which is 70 km from Chiang Mai, supplied woven silk to the court.[24]

Other distinct artisan villages and localities can be traced in parallel with the conscious government-led development of Chiang Mai as a domestic and later international tourist destination. Chiang Mai was connected to Bangkok by railroad in 1927 when it was totally integrated into the Thai government structure. This rail connection and subsequent highway and air connections modernized this century old hub of trade, bringing visitors from all over the world. The early tourists in the post World War II era were treated to souvenirs of silver bowls and lacquer boxes from the remaining old craft neighborhoods south of the old city walls. Foreign missionaries 'discovered' the viability of wood carvings like those that were still being done for the temples as souvenirs, encouraging experimentation with local wood carving skills which are now used to carve everything from Mickey Mouse to cowboy statues along with interpretations of traditional subjects. What started as a cooler, romantic getaway for the wealthier Bangkokians evolved into the 'hidden gem' in Thailand for the adventurous international traveler. Chiang Mai did not remain 'hidden' for long, and by the 1970s it became apparent that steadily increasing numbers of international tourists could be attracted.

This realization in the 1970s impacted the types and quality of crafts being produced locally. The hilltribes were 'discovered' and transformed into tourist attractions and craft producers.[25] The local Lanna culture was reinvented and sometimes invented to provide tourist attractions and souvenirs.[26] The mass outlet for the wholesale distribution of this culture was the plaza outside the Tapae Gate. It was a natural gathering point with space for food vendors, souvenir vendors, and other activities. As the popularity of this location burgeoned, the concentration of activities became a nuisance and they were pushed out of the area. In 1975, this gathering was transferred to an empty lot in front of the Chiang Inn hotel. Without any access to electricity the activities were lit by candles. This was the birth of the night market as it is today with a building on this sight which houses many permanent outlets and the snaking along the streets in both directions out of this building of both formal and informal booths for selling local and imported ware mainly for tourists.

At the same time the night market was born, so was the revitalized umbrella village of Bo Sang[27] about 8 km east of Chiang Mai, out San Kampaeng Road. By the mid-1980s the Thai government formalized the integration of the handicraft economy with tourism development in Chiang Mai in the second five-year tourist plan by explicitly encouraging export production of handicrafts, craft related festivals, and the designation of San Kampaeng Road for only craft production and other tourism related commercial activities where the trend had already been underway since the 1970s.[28] This has resulted in a 15 kilometer strip of sleek sales outlets, some which still house the production of the crafts. A similar craft road

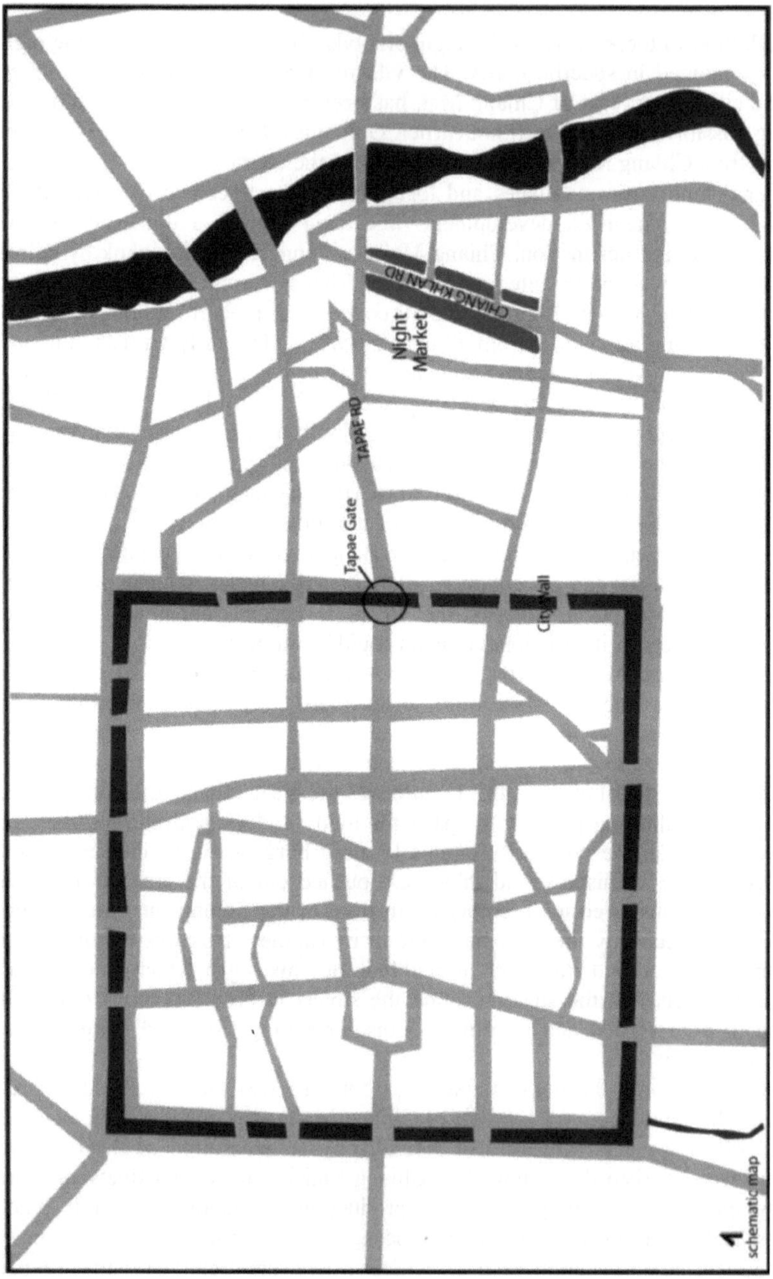

Map 2: Tapae Gate and Night Market

occurred as an extension of the old Wulai Road. Known as Hang Dong, for the main district town on the road, this corridor extended to Ban Tawai, which was transformed into a wood carvers village starting in the 1960s and became a popular tourist destination by the 1980s with the help of government advertising. Ban Tawai, about 15 km west of Chiang Mai was a small agricultural village. It seems that the transformation into a crafts village was serendipitous, in that several members of the village learned woodworking by working at an antique shop in Chiang Mai. They then took these skills and started production of wood carvings at home, transferring the skills to family and neighbors.[29] Lesser celebrated craft villages near Chiang Mai with traditional roots are Muang Koong known for pottery, and Ban Pha Bong in the district of Saraphi known for basket weaving (Cohen, 2000, 225).

3.3 Why Chiang Mai?

Several factors become apparent to answer "why Chiang Mai became a craft center" and "why they make the crafts that they do". Chiang Mai experienced the forces of globalization from its outset in the 15th century. This globalization was possible due to the relative location of Chiang Mai as well as the transportation links that have evolved from caravan trails to an international airport. Talent, ideas, art forms, products, and raw materials all flowed in and out of this city providing the necessary wealth and impetus for the development of fine crafts. Innovations on traditional objects and art forms happened slowly at first, but by the fifth century of this process the innovations were speedy, matching the ever quickening pace of global connections and expectations. In addition to this 'outside pull' there was always an 'inside push' as well. In the earlier days, the wealth of the kingdom and the wealth of the traders provided local demand for high quality crafts. In the past century, the Thai government provided the nudges necessary to facilitate the continuing growth of these activities, albeit in new and more commercialized forms.

4. HANDICRAFTS IN CHIANG MAI TODAY

This historical overview provides the context for understanding the handicrafts of Chiang Mai today. Building on this background, this section provides a brief discussion of the supply, demand,and distribution of these handicrafts, as well as a look at the key actors involved. The objective is not to fully describe this sector, but rather, to communicate a sufficient conceptualization of the sector to facilitate the analysis of its role in sustainable development of the city and its region.

4.1. Suppy-Side: What Do They Make and How Is It Made?

Today Chiang Mai produces a remarkably wide range of handicraft products, many a continuation of or spin-off of traditional crafts and production skills. These include gems, silver jewelry, silverware, hilltribe crafts and embroidery, silk and cotton fabric and garments, baskets, celadon ceramics, furniture, lacquerware, wood carvings, parasols, and saa paper items such as stationary and picture frames.

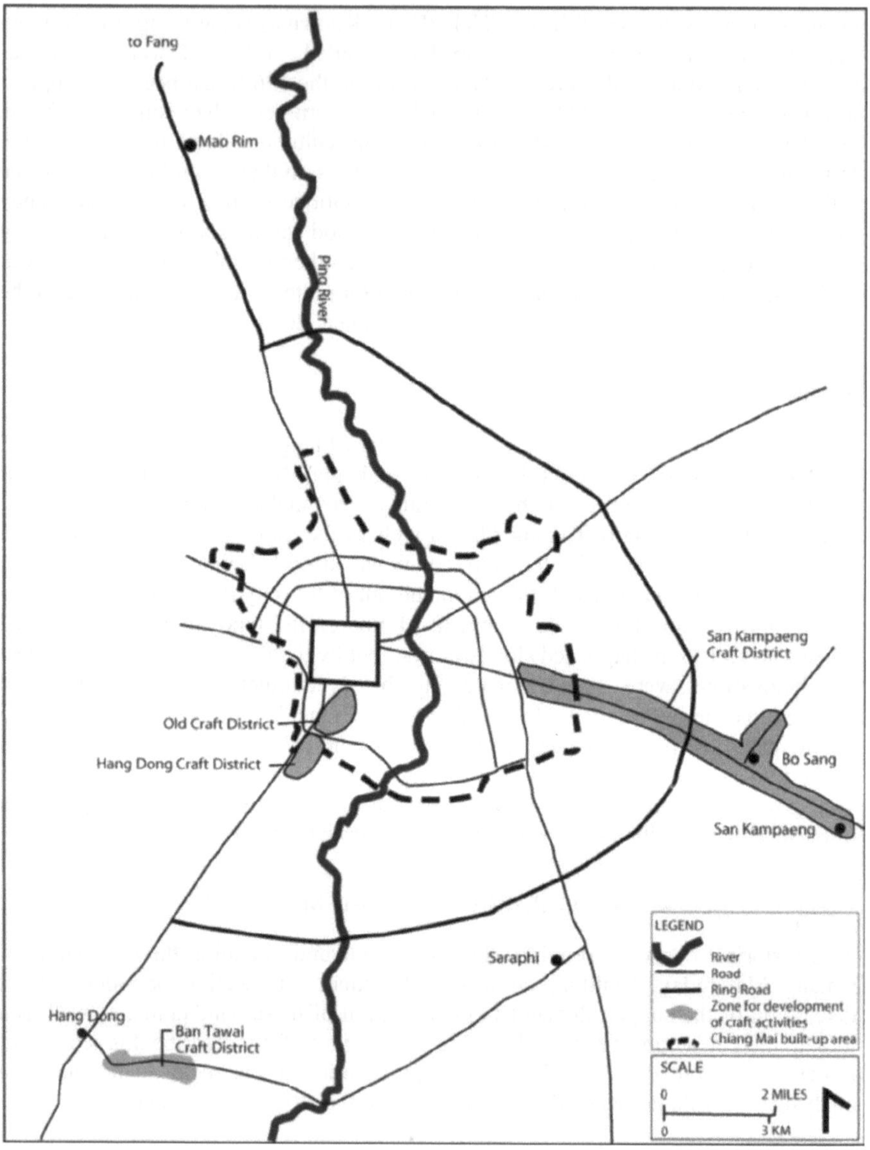

Map 3: Craft Districts

Through the discovery of the range of craft products and observations of where and how many of them are made, a variety of categories became apparent. First are the items that are produced with the skills and techniques handed down through generations and that have some utilitarian purpose or cultural use; such as some of the cotton and silk textiles, some clay pottery, and some basketry items. Second are the items with the same purpose as above, but the production techniques have evolved in such a way that the quality of the end product is often less. These techniques may include replacement of raw materials to decrease the cost of production, to make production quicker, or to replace resources that are rare. Examples include replacement of teak with softer woods, replacement of bamboo with soft wood as the base for lacquered items, or replacements of cotton and silk with synthetic material. Another evolution of technique is the replacement of hand work with machine work, such as seen in the textile production. The third category contains items that are produced with the traditional skills and techniques, but the products are modern items that may or may not be of Thai origin. Examples include the use of saa paper in making stationary and picture frames. Fourth are the purely tourist items; souvenirs or trinkets that may have evolved from traditional crafts, such as the large paper umbrellas or hilltribe items, or items that have no connection to the traditional craft industry, such as t-shirts, watches, and designer clothing knock-offs. The fifth category comes from large factories making export items for other Thai markets as well as the international tourist trade market.

Production techniques and arrangements are as varied as the crafts themselves, although there are two predominant models: 1) home-based, or 2) factory-based. In the first case, production is primarily through piecework. In rural areas, this takes place during the "slow" season for agriculture. In urban areas, the work is done by slum residents for whom it is a primary—if not sole—source of income. In the second group, factory-based producers may be paid by the piece or by the hour, although it appears that piece-work is the norm. Home-based work is either done in villages and neighborhoods by individual artisans or in community groups. In the case of community groups, members may or may not come together for the production phase itself. If not, they may only meet to turn in their individual production. In the case of factory-based work, the work may take place in backyard factories ("cottage industries") or in large-scale factories. In these settings, workers may be paid by the hour or by the piece.

4.2. Demand-Side: Who Buys Handicrafts?

Traditionally, the crafts of the Chiang Mai region were driven by three broad sets of demands: the daily household and ceremonial needs of the general population, the desires of the court for items of prestige and of ceremonial significance, and the demands of regional and international trade. This diverse mix of forces shaped an arena for a wide array of handicrafts and a range of quality of workmanship and materials. Chiang Mai's economy was never an isolated, subsistence economy with each household making its own goods; but rather a globally influenced, well articulated economy based on specialization of skills, often organized by village.[30]

Today handicraft products in Chiang Mai are driven by a broader, more complex set of demands. At least five sources of demand can be identified. First, the tourists to Chiang Mai, both from other parts of Thailand and from abroad, come for shopping. This group demands at least two types of products – souvenirs of Chiang Mai, usually something that they perceive as representing the local culture and other handmade decorative objects that are not necessarily cultural in nature, but are different from what can be found at home and are perceived as being a good buy. Second, handicrafts from Chiang Mai are distributed to other cities in Thailand, especially Bangkok, to provide the same two types of goods for the local tourist market. Third, these handicrafts also are sold in tourist markets around the world, sometimes passed off as 'local' handicrafts.[31] Fourth, domestically there is a growing interest in local cultural items. For example, locally produced traditional textiles and jewelry are popular items with some women in Chiang Mai. In addition to the value placed on culture, there is also a growing interest in quality local art. And fifth, there is a world-wide non-tourist market willing to buy affordable and unique handmade decorative items. For example, in the U.S. there are many import shops, some large national chains and other independent operations, importing furniture, textiles, and other home furnishings and decorative items, many with no identifiable connection to Thailand's culture. This range of demand requires not just a wide array of products and the flexibility to respond to changes in these demands, but also a level of sophistication to participate in these far-flung economies.

4.3. Distribution: Where Can You Buy Handicrafts?

Chiang Mai handicrafts are distributed for sale locally, as well as regionally in tourist towns, such as Mae Hong Son, further Thai tourist destinations such as

Phuket, Bangkok, as well as abroad. Within Chiang Mai, the distribution of handicrafts has a wide range of outlets, including individual shops, showrooms, the Night Bazaar and nearby sidewalks, large factory-showrooms along highways, and in "traditional" villages.

Shops are scattered throughout the central area of the city, with the greatest concentration near tourist destinations. Some of these shops are outlets for family-produced crafts, while others are filled with items bought by the shop owner which are of the quality and design that he/she cares to sell. Shops can be small and specialized or large and striving to be a one-stop tourist shop where a large range of crafts can be found. Most of the shops seem to cater to middle- to lower-income buyers, hoping to catch a majority of the market. However, high-end boutiques are becoming more apparent.

Showrooms are also scattered throughout the city. In addition to two government-supported showrooms (that of the Department of Industrial Promotion and that of the Department of Economic Promotion), the Northern Handicrafts Manufacturers and Exporters Association maintains a shop/showroom, as do several of the NGOs (particularly with regard to hilltribe crafts). Most of the show rooms are not catering to the tourist shopper, but rather to the buyer who is looking for bulk purchases for resale in other regions or abroad.

The informal gathering of sellers of crafts and souvenirs on the grounds in front of the Chiang Inn hotel in the 1970s has evolved into the more formal Chiang Mai Night Bazaar Shopping Centre on this site. Across the street the idea was extended and a food court was added at the Galare Food and Shopping Centre. These two structures house formal shops as well as booths that operate during evening hours, from around 5:30 p.m. to midnight. The Chiangmai Night Bazaar houses over 75 shops and over 200 booths on three floors. These are both private sector undertakings, so booths are rented by vendors. These two focal points also attract itinerate sellers who hang around with items to sell. On the sidewalks near the Night Bazaar buildings are hundreds of booths for the same similar items available inside the market buildings. Apparently, the vendors make arrangements with the individual shop owners in front of whose businesses they put up their booths. They are also expected to pay 10 baht per square meter (about US$0.25 at the time of the research) to the Municipal Government for cleaning.[32] This combination of food, entertainment, high-end shops, one-stop shops, specialized booths, itinerate sellers, and nighttime street life creates the primary tourist draw to Chiang Mai.

Cohen (2000) has aptly dubbed the government designated handicraft routes emanating east and west of the city "Ribbon Development". These two highway-type roads, San Kampaeng Road and Hang Dong Road, have become multi-kilometer stretches of often discontinuous development of a large number of craft industries and shops. The Board of Investment permits and promotes the construction of large craft enterprises that can receive busloads of tourists in their showrooms, hence large parking lots and sprawling grounds for each individual property. The government incentives have also fueled land speculation along these roads.

As discussed in the previous section of this chapter, there are a few traditional craft neighborhoods and villages that have maintained that characteristic until today. The government has also targeted some of these neighborhoods and villages as well as some new areas which are conveniently located near or on the designated tourist routes to be sites where incentives are given for craft shops and workshops. It is not clear how much of the new investment in these areas, as well as along the routes, is from Chiang Mai or from outside the region.

4.4. Who Are the Key Actors in the Sector

Our discussion of the key actors in the sector is divided into three sub-sections. First are the individuals, both workers and employers, next is the government, and last are the non-governmental agencies.

4.4.1. Individuals

The individuals range from individuals who work on a piece rate or as independent artisans who sell their production, to small or large businesses that employ individuals either in-house or as piece workers. From the first group, we visited a slum where women produced lacquer boxes. Other than gathering to work together and word-of-mouth, there was no form of organization among these producers. The mid-size owner/producers were often found either in the formal or the informal "Night Bazaar," again with no apparent association or organization. The larger producers often had autonomous workshops and are members of the Northern Handicrafts Association, an official subgroup of the Chamber of Commerce. This association is a membership organization that maintains a showroom near the Night Bazaar. The association works to represent the interests of the larger businesses at the local and national levels.[33]

The complexity of this sector can be seen by envisioning women artisans we observed over the course of the field research. In each case, the women were seated on the floor or on the 18-inch high benches common in Thailand, legs tucked beneath them, working on the craft item in question. They were involved in producing five different items—clay pots, lacquerware, dolls, umbrellas, silver, and saa paper frames —and they reveal five different relations of production. Despite the strikingly similar physical appearances, their participation in the handicraft sector is not the same.

The woman who made the clay pots lived in a village near Chiang Mai and used to make them for everyday use by herself and her neighbors. An older woman, she now made them infrequently, but in the past she almost certainly made them as neighbors requested them, in time left by her other household tasks, as a complement to her family income. In contrast, the women who were etching lacquer boxes in a slum in Chiang Mai were paid by the piece, measured in inches etched in the case of lacquerware. They were provided all inputs by a middleman and the amount they earned was quite small, barely a complement to their household income, considering the number of hours devoted to the activity. The women who made dolls for sale to tourists were in a small village near Chiang Mai and were self-employed. They bought their own supplies and sold their production to a middleman who took it to Mae Hong Song for sale. The women seen making umbrellas at the "umbrella village" were probably paid an hourly rate to work in the area of the shop where tourists could see their craft techniques; they were the public face of the many nearby artisans who work at home on a piece rate. The women making silver were located in the factory and probably paid by the hour to pound the designs into the silver jewelry and bowls. This employer-employee relationship probably emerged because of the necessity of working near the ovens and equipment. In the case of saa paper frames, the women were members of an association. They collectively purchased materials and, together with their NGO sponsor, calculated a piece rate for each activity, based on a reasonable hourly (or daily) wage. Their incomes were divided between the individual artisans and their community

4.4.2. Government

There are two key national government agencies that work with handicrafts promotion in the Northern Region of Thailand: the Department of Export Promotion and the Department of Industrial Promotion, which work primarily with small- and large-scale producer/employers in both rural and urban areas. Both organizations maintain showrooms where producers can display their products to potential buyers. In addition, the municipal government's community development department and an integrated rural development project work with community-based production.

4.4.3. NGOs

Community development, income generation, women's development, or political work are the focus of the bulk of the NGOs working with the handicraft sector in the region. The YMCA Foundation's work focuses on craft production that will provide a reasonable wage by marketing through alternative trade organizations. WEAVE (Women's Education for Advancement and Empowerment) is another organization that markets through alternative trade organizations. It is affiliated with the Bangkok-based ThaiCraft organization, which holds sales several times a year, primarily for expatriate residents of Bangkok. WEAVE works primarily with Burmese refugees in Northern Thailand. Other organizations work with hilltribe families to supplement their incomes (Thai Tribal Crafts) or with women's groups (Grassroots HQ).

5. ANALYSIS: APPLYING THE SUSTAINABLE DEVELOPMENT FRAMEWORK TO THE HANDICRAFT SECTOR IN CHIANG MAI

We return to our proposed analysis framework and discuss the handicraft sector in light of our conceptualization of sustainable development. The themes and issues are categorized by the four components we previously described economic structure, social equity, environmental relationships, and cultural identity and heritage.

5.1. Economic Structure

The above section has described the organization and characteristics of the handicraft sector in Chiang Mai, so we will not repeat the details already provided. However, four themes deserve further discussion in our quest to evaluate the sustainability of this activity.

The first of these themes is that of the role of the handicraft sector in the economy of the city. Because of its somewhat cross-sectoral nature in terms of traditional economic accounting, a potentially large unaccounted for labor contribution, and diverse and sometimes amorphous methods of distribution, reliable measures of inputs, outputs, income generation, and sales are not readily available. However, we can characterize the role of this sector in the economy in more general terms. For most of the sector's five hundred or so year history the products have had three purposes: to meet local needs, both day-to-day as well as ritualistic (ceremonial, religious, or life stage events such as weddings); to build the physical wealth of the city and its inhabitants, as long-lasting art and artifacts both public and private; and to generate more wealth through trade. In the last fifty years, the latter role has come to dominate. The handicrafts of Chiang Mai have become inextricably linked with the tourism sector of the region as well as of the country. This shift in focus has shifted the products from necessities or items with a cultural use or meaning to commodities meant to feed into an increasing consumeristic society with a shift in what they represent. In addition to the myriad of local outlets, these items flood the shops and markets of Bangkok as well as other Thai cities. The Chiang Mai crafts have also been observed in tourist shops around the world.[34]

This widespread distribution of these items certainly points to the potential for income generation not only to the individuals involved in the production and distribution, but also to the city and region, both directly and indirectly. Taxation on tourist related sales seems to exist, hence bringing direct revenues to the city. This may only be able to be collected officially in high-end stores and hotel shops, but informal ways of taxing these activities also exists with both official and unofficial fees on night market activities. As more outlets for these products emerge demand for increased production follows. So far this demand seems to be mainly met by increased labor input, but at some point a more mechanized mass production may evolve which will change the structure of who will make money from this sector.

The third theme is that of scale or the issue of local vs. global in the sector, both in terms of demand and of ownership. It is obvious that these handicrafts are increasingly produced for a global market. What is less clear is the issue of ownership of the firms, which determines where a majority of the profits will flow. It appears at this point that most of these firms are locally owned, but we were given intimations that some of the ownership is shifting to Bangkok and the potential exists for the multinationalization of some of this production, reducing the local control and benefits. There is already growing concern over the fairness of these relationships with NGOs stepping in to explore Alternate Trade Organization models for fairer oversight of the distribution of benefits (ie. wages).[35]

The last economic theme is that of long-term viability of this sector. At first glance you can point to the five hundred year history of this activity and assume that it must be firmly integrated into the lifestyles and livelihood strategies of the region. And there is a tradition that has been able to reinvent itself many times as local and world conditions have changed. However, tourism creates a double-edged sword: it creates an easy market for products, but the market distorts the purpose and quality of production. In addition, some of the transformations are not necessarily made with a good understanding of what foreign tourists want and like. With the globalization of production/marketing, products will be less distinctively Thai. There will be more products like the rest of the world, which may ultimately mean that the market will diminish (though probably not disappear). In addition, tourism relies on stable world politics and a strong economy so that people will both travel and have money to spend on non-necessities. Basing an economic sector on this can be precarious and may experience unexpected fluctuations. It is not clear how robust the current direction of this sector is or if there is sufficient flexibility in the sector to react to the ever increasing changes in global demand for such products.

5.2. Social Equity

Regarding the social sustainability of the handicraft sector the main concerns revolve around distributive and welfare issues. The distribution issues are twofold: the distribution of economic benefits and the distribution of negative externalities such as environmental or health problems. The former will be a function of the structure of the labor market, the structure of ownership, and the structure of trade relations. In our study we were exposed to multiple worker-firm relationships, some which lend themselves to more exploitation than others: the more 'hidden' the worker, the more probability that unfair practices are possible. Probably the most 'hidden' are the home workers that are paid by the piece or by the batch. Such working arrangements offer little leverage to the laborer to demand better wages or other benefits. But such a scheme allows the workers, usually women, to combine child care and home care with their income generating activity due to the flexibility of timing of work. A much more visible form of labor are the workers who demonstrate production in the sleek tourist showrooms. Of course, there is no obvious correlation with these structures and income levels, but certainly working conditions are more scrutinized as the workplace is made more public.

Ownership affects equity of economic benefits in several ways. As discussed in the section above, an owner of the firm who is farther away (like in Bangkok or even in another country) will extract profits away from the neighborhood, city, and region. Also, with such distance labor one can more easily become a faceless entity to the employer, reducing the willingness to accommodate the varied needs of the workers with flexibility and compassion. In such situations workers may be forced to choose between work and other family obligations. Ownership is also closely related to trade relationships. A local workshop owner in a village will have a more difficult time identifying, and probably dealing with, importers abroad or even in Bangkok, unless she has access to technology, expertise, and experience to facilitate this type of transaction. For this reason, without technical assistance, many local owners are at a disadvantage in getting direct access to larger profits, as they must rely on a chain of middlemen to get their product to the same ultimate location in Bangkok or overseas.

We found little, if any, evidence of any recognition of environmental or health issues related to this sector. Handicrafts are viewed as small, light manufacturing activities, not deserving governmental attention in the areas of environmental standards or occupational health. However, this does not mean that there are not occupational health issues or environmental justice issues. The informality (ie. lack of regulatory oversight) of a majority of the handicraft manufacturing operations, many growing in size of number of employees, coupled with lack of awareness of manufacturing and health linkages is sure to be harboring unhealthy practices. In fact, through our interviews there were employers who suspected that their operations were causing environmental damage with potential health implications but they implied there were no accessible mechanisms to help them deal with such issues.

Because of the characteristics of the handicraft sector – can be done at home or on a small scale, often with little equipment, is often based on traditional skills, and produces a readily sellable product – it has been increasingly viewed as a feasible community development strategy, both by governmental and nongovernmental entities. Both the Municipal and District Community Development Departments have been amenable to integrating handicraft business development into community development projects. These schemes are viewed as a means to achieve "participation" by citizens in the development of their community (but they still contain elements of government controlling people). NGOs see these projects as a vehicle to empower people in market negotiations, such as through the introduction of alternative trade organizations (ATOs). Handicrafts have long been a tool of 'developing' the hilltribes. Some of these efforts have been genuine community development and cultural heritage preservation efforts, while other efforts are simply (potentially questionable) income generation opportunities such as schemes to use hilltribe members to market Bangkok-produced factory items.

5.3. Environmental Relationships

The discussion of environmental relationships of the handicraft sector in Chiang Mai revolves around two organizing principles – those aspects related to the natural environmental and those aspects related to the built environment.

The natural environment provides both the source and the sink for the inputs and outputs of any economic sector. Handicrafts, because of their traditional origins tend to be more directly related to natural resource base than more modern sectors. In the case of handicrafts in Chiang Mai, it is not obvious how closely the local or regional availability of natural resources led the development of this sector. We have found anecdotal evidence that in the past both cotton and silk were locally produced, and these are now imported for the local weaving industries. Wood was always a local product as well. Even by the 1960s when carving started taking off in Ban Tawai, teak was readily available in surrounding forests. However rapid deforestation[36] in the area made teak more expensive and finally government regulations banned the cutting of teak in 1989, so cheaper woods, such as raintree, were substituted. And now enough raintree wood, especially of the sizes needed, is not available locally and must be imported from as far away as 1000 km (Cohen, 2000, 226-7). It is known that silver and other precious metals and gems were not local, but imported first as coins, and later as bulk metal and stones. Rattan and bamboo were once abundant in the Ban Pha Bong region, but demand depleted the stock, requiring importation of cheaper substitutes (Cohen, 2000, 278). Other raw materials such as lacquer resin, dyes, and clay may have been local products, but seem to be imported now as well. Regarding other natural inputs, such as energy and water, the demands for these inputs are not excessive since most of the operations are still relatively labor intensive using little mechanization.

The inputs to this sector seem to be relatively removed from impacting the local and regional environment. On the other hand, the outputs of the various handicraft production processes are causing water and air pollution, as well as solid waste. It is likely that most of these polluting incidents at an individual level are small and inconsequential. However, there are many of these activities polluting in the same way and the accumulated effect of each of these activities probably is significant. There is no official interest by the government to deal with small polluters; hence, no incentives exist for these activities to even consider environmental issues. The same goes for the health issues of workers that may result from processes and materials that are not environmentally sound. So not only is the surrounding environment suffering, but the workers may be as well. This inattention to small enterprises has been causing bitterness in the larger businesses which do fall under government environmental regulatory requirements. They complain that the smaller firms will have an advantage over them. Finding a solution to this issue of pollution and the potential inequities will be a challenge. This will require dealing with registered vs. unregistered firms and small vs. large firms. Of particular concern are those production processes that utilize electroplating, use of dyes, or chemical washes. Textile and paper making processes can also be environmentally harmful. As volumes of production increase, techniques of production become more mechanized, and input materials change, possibly towards more manufactured instead of natural materials. The environmental issues will only expand if foresight does not lead to action now.

In addition to the impacts on the relationship with the natural environment, the built environment has been affected by the handicraft sector as well. This has mainly been manifested through land use patterns that have been both mandated and have more or less spontaneously evolved to accommodate the activities of this sector. These resulting patterns then have implications for both the natural environment and for the characteristics of the businesses themselves. The primary land use pattern connected with the handicraft sector is the relocation of the sales outlets to two main roads out of the city (San Kampaeng and Hang Dong as already discussed in section 3 above and illustrated in Map 3). The concept of physical agglomeration of these like activities has its benefits. People interested in seeing handicrafts being made or in buying them can find these establishments in clearly identifiable areas of Chiang Mai. In addition, having these activities in close proximity can aid in the more efficient and effective implementation of solutions to issues such as environmental pollution or waste management. Unfortunately the ribbon configuration also poses some problems. These developments are many kilometers in length, with large buildings or compounds on large properties fronted by large parking lots to accommodate tourist buses. Such a configuration, as opposed to a compact village-like development pattern, requires motorized transportation, usually buses, sometimes tuk-tuks or taxis, to move people from place to place. Walking from establishment to establishment is not desirable or even feasible in most cases. These roads have become like the worst of American strip development with visual blight, no sidewalks, and streets that are impossible to cross. The environmental impacts of this type of development include increased air emissions from the increased reliance on vehicular transport and increased problems with storm water runoff as more of the land is covered by impervious surfaces and vegetative cover is lost. This storm water runoff both increases pollution in surface water sources by carrying oils and chemicals from the streets and other surfaces with it, contributes to more frequent and more severe flooding and reduces the water available to recharge groundwater aquifers. A non-environmental impact of this type of development is that it tends to favor large establishments over small. Only those firms with access to land on these roads and with the size to accommodate the bus-based mode of tourism that is frequenting these locations will be able to build and survive. The smaller producers and sales outlets will remain off the beaten path and not be able to compete with the larger ribbon-based firms and stores.

5.4. Cultural Identity and Heritage

There is a tension between preserving craft items and letting market-driven evolution take over for several reasons. One is that many crafts are utilitarian in origin and our modern lives do not have the same material needs, hence the items become obsolete for more than reminders of days gone by. Crafts that were driven by their status appeal are also era-specific, as the icons of status are shifting ever faster in this age of global marketing. In addition, unless traditional skills can gain a market value, the shift to profitable activities is inevitable.

Part of the effect of increasingly global markets is the shift to more globally homogenous goods. Simultaneously, especially when a good portion of that global market demand comes directly to the locale as tourists, there are pressures to "mythologize" the culture - finding objects that will satisfy the desire for a piece of northern Thai culture as conceptualized by outsiders and reinterpreted by insiders. This cycle of created and recreated culture often ends up being depicted in mass marketed increasingly less expensive items. As the uniqueness and quality of these representations decline, the potential contribution to cultural identity and heritage preservation, in turn, declines. The modern visitor to Chiang Mai, who increasingly has seen other parts of Thailand as well as other tourist centers around the world, wants to find 'authentic' cultural objects from Chiang Mai; however, cost is still a barrier. The feasibility of high quality, authentic crafts that fit the bargain basement atmosphere of most of the craft marketing, especially at the Night Bazaar, is low.

Local, regional and national level efforts try to reconcile some of the tensions between heritage preservation and demands of the global market place. Locally the Cultural Council of Chiang Mai Province, presided over by a member of the Chiang Mai royal lineage, Princess Chao Duangduen Na Chiang Mai, is leading efforts to preserve traditional values and a 'way of life' and their representation through clothing, dance, food, and festive artifacts, such as the paper hot air balloons, but little emphasis is on artisan skills. A variety of other NGOs are striving to maintain the value of the traditional techniques by finding modern form for their expression so that the crafts have an economic reason to exist.[37] At the national level, a heroic effort is led by Queen Sirikit, through her SUPPORT initiative, which has a clear emphasis on both craft as artifact and craft as livelihood. This program is successfully reviving and preserving some of the most intricate art forms of the Thai culture, although these efforts have not been in Chiang Mai to date.

6. THE POTENTIAL ROLE OF THE HANDICRAFT SECTOR FOR A SUSTAINABLE CHIANG MAI

This analysis using our proposed sustainable development framework reveals the multidimensionality of an economic sector by starting to connect an economic activity to its web of environmental, social, and cultural inputs and outcomes. Becoming more cognizant of these interconnections may not offer immediate answers to the question of urban sustainability, but it provides a basis for realizing the many facets of the sector, which in turn may lead to new strategies for guiding a long-term balance among environmental quality, social equity, cultural sensitivity, and economic prosperity. Our analysis could only scratch the surface of the intricacies of this sector and the interrelationship it has with long-term livability in the region. But it suggests the need for a more holistic view of urban management and economic development initiatives.

Handicrafts may or may not be the most important or promising economic sector in the city and region of Chiang Mai, but it is probably the sector for which the city and region are most known. Some form of handicraft sector will remain and, by its nature, continue to evolve in this region. But the question is 'will this sector follow one of the many possible unsustainable trajectories or will it fall within the range of what are probably fewer sustainable trajectories?' Left on its own, to react to the current incentive/disincentive structures locally, nationally, and globally, the handicraft sector will most likely encounter many pitfalls which would be better avoided. Elements of unsustainable trajectories would include continuation of the ribbon development pattern instead of a compact village-like development. The 'craft villages' and 'craft neighborhoods' have fewer and fewer elements of the village or neighborhood with little pedestrian accessibility, lack of mixed uses, and potential for implicit exclusionary practices. There is also a trend towards larger and thus fewer enterprises, both in terms of sales outlets and in production. The required flexibility of the sector may be compromised with this structure. In order to maintain this flexibility, these larger enterprises may rely on exploiting sub-contracted labor who will have fewer options to choose from. The rapid commercialization and turnover of products to meet local and export demand may lead to items with little, if any, local content. These may eventually lose the interest of an increasingly more sophisticated and well-traveled tourist population. In addition, this rush to meet this perceived global demand may squeeze out any niche for art or ethnic crafts.

To avoid some of the ill effects of these possible trajectories, some interventions which will capitalize on the more sustainable characteristics of the sector will be necessary. At least six such areas that can be built upon emerge from our analysis of the handicraft sector. First, the sector has an inherently strong relationship with the cultural heritage base of the region, both in the skills and artisanship needed to create the crafts and in the products themselves. Revival of traditional craft-making techniques can be of value in maintaining cultural identity. Additionally, the fine execution of traditional techniques is highly valued in selected world markets, making high quality hand crafted items and the artistic expressions that may evolve using these techniques highly sought after collectibles.[38] Queen Sirikit has been using this principle to underlie her SUPPORT project. There is potential to expand financial and technical aid to increase the types of handicrafts that can be elevated to cultural art pieces. With respect to the role of the products themselves in cultural preservation, wider local markets should be cultivated. Princess Chao Duangduen Na Chiang Mai is leading the effort to make traditional weaving and jewelry fashionable among the residents of Chiang Mai. Other traditional items may also be in demand if it was known that such items could be available.

Second, the sector is labor intensive. This characteristic bodes well for continuing to provide income generation possibilities full-time, part-time, or seasonally as needed. However, efforts should be made to ensure that these opportunities provide fair compensation and benefits as well as a healthy workplace environment.

Third, the handicraft sector provides both urban and rural labor opportunities. It is a sector that can and does take place in urban neighborhoods, urban commercial areas, rural villages, as well as rural centers. The local sale of these items also occurs in both urban and rural locations. These activities can and do provide the nuclei for other income generating activities related to tourism as seen in the burgeoning development of crafts villages such as San Kampaeng and Ban Tawai. With appropriate planning and design, these revived villages can be viable places to live and work, as well as attractive visitor centers.

Fourth, the small scale and relatively low-tech approaches to craft making make this a less polluting industry than most other manufacturing. If resources could be dedicated to informing and training these small entrepreneurs about environmental issues, simple changes in raw material choices and manufacturing process refinement could stop much of the polluting that is currently occurring.

Fifth, with respect to the handicrafts that rely on renewable natural resources such as wood, rattan, cotton, and other plant materials, the maintenance and improvement of these resource bases could be tied to the craft making and buying experience. Community-based forestry, revival of cotton cultivation, and reintroduction of basketry grasses and bamboos could be directly integrated with a comprehensive approach to handicraft development.[39]

And sixth, the handicraft sector produces goods that can be exported globally, linking Chiang Mai into the inevitable global market. Yet, at least to date, it appears that this type of economic activity is less likely to be absorbed into a multinational ownership structure. Production and distribution is still best suited to local control. Efforts should be made to ensure that this characteristic is maintained and enhanced.

Local and national government, private enterprises, and non-governmental entities should all strive to capitalize on the unique and flexible characteristics of the handicraft sector. This sector has the potential to link economic opportunities, cultural preservation, environmental quality and resource base revival, as well as social development.

NOTES

[1] This kingdom is referred to as Lan Na Tai, Lanna Tai, and Lanna Thai in the literature. We have chosen to use Lanna Tai. Lanna, combining Lan Na into one word, appears to be the most common spelling, and Tai is used because it refers to the ethnic groups and languages of that region as opposed to the Thai designation used for all of Thailand.

[2] See Porus Olpadwala and William w. Goldsmith. (1992). The Sustainability of Privilege: Reflections on the Environment, the Third World City, and Poverty. *World Development, 20*, 627-640, for insights into some of the controversies around defining a sustainable urban system.

[3] "Sustainable development is development that meets the needs of the present without compromising the ability of future generations to meet their own needs" (World Commission on Environment and Development, 1987, 43).

[4] We refer to handicrafts as an economic sector even though it is not considered so by the International Industrial Classification System, where the production and retail activities related to handicrafts would fall under a variety of classifications. For our purposes handicrafts represent a set of economic activities that are related through their intention as products. In other words, these are products that historically were the work of skilled artisans who created works of art or utilitarian items by hand. Many of these products have evolved into items for the local and global tourist markets while others are considered traditional arts of value to the heritage of Thailand. Today the products are not necessarily made by hand or even of traditional origin but are perceived of as Thai crafts.

[5] Skepticism abounds in the literature. One such example which speaks specifically to cities is: Scott Campbell. (1996). Green Cities, Growing Cities, Just Cities? Urban planning and the contradiction of sustainable development. *Journal of the American Planning Association, 62,3*, 296-312. A broader critique can be found in: Sharachchandra Lele. (1991). Sustainable Development: A Critical Review. *World Development, 19, 6*, 607-621.

[6] For example, in Mark Roseland. (1998*). Toward Sustainable Communities: Resources for Citizens and Their Governments*. Gabriola Island: New Society Publishers, one of the better attempts at making sustainability a workable concept for community development, only one of the nine chapters allocated to defining the building blocks of a sustainable community deals with economic development. The social component is talked about in a chapter on housing and the remaining seven chapters deal with environmental management and land use choice.

[7] Such as in Mohan Munasinghe. (1993). *Environmental Economics and Sustainable Development.* Washington, DC: The World Bank.

[8] See Wheeler, 1996 for one attempt at summarizing the literature on sustainable urban development.

[9] For good examples see Michael Redclift (ed). (2000). *Sustainability: Life Chances and Livelihoods.* London: Routledge and Koos Neefjes. (2000). *Environments and Livelihoods: Strategies for Sustainability.* London: Oxfam Publishing.

[10] We prescribe to Geertz's definition of culture, where culture is shared meaning (Geertz, 1973). Of course, culture is evolving and anything "new" can also become a source of identity. In this work we emphasis the historical aspects of culture and tradition because of the locally-expressed concern with historical elements of the sector.

[11] The historical evolution that is portrayed in this section was pieced together from several in-depth interviews with Professor Vithi Phanichphant, from the Faculty of Fine Arts at Chiang Mai University supplemented and cross-referenced with the few historical and academic works that could be located. In general, finding documentation of the culture and crafts of the Lanna Tai Kingdom is difficult, as most of the research and documentation has been done from the point of view of the Siamese Kingdom and therefore not distinguishing or recognizing the richness of culture and art originating in the North.

[12] The Lawa tribes of Mon-Khmer were the inhabitants of this region before the Lanna Tai Kingdom. Vestiges of their culture and crafts were absorbed into the Kingdom, including cotton weaving techniques and designs and architectural elements, such as the signature wooden carved roof pieces that today still characterize 'traditional' northern Thai houses (Naenna and Phanichphat, 1993).

[13] These economic benefits most likely contributed to the fact that Chiang Mai remained the most important of the Northern Tai kingdoms. Bowie's work (1988, 1993, 1996) seems to imply this.

[14] Details of the trading routes are not well documented. In fact Forbes (1987, 3) who specializes in studying Yunnanese overland trade with the bordering Thai, Burma, and Lao lands has found little descriptive information of these routes before the 19th century. Evidence does exist that caravan routes linking these regions have existed since the T'ang Dynasty (618-907 AD) (Yi-sein 1966 as cited in Forbes 1987, 3).

[15] Silk was imported from the nearby Nan Province for the royal court (Naenna and Phanichphat, 1993). Eventually the local weaving skills were used to weave their own silk with thread imported from China and Burma.

[16] Bowie (1993, 185) also found evidence in archival records that raw cotton was a significant export via overland Chinese caravans going back to Yunnan.

[17] British visitors to Chiang Mai in the late 1800s wrote about a thriving market town lined with shops. More than 100 of these shops were selling cotton goods – both local products and imported goods of British manufacture, including cotton piece-goods, yarns and threads, as well as hardware items (Bowie, 1993, 190).

[18] This pottery was described as traditional stoneware that was by this time being produced using Asian and Western patterns. The production was rated by the ILO as being more developed than elsewhere in Thailand (ILO, 1959, 41).

[19] Scholars note that war in this part of the world was not about acquisition of land as much as about acquisition of people. Much of the population that was gained were given the status of slaves. Bowie (1996) estimates from archival research that at least one half of the population of Chiang Mai were war captives/slaves during the 19th century. In addition Bowie (1993, 188) found informants in San Kampaeng who tell of their forefathers who were captured during the various wars specifically for their skills in weaving and were settled in San Kampaeng with the express purpose of providing weaving skills to the kingdom.

[20] Mabry and Mabry (1981, 248) point out that the occupation of artisan was not one of high status or one that was desirable. This is linked to both the quality of life and income. Artisans were removed from their homes under the guise of corvee obligation to service the royal courts, separating them from friends and family. In addition, income was higher for farmers than for artisans. Due to this situation the native populations had no incentive to excel in artisanship, hence the need to find these skills in outsiders.

[21] Teak, the primary wood of the region, was a powerful natural resource of the area. The intricacies of the control of and use of this resource by upland tribes, Northern princes, and foreign concessionaries are explored in Jason A. Ramsay (1971). *The development of a Bureaucratic Polity: The Case of Northern Siam*. PhD dissertation. Cornell University. The political economy of the upland-lowland interrelationship based on natural resources and production needs, especially with teak and cotton, are also dissected in Hjorleifur, R. Jonsson. (1996). Rhetorics and Relations: Tai States, Forests, and Upland Groups. *In State Power and Culture in Thailand*, edited by E. Paul Durrenberger, 166-200. New Haven: Yale University Southeast Asia Studies.

[22] This ethnic group immigrated to Chiang Mai province between 1782 and 1815 and settled the district of Baan Khoen. Their lacquerware (also referred to as Kruang Khoen), bamboo basketry covered with a black lacquer and inscribed with designed of contrasting colors, was made into housewares such as trays, cases, and bowls. Their technique has remained the same until recently, with fakes consisting of wooden shapes, with a thin layer of lacquer and painted designs, flooding the market (Leesuwan 1981, 53-60).

[23] Bowie (1988, 1993) studied accounts of 19th century missionaries and other foreign observers in Chiang Mai to learn the details of the role of textiles to the Northern kingdom. During the 1800s textiles were the chief course of income to Chiang Mai. Much of the product went to the British via British Shan.

[24] Bowie, 1993, confirms the roles of San Kampaeng and Hot in weaving. Wild native silkworms were cultivated in Hot for a period, but that could not be sustained. So for the most part the raw materials for silk weaving were always imported from NE Thailand, Laos, Mandalay, and Yunnan.

[25] The hilltribes were relatively recent in-migrants to the Chiang Mai region, coming from Laos and Burma in the late 1800s. Crafts of these tribes were mainly of items of utilitarian needs such as clothing, blankets, bags, baskets, and tools, and for the more wealthy silver items. Large or breakable items were not compatible with the lifestyle of the tribes and religion for the most part did not include ceremonial or representative objects. Also, there was little surplus of crafts (i.e. crafts were not used for exchange outside of tribal relations) (Campbell, 1978). So there was not a history of crafts from these tribes playing a role in the region until tourism provided an impetus to create or transform tribal handicrafts.

[26] This is not unlike the creation of the culture of Santa Fe as documented in Chris Wilson. (1997). *The Myth of Santa Fe: Creating a Modern Regional Tradition*. Albuquerque: University of New Mexico Press.

[27] Cohen states that Bo Sang was a center of umbrella production from saa paper since the 1700s or so. (Cohen, 2000, 179). Professor Vithi also confirmed that such umbrellas have been used as gifts for monks for many generations.

[28] In addition 1988-1989 was designated the National Arts and Crafts Year in Thailand.

[29] Cohen (2000, 225) confirmed this through interviews in the village.

[30] Bowie's (1988, 1993, 1996) research provides evidence of this.

[31] Items produced in Chiang Mai have been spotted in the tourist markets of Athens.

[32] About three years before the research was conducted there was a great controversy about a "mafia" collecting payments for the "right" to locate a booth on the sidewalk. When one vendor refused to pay, he was murdered. Much of the Municipality's real or apparent efforts to "regulate" the sidewalk vendors stem from either improving the situation or covering for ongoing corruption.

[33] This membership organization is included as an "individual" rather than an "NGO" because its activities are an extension of the activities of the individual members. NGOs, in contrast, are generally "outsiders" who come to work in behalf of a group.

[34] These crafts have been seen in the markets of Plaka in Athens, the tourist market center for Greece.

[35] For example, the work of the YMCA for Northern Development Foundation.

[36] In 1938, 72% of Thailand was forested (England 1997, p. 60) and by 1990, 22% was forested (Hurst, 1990, p. 81).

[37] Such as Thai Tribal Crafts, WEAVE, Y-Development Cooperation, and Grassroots HQ.

[38] The revival and artistic enhancement of traditional native American pottery techniques in the pueblos of the southwest United States in the past century is a prime example of the recreation of cultural identity and pride while increasing opportunities for income generation. Prices that can be obtained for high quality native American pottery is increasing yearly, with hundreds of dollars per piece quite common, and thousands of dollars per piece becoming more common.

[39] The USEPA, through its now defunct Sustainable Communities Grants program, has experimented with funding community projects which reinforce the revival of a natural resource base, economic base, and cultural heritage linkage, such as the replanting of native grasses for traditional basket making in the Western United States.

REFERENCES

Beatley, Timothy and Kristy Manning. (1998). *The Ecology of Place: Planning for Environment, Economy and Community*. Washington, DC: Island Press.

Berke, Phillip R. and Maria Manta Conroy. (2000). Are We Planning for Sustainable Development? An Evaluation of 30 Comprehensive Plans. *Journal of the American Planning Association 66, 1*, 21-33.

Bowie, Katherine A. (1988). *Peasant Perspectives on the Political Economy of the Northern Thai Kingdom of Chiang Mai in the 19th Century: Implications for understanding of peasant political expression.* Ph.D. dissertation. Chicago: University of Chicago.

________. (1993). Trade and Textiles in Northern Thailand: A Historical Perspective. In Songsak Prangwatthanakun (Ed.), *Textiles in Asian: A Common Heritage*, (pp. 180-213). Bangkok: Office of the National Culture Commission, Ministry of Education.

________. (1996). Slavery in Nineteenth-Century Northern Thailand: Archival Anecdotes and Village Voices. In E. Paul Durrenberger (Ed.), *State Power and Culture in Thailand*, (pp. 100-138). New Haven: Yale University Southeast Asia Studies.

Cameron, John. (1997). Labour force analysis as a means to understand the livelihood dimension of sustainability. In Richard Auty and Katrina Brown (Eds.*), Approaches to Sustainable Development* (pp. 50-82). London: Pinter.

Campbell, Margaret. (1978). *From the Hands of the Hills*. Hong Kong: Media Transasia Ltd.

Campbell, Scott. (1996). Green Cities, Growing Cities, Just Cities? Urban planning and the contradiction of sustainable development. *Journal of the American Planning Association 62*, 296-312.

Chambers, Robert. (1992). Sustainable Livelihoods: the poor's reconciliation of environment and development. In Paul Ekins and Mandfred Max-Neef (Eds.), *Real-life Economics: Understanding Wealth Creation*, (pp. 214-230). London: Routledge.

Cohen, Erik. (1993). The Heterogeneization of a Tourist Art. *Annals of Tourism Research, 20*, 138-163.

________. (2000). *The Commercialized Crafts of Thailand: Hill Tribes and Lowland Villages.* Honolulu: University of Hawaii Press.

England, Phillippa. (1997). UNCED and the implementation of forest policy in Thailand. In Philip Hirsch (Ed.), *Seeing Forests for Trees: Environment and Environmentalism in Thailand*, (pp. 53-71). Chiang Mai: Silkworm Books.

Forbes, Andrew D. W. (1987). The "Cin-Ho" (Yunnanese Chinese) caravan trade with North Thailand during the late nineteenth and early twentieth centuries. *Journal of Asian History 21, 2*, 1-47.

Gallagher, B.M. (1973). Craft Villages of Northern Thailand. *Arts of Asia, 3*, 48-55.

Geertz, Clifford. (1973). *The Interpretation of Cultures.* New York: Basic Books.

Getz, Donald. (1993). Tourist shopping villages: development and planning strategies. *Tourism Management 14, 1*, 15-26.

Government of Thailand. (1998). *Investment Plan for Chiang Mai Province.* Executive Summary. Bangkok: Government of Thailand.

Hallett, Holt S. (1890). *A Thousand Miles on an Elephant in the Shan States*. Edinburgh: William Blackwood and Sons.

Hurst, P. (1990). *Rainforest Politics*. London: Zed Books.

International Labour Office. (1959). Report to the Government of Thailand on Handicraft Activities at the Fundamental Education Centre in Ubol, Part I. Geneva: ILO.

Jonsson, Hjorleifur R.. (1996). Rhetorics and Relations: Tai States, Forests, and Upland Groups. In E. Paul Durrenberger (Ed.), *State Power and Culture in Thailand* (pp. 166-200). New Haven: Yale University Southeast Asia Studies.

Le May, Reginald. (1926). *An Asian Arcady: The land and peoples of Northern Siam.* Cambridge: W. Heffer & Sons Ltd.

Leesuwan, Viboon. (1981). *Thai Cultural Heritage.* Bangkok: The Office of the National Culture Commission, Ministry of Education.

Lele, Sharachchandra. (1991). Sustainable Development: A Critical Review. *World Development 19,* 607-621.

Lubeigt, Guy. (1994). Traditional and recent aspect of the urban development of Chiang Mai, Thailand. In Marc Askew and William S. Logan (Eds.*), Cultural Identity and Urban Change in Southeast Asia: Interpretative Essays*, (pp. 117-134). Geelong, Victoria, Australia: Deakin University Press.

Mabry, Mary Connelly and Bevars D. Mabry. (1981). The role of the arts in developing countries: Thailand, a case study. *Ekistics 288, May/June*. 247-250.

Meadowcroft, James. (1997). Planning, democracy and the challenge of sustainable development, *International Political Science Review 18*:, 67-90.

Munasinghe, Mohan. (1993). *Environmental Economics and Sustainable Development.* Washington, DC: The World Bank.

Naenna, Patricia Cheesman and Vithi Phanichphant. (1993). Tai Lineage in Textile Art. In Songsak Prangwatthanakun (Ed.), *Textiles in Asian: A Common Heritage* (pp. 40-49). Bangkok: Office of the National Culture Commission, Ministry of Education.

National Research Council. (1999). *Our Common Journey: A TransitionTowards Sustainability.* Washington, DC: National Academy Press.

Neefjes, Koos. (2000). *Environments and Livelihoods: Strategies for Sustainability.* London: Oxfam Publishing.

Olpadwala, Porus and William W. Goldsmith. (1992). The Sustainability of Privilege: Reflections on the Environment, the Third World City, and Poverty. *World Development, 20*, 627-640.

Parnwell, Michael J.G. and Suranart Khamanarong. (1990). Rural industrialization and development planning in Thailand. *Southeast Asian Journal of Social Science 18, 2*, 1-28.

Penth, Ken. (2000). *A Brief History of Lan Na: Civilizations of North Thailand.* Chiang Mai: Silkworm Books.

Pye, Elwood A. (1988). *Artisans in Economic Development: Evidence from Asia.* Ottawa: IDRC.

Ramsay, James A. (1971). *The development of a bureaucratic polity: The case of northern Siam.* PhD dissertation. Cornell University.

Redclift, Michael (Ed). (2000). *Sustainability: Life chances and livelihoods.* London: Routledge.

Roseland, Mark. (1998). *Toward Sustainable Communities: Resources for citizens and their governments.* Gabriola Island: New Society Publishers.

Souvenir Guide to Chiang Mai celebrating the 700th anniversary of its founding 1296-1996.

Warren, William and Luca Invernizzi Tettoni. (1996). *Arts and Crafts of Thailand.* San Francisco: Chronicle Books.

Wheeler, Stephen. (1996). Sustainable Urban Development: A literature review and analysis. Monograph 51. Berkeley: Institute of Urban and Regional Development, University of California at Berkeley.

Wilson, Chris. (1997). *The Myth of Santa Fe: Creating a Modern Regional Tradition.* Albuquerque: University of New Mexico Press.

World Commission on Environment and Development. (1987). *Our Common Future.* Oxford: Oxford University Press.

Yi-sein, Chen. (1966). The Chinese in Upper Burma Before A.D. 1700. *Journal of South-East Asian Researchers, II*, 81-134.

INTERVIEWS

Ban Maew Village in San Kampaeng District.
Chiang Inn Plaza.
Chiang Mai Chamber of Commerce:
  Mr. Thavatchai Taeriyaphirom, Deputy Secretary General
  Mr. Danai Leosawathiphong, President
  Mr. Rachan Veeraphan, Chairman, Tourism and Hotel Industry Committee.
Chiang Mai Municipality, Department of Social Welfare, Office for Community Development:
  Mr. Chalor Bunthong, Director
  Mr. Somsak Upapan
  Mr. Jedsada Phromkhan.
Cultural Council of Chiang Mai Province:
  Princess Chao Duangduen Na Chiang Mai, President.
Department of Export Promotion.
Department of Industrial Promotion, Region I:
  Ms. Pimpan Areeboonsap, Director.
Department of Thai Art, Faculty of Fine Arts, University of Chiang Mai:
  Professor Vithi Phanichphant.
Dong Luang Village and Nong Buah Village in Pasaang District.
Grassroots HQ.
Imboon, Consumer and Farmer Producers Cooperative.
Mae Chaem District.
Northern Development Foundation:
  Ms. Karan Aquino.

Northern Handicrafts Manufacturers and Exporters Association:
  Ms. Victoria Nimmanahaeminda, Vice President International Relations
  Ms Natha Ojarasporn, Director, Public Relations.
Rakang Community: Mr. Jaroen, Community leader.
Sanpatong Handicrafts Center.
Thai Tribal Crafts.
WEAVE, Women's Education for Advancement and Empowerment.
Y-Development Cooperation Co., Ltd. (July 16 and 22, CC, JL):
  Ms. Nalinee Pussateva, Export Manager.
YMCA for Northern Development Foundation:
  Ms. Patcharin Aviphan, Director
  Ms. Regine Norden, Environmental Education Advisor
  Ms. Jiravan Sodawat, Head, Environmental Education.

CHAPTER 7

MICHAEL ROMANOS

THE CITY OF A THOUSAND SMILES: SUSTAINABILITY AND THE PURSUIT OF TOURISM IN CHIANG MAI

Abstract Tourism has been emphasized extraordinarily in the development plans of Thailand in the last thirty years, with some significant results. Its contribution to the GDP has been considerable, and public investments in its infrastructure have also benefited many other sectors of the economy and the development of the provinces, while private investments have stimulated local economies. But tourism has also raised many questions about its ability and appropriateness to support long-term sustainable development, especially at the local level. This chapter reports on an exploration of a range of issues associated with tourism in Chiang Mai and its impact on the local economy, quality of visitors, physical infrastructure, environment, public health, and local culture. It also assesses the tourism sector's sustainable development potential for Chiang Mai and identifies some courses of action that can be taken in order to shape future local tourism policies.

1. INTRODUCTION

Since the 1960s, largely because of the increased affluence of the citizens of the industrialized countries, and the pursuant demand for more leisure and recreation, tourism has evolved as a major social activity and an important economic sector (Bosselman et. al.,1999; de Kadt, 1984, x). For countries with developing economies like Thailand, which were entering their post-industrial-revolution, post-colonialism stages of their development in the 1960s, and were feeling the pains of transformation from resource-based to service-based economies, tourism provided a great opportunity for achieving some of their economic growth and development objectives quickly and relatively inexpensively. Tourism is an export sector, unique in the sense that its products are consumed by their buyers at the place of production. Natural beauty, cultural expressions, art and archaeological sites, food, and the exotic environment provided by many developing countries and regions, were ready products to make available to the growing middle class of the West without too much need for product processing, packaging, and marketing. In addition, the infrastructure and service needs required for the accommodation of tourism demand were seen by both governments and the public as public investments also fulfilling broader needs of the local populations. Thus, tourism expenditures were considered as beneficial to both the expansion of the new tourism industry and the overall development of the country.

M. Romanos and C. Auffrey (eds.), Managing Intermediate Size Cities, 127-151.

In Thailand, much of the infrastructure that supported tourism was donated. Already from the 1950s on, the U.S. extended considerable economic and military assistance to Thailand, estimated at over $1 billion between 1950 and 1970. In the 1960s and early 1970s, while the war was raging in Vietnam and neighboring countries, American soldiers were stationed in Thailand or were spending R&R time there, and the U.S. government was investing in infrastructure, especially road projects, which helped open regional markets and enhance agricultural and industrial activity in regions outside Bangkok (Dixon, 1996, 30; Elliot, 1978). Thus, by the end of the Vietnam war, Thailand found itself in a position of well publicized tourist attractions, and with a reasonable tourism and transportation infrastructure in place. Inevitably, after the secession of hostilities and the end of the oil crisis by the late 1970s, Thailand emerged as a prime tourist destination in Southeast Asia. In addition to the cosmopolitan image of Bangkok, several other destinations became well known for their natural beauty or their cultural treasures and resources. Chiang Mai was one of them.

For Chiang Mai, receiving large numbers of visitors is not a new experience. Historically, the city has been an administrative, commercial, and cultural center as the capital of the Lanna kingdom, and the entire region is home to a large and diverse number of tribal groups with a rich culture, who use Chiang Mai as their central market and service center. The region's mild and pleasant climate in the highlands of the north, and its magnificent natural setting, have also contributed to the attraction the city exerts on large numbers of domestic visitors. In recent years, Thai visitors have not only multiplied, but have sought to establish second residences in the city. A significant industry has sprouted locally for the construction of condominiums, most of which are purchased by Bangkok residents planning to visit often and eventually to retire in Chiang Mai. Thus, domestic tourism has been a long present phenomenon in this city. But the composition of Chiang Mai's visitors has changed dramatically in recent years, as large numbers of tourists are visiting by the busloads, and that change has brought about many controversies as to the net effects of tourism on the region, the city, and the populations inhabiting both.

It is the purpose of this chapter to explore this controversy, by focusing on a number of issues that have been identified by the literature as well as by the local residents and their government and NGO representatives through our surveys. While common wisdom still argues that tourism has brought economic prosperity to the city, that same common wisdom blames most ills of the city, from traffic congestion and pollution to the use of drugs and the incidence of HIV to the tourism sector and its local activities. Because tourism is such an important economic sector in Thailand, and because all cities of the country, from Bangkok down to the smallest of the intermediate size cities, consider tourism as their best chance for rapid economic growth[1], it is important that the implications of this over-reliance and trust in the sector be examined closely. This is needed not only because it may help identify flaws in this logic and make local governments and citizens alike reconsider the importance of the sector and the priority they give to its investments, but also because the future of the sector may be critical for the future sustainable development of the city and its region.

2. TOURISM IN THAILAND AND CHIANG MAI

2.1. Tourism in the Framework of Thai Development Policy

Thailand has made tourism one of its highest priorities. The Sixth Five Year National Economic and Social Development Plan covering the period 1987 to 1991 placed tourism as one of the top sectoral priorities for public infrastructure investment allocations[2]. Its Seventh Plan, covering the period 1992-1996 continued the emphasis, projecting a rise of 25% in the numbers of tourists entering the country during the period (Dearden, 1996, 204; *Bangkok Post Economic Review*, 1991). The Tourism Authority of Thailand (TAT), the official government agency coordinating the sector at the national level, has devised marketing approaches and campaigns that promote all regions of the country to all types of visitors, and especially mass tourism. Its publicity campaigns ("The land of smiles", "Visit Thailand Year 1987") created high visibility for the country not only among Asian countries but in Australia, Europe, and the Americas as well. The growth of tourism that resulted from these drives has been spectacular (Hewison, 1989). The country passed the one million tourist arrivals mark in 1973, the two million mark in 1981, and the three million mark in 1987. That number had exceeded 5 million by 1990. The years 1991-92 saw a sharp overturn in this growth, down to 3.6 million visitors, illustrating the volatility of the sector to external and internal political and economic events. But the sector recovered fast, and had again reached 5.7 million visitors by 1993, 7.2 million by 1996, and over 8 million by the end of the century (Bell, 1996, 55; TAT, 1996; Santikarn, 1994; TDRI, 1994). Tourism gross revenue was already about $1 billion in 1985 and increased to $3 billion by 1988, to $7.7 billion by 1995, and down to $6 billion by 1998 (in current prices), fluctuating between about 11% and 15% of total exports of Thai goods and services for that period. In the 1990s, this represented between 5.1% and 5.5% of the country's Gross Domestic Product (United Nations ESCAP, 2000; Muscat, 1994, 197)[3].

What attracts visitors to Thailand is a combination of factors. In tourist surveys Thailand ranks very high in quality of service, quality of food, friendliness of its people, nightlife and entertainment, and cultural and archaeological attractions. It ranks average in scenery and physical attractions; but ranks at the bottom of Asian countries in terms of its environmental and sanitary conditions (Santikarn Kaosa-ard, 1994; TDRI, 1994).

2.2. Chiang Mai as a Tourist Destination

The ancient city of Chiang Mai, designated by King Mengrai in 1292 as the capital of the Lanna kingdom, became a major center of Theravada Buddhism between the 13th and the 16th centuries. Its old town, a square mile of densely built streets enclosed by an impressive brick fortification and surrounded by a moat, contains many significant temples (wat), about the same number as Bangkok, in addition to some fine examples of residential Lanna architecture. Many other temples are situated in the greater urban region and the surrounding forests and hills, forming a system of approximately 300 Buddhist monuments, and making the city one of the premier religious centers in Southeast Asia. Northeast of the old city are the old colonial suburbs with distinctive mansions. The mountains surrounding the city are covered with dense jungle, are dotted with many tribal villages, and are home to some magnificent monasteries and temples. In addition, Chiang Mai has traditionally been a major center for arts and crafts, and is famous for its shopping along Tha Phae Road, the Warorot Market, and the Night Bazaar. With so much cultural and historic heritage concentrated in a modest size city, it is no wonder that Chiang Mai has become a top destination within Thailand, for both domestic and international tourists.

That attraction translates into some revealing statistics. Around the period of the Joint Study, the number of visitors to Chiang Mai province increased from 3 million in 1996 to 3.32 million in 1999. The share of foreign visitors during that period of time also increased, from 32% in 1996 to 39.45% in 1999. Total revenue from domestic and international visitors in 1998 was 21 billion baht, or $525 million, up from 19.3 billion baht or $482 million in 1997 (Thailand Provincial Investment Gateway, 2000; Associated Press, 1999; TAT, 1998). Visitors coming to Chiang Mai in 1998 stayed for an average of 2.85 days (domestic visitors) and 3.52 (international visitors). Even though we do not have detailed data for these tourists, we know that the majority, about 63 percent, came from the countries of East Asia, about half of them were returning visitors, about 55 percent of them traveled independently, 88 percent of them were on holidays, and they all spent money mostly on shopping (38%), accommodations (20%) and food (15%) (TAT, 1996a). Among the attractions of the Chiang Mai region, these tourists rated Thai culture and the regional landscape and scenery as the two most important attractors to the area (Butler et. al., 1999).

The tourism superstructure of the city in 1997 consisted of 223 hotels and guest houses with a total of 14,344 rooms. A large number of private tourist agencies have been operating out of the city, and an equally large number of restaurants, bars, bakeries, pastry shops, and fast food shops have been serving visitors and locals alike. Detailed data for these establishments or for the labor force occupied in and by them were not available during the period of the study and are still fragmented and unreliable. More information was available for trekking companies. Chiang Mai only had 20 or so trekking companies in the early 1980s, but their number had exploded to approximately 200 by 1992. The main service of these companies is to organize trips to the hill tribe areas. The guides they use are mostly uneducated, untrained, and uninterested in the cultures of the tribes. Much of the information they provide to the tourists is simplistic or outright wrong, but they play a very controlling role as intermediaries between the visitors and the mountain villagers (Toyota, 1996; Putsatee, 1992). No information exists on the expenditures and revenues of these companies, but estimates show that they make a comfortable profit (Dearden 1996, 209). So do the headmen of the hill tribe villages and their families, because they are the ones collecting fees from the trekking companies and organizing the accommodations in the tribal villages.

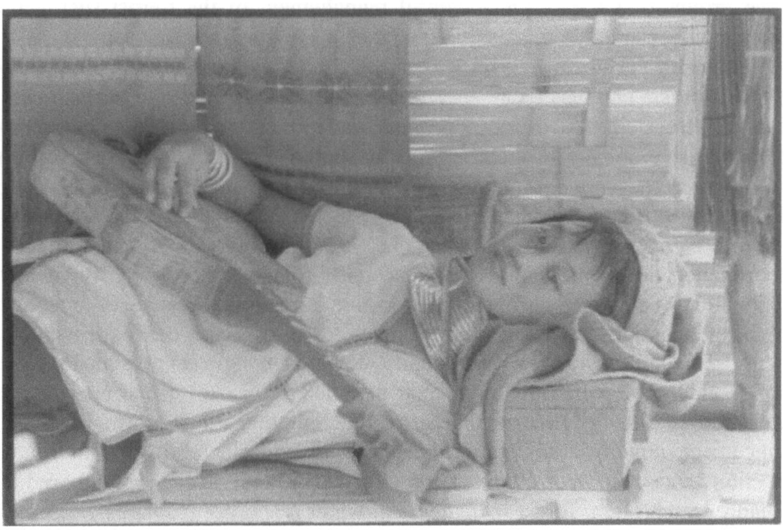

3. SUSTAINABLE TOURISM VS. TOURISM FOR SUSTAINABLE DEVELOPMENT

While it is not the purpose of this chapter to elaborate on definitions and concepts of sustainability, there is a distinction that I want to make, between sustainable tourism and tourism for sustainable development. The first concept, that of tourism sustainability, focuses its attention on the reasons for the tourist activity, and desires the preservation of the places, activities, and conditions that have attracted visitors there, so as to ensure that the reasons for this attraction will continue to exist in the future and thus visitors will continue to come. While the objective of sustainable tourism is basically an economic profit one, it nevertheless has good side effects in most instances. To the extent that visitors come to a destination like Chiang Mai in order to see and enjoy the various cultural and historical sites, the native architecture, and the local ethnic groups and hill tribes, it makes perfect sense to make special efforts to preserve these places, people and manifestations of culture, so that the income stream generated by their visitors will not diminish in the future. That means that investments, both public and private, are likely to be made to ensure the conservation and protection of these resources.

But conservation alone will not guarantee the sustainability of the tourism sector in a region like Chiang Mai. A balance must be found between it and the needs for tourist development (Ballinger, 1996, 51). For one, many of the tourist destinations in the region are living and working places, like the temples and monasteries as well as the tribal villages and rural landscapes. As such they are being used, and usage creates needs for improvement, maintenance, new structures, land use changes, and life style evolution. Also, tourism facilities, businesses, services, and accommodations need to grow, to improve, and to meet the changing needs of diverse groups of visitors. Only careful management of the tourist sector and its resources will ensure that this difficult balance may be achieved.

Tourism for sustainable development, on the other hand, signifies not only the role that tourism can play as an economic sector, but also the role tourism can play as a dynamic system of activities, products, and human interactions in the long-term sustainable development of a community or region. This is a tall order, for even if the tourism sector were completely dedicated to the achievement of that balance, it would not be sufficient to ensure long-term system sustainability. Communities and regions are open systems, affected by a multiplicity of decentralized and uncoordinated decisions, and while some parts of the system may move towards sustainability, others may go in the opposite direction, because of uncontrolled growth, or unanticipated exogenous changes, or unexpected consequences of new policies. Development of a community towards long-term sustainability is a complex enterprise, which requires clear parameters for: (a) economic growth and distribution of the benefits of that growth – employment opportunities, income enhancement, poverty eradication, and public revenue increase, (b) environmental protection and resource conservation – land conservation, sprawl reduction, farmland preservation, forest management, water resources management, provision of environmental services, natural resource conservation and management, elimination of pollution, environmental justice, (c) cultural conservation and the preservation of historical continuity – buildings and architectural complexes, urban components of particular significance, historic and archaeological sites and monuments, art objects, traditions and ceremonies, artistic expressions of theater, music and dance, customs and mores, and (d) social equity, social justice and the presence of civil society – equitable distribution of opportunities and rights, a just legal system, absence of government corruption, equal representation, and participatory mechanisms. This last one is the critical ingredient in our sustainable development framework. Development cannot be sustainable in the absence of accountable democratic institutions.

4. TOURISM FOR THE SUSTAINABLE DEVELOPMENT OF CHIANG MAI: A ROSTER OF ISSUES

Consistent with the exploratory research purposes of the Joint Project, this section discusses a range of issues identified through the structured interviews with tourism officials in Bangkok and Chiang Mai, and with a wide spectrum of tourist agencies, operators, NGOs, and concerned citizens[4]. I call them issues, rather than problems, because their foundation is a set of questions, rather than already identified crises. Most of these issues are not clear-cut. Even though the literature on the impacts of tourism tends to classify effects as positive or beneficial, and negative or detrimental, a growing number of researchers considers this "good" or "bad" approach too simplistic to describe the complex dynamic process of social and cultural changes that occur in traditional societies as they come into contact with urban and western visitors as a consequence of tourism (Toyota, 1996, 226-7). There is no clear and undisputed evidence of what the answers are at this point. However, both the issues and the ways in which they can be addressed have a lot to do with values and value judgment. Similarly, potential answers are dependent on the goals and priorities of the city of Chiang Mai. If these goals and priorities change in the future, so will the possible responses and their feasibility for the sustainable future of the city.

4.1. Tourism's Impact on Chiang Mai's Economy

In the early 1990s, tourism-generated revenue (export income) in Thailand was equal to two-thirds of the revenue generated by agricultural exports[5] and roughly equivalent to the revenue generated by textile and garment exports country-wide. By the late 1990s, it had become the largest single export in the country. Thus, tourism became a leading source of foreign exchange, which was critical for the importation of capital goods to support the industrialization and modernization of the Thai economy.

At the provincial and local level, tourism is considered a profitable economic sector, the inputs of which, at least initially, are inexpensive because they are the local natural and cultural resources. Their combination with an improved physical infrastructure and quality services provides a successful model of local and regional development, and therefore an effective means of creating jobs and increasing local income. But the actual impacts are questionable. In Chiang Mai, tourism has opened up many new employment opportunities, and has caused widespread in-migration from the surrounding rural countryside as well as from farther away provinces in search of jobs (McLaren, 1998, 72-7; de Kadt, 1979, 42-7). But many of these jobs are of low quality and pay, because they are mostly unskilled labor. Managerial and technical positions have gone to Bangkok professionals, and the larger tourist enterprises either belong to investors from Bangkok or are operated by outsiders. So in terms of upward mobility opportunities provided to the local labor force by tourism, they to date appear limited[6]. The issue becomes even more problematic when we consider large hotels and resorts financed, owned, and operated by large national and international conglomerates. For their supplies, these facilities rely almost exclusively on the national, rather than the local market, on the justification that the local market – fruits and vegetables, meat, other foodstuff, maintenance services, local outsourcing – are unreliable and of varying quality. As a result, the multiplier effect of these large units on the local economy is negligible, since their economic linkages with the community are weak.

As the distribution of tourism investments is uneven, with major investments owned and controlled by outside interests, so is the distribution of financial benefits to the local economy uneven as well. This distribution favors large enterprises and operators, but gives few direct benefits to local people. Thus, at the end, the weak role of tourist activity in the local economy leaves some of the most critical local development problems unresolved. A case in point is the poverty concentration in the 38 informal settlements inside and around the Chiang Mai urban area. Most of these settlements lack sanitation, and have no running water, no garbage collection, and no access to public transportation for their residents and their mobile businesses. Granted, these informal settlements have not been caused by tourism. But surveys indicate that the majority of the slum residents came to Chiang Mai in search of employment, and tourism is the largest single revenue generator in the city. Even though we have no data available regarding the sectoral distribution of employment in the informal settlements, we know from our surveys and those of CASE that many of the residents work in marginal jobs in hotels, restaurants, and in the Night Bazaar. That segment of the local labor force is deliberately kept poor because of the extremely low wages paid. So many other activities and employment are related to tourism, directly or indirectly, that the sector at least indirectly bears responsibility for the inability of the municipal government to address and resolve the residential infrastructure and services problems of these residents (CASE, 1998). The point is that tourism, to date, has not been able to improve the living and working conditions of poor urban residents. The only relation that can exist between slums and tourism is indirect: tourism generates local municipal revenues, and these revenues can be used to improve conditions in the slums. No evidence exists that this chain has occurred yet in Chiang Mai.

So, the question of how the tourism sector affects the conditions of equity of those at the bottom of the social scale in the urban region of Chiang Mai, remains unanswered. We know that tourism brings considerable revenues to the tourism business of the city, and we know that approximately two-thirds of that business is controlled by non-Chiang Mai interests (our survey). We can surmise that more than 50% of the people employed in the sector are non-Chiang Mai residents (our survey)[7]. Finally, we know that annually the domestic and international visitors to Chiang Mai generate approximately \$525 million in gross receipts (Thailand Provincial Investment Gateway, 2000). Based on the above, government revenues from fees, permits, licenses, and "hospitality" taxes on tourism enterprises should build up to a substantial amount of funds. But the majority of this revenue, even though collected by the municipality, is forwarded to the central government. No separate entries appear in the municipal annual budget identifying the magnitude and the uses of these funds. As a result, we cannot estimate the direct and indirect effects of these revenues on the municipal finances and, by extension, on the development programs of the city.

One way that tourism brings benefits into the local community is through the market for arts and crafts. Visitors create demand for handicrafts, and Chiang Mai has had a long tradition and fame for its woodcarvings, paper umbrellas, celadon porcelain, and other art forms. (See Chifos and Looye, Chapter 6, of this book for more on the Handicraft Sector.) This demand has caused a resurgence of traditional crafts and art forms, and is thus credited with the prosperity of the sector. Besides the economic benefits of this resurgence, demand for local art forms has created incentives for some artists to enhance and improve their work, and thus evolve into real artist-creators, rather than simply arts and crafts people. On the other hand, visitors' demand for arts and crafts may, and has in many places, cheapened the quality of the handicrafts, commercialized them, and taken away their artistic and cultural significance and identity. This loss of quality does not appear to be an official concern at this point. On the contrary, both the government – through its Thailand Tourism Authority – and the private sector have gone to extremes to promote the sale of handicrafts. The Tourism Authority has even created entire staged villages by the roadside to encourage the sale of handicrafts[8]. More on this is discussed in this chapter's Section 4.6 on culture.

But while we do not know exactly how much share of the benefits from tourism the city and its economy derives, we know that the city, and its local population, are the recipients of all the costs of the sector's operations. That is because every cost associated with tourism incurs in the city, and thus it directly and indirectly impacts the locals. Such costs include: new construction of hotels and tourist services, more traffic and the resultant congestion and air pollution[9], more crowded urban conditions, and more adulteration of the local culture. Other effects are on the overloading of the physical infrastructure and on the threats to public health. These issues are discussed in the sub-sections that follow.

4.2. Quality versus Quantity of Tourist Visitors

The desirability of a shift from "quantity" – mass tourism generating little revenue for the local economy on a per capita basis – to "quality" tourism is an issue that dominates discussions about public policy and market initiatives in the tourist industry. The trend, not only in Thailand but around the world, is that more and more people want to partake of the leisure and recreation opportunities offered, and in response, large syndicated travel operators are organizing package tourism programs in which, because of the large numbers of buyers, the per person cost is reduced, making participation in the programs more affordable. But this results in mass tourism brought into a city or region by the bus-load, and it is exactly the kind of tourism mostly criticized as the least desirable, because of its superficiality and its lack of connectivity with the place visited. Mass tourism now comes in pre-paid, all inclusive package deals, in which everything, from snacks to local transportation and museum fees are part of the package. It is the kind of tourism that leaves very little in the local community, while most of the revenue goes to the large package operators.

"Cheap" tourism, when it is not of the bus-load type, is often associated with negative situations, such as unruliness, crime, and drugs. Chiang Mai does not have as much of that problem as other Thai tourist destinations like Pattaya, mainly because it is located away from the sea and thus tends to attract a different type of tourism, at least partially based on cultural visits and interests. Even so, the trekkers' areas in the old city of Chiang Mai are rumored to be places where drugs, prostitution, and other unappealing activities take place.

Quality tourism, on the other hand, while considered the desirable kind, has not been defined adequately, and, at least in Thailand, very little has been done to actually encourage it. If quality tourists are the very wealthy, Chiang Mai would have to develop its tourism infrastructure so that expensive hotels and resorts will be available for the visitors. Wealthy tourists tend to look for high luxury, at least equal to what they have available at home, both in terms of hotel accommodations and restaurants, and in terms of recreation and shopping. Tourist destinations catering to such visitors – from Santa Fe in New Mexico, to Capri in Italy, to Santorini in

Greece – have attracted art galleries, opera and symphony, high quality transportation, and all sorts of frills to satisfy the desires of a very discriminating clientele. At the same time, it is not clear that attracting wealthy visitors would generate significantly higher local revenues. Wealthy tourists spend more on a daily basis, but much of that is for hotel accommodations and meals, and in most cases these facilities are owned by large international conglomerates who siphon revenues from outside the community.

If quality tourism means middle class family tourism, because it is associated with family values and is considered not to cause crime, then the focus of both the public and the private sector regarding priority tourism investments will have to shift. This class of tourists will stay in local hotels, will eat in local restaurants, and will visit the miscellaneous sites promoted by the city as tourist destinations. They will shop in the local stores, and their purchases are not likely to be too demanding or too expensive. They, in fact, will spend more that the average wealthy tourist, because they tend to stay in the city longer (TDRI, 1994). But they will require family-type entertainment, public transportation, safe streets, and clean air and water. The continuing orientation of Thailand, and of Chiang Mai, towards seedy entertainment and sex tourism, is not likely to make the country and the city more attractive to this class. And the environmental problems of the city, discussed below, would act as deterrents and disincentives for longer stays.

Alternatively, quality tourism could be considered represented by the visitor who comes to learn, to explore, to understand local cultures, and to get close to the local people. This kind of tourist comes in small numbers, tends to stay in local establishments and, while staying longer on the average, also tends to spend less. Ecotourism has been one of the options promoted for this kind of sophisticated traveler. Chiang Mai has two ecotourism agencies, both organizing small group and individual treks including mountain climbing, tribal visits, rafting, nature trails, and species identification. While successful, these operations tend to cost a lot, because of all the individualized plans and preparations, and leave little profit, because of the small number of people they accommodate.

Finally, quality tourism could be defined as tourism that would have the most positive impact on the local economy and community. The idea here is that wealthy tourists may spend more but may not necessarily contribute more to the "local" component of the local economy, since they tend to stay in resorts and luxury chain hotels, while middle class visitors tend to stay longer, chose accommodations in the community, and spend there. Each of these types of visitor would require different investment levels of infrastructure and accommodations, and accessibility to entertainment and destinations[10].

4.3. Tourism's Impacts on Chiang Mai's Infrastructure

As noted earlier, the emphasis on tourism given by the Sixth Plan was not followed with substantial public sector investments during the 1990s, other than promotion and publicity. Most of the development of the tourism industry during the decade was due to the initiative of the private sector. As a result of that initiative, new businesses were established, hotels and resorts were built, and the tourism industry in the country developed contacts and established linkages with the international network of operators. As a result, the sector built up its strength and sophistication during the decade. This development model, however, may not be sufficient in the future. Visitors, who in the 1980s and early 1990s came to Thailand to enjoy the natural beauty and archaeological sites of the country and were not particularly demanding about services and accommodations, are much more discriminating today, and demand quality infrastructure, facilities, and services that will require major public investments. In the absence of these investments, the quality of tourism in the country is likely to deteriorate even more, because as the quality of service declines, so will middle class preference for locations in Thailand dwindle.

Chiang Mai's physical infrastructure needs considerable improvement regardless of tourism sector needs. The city has a serious waste water treatment problem, the result of which is contamination of both ground water and the surface waters of the Ping river and the city's canals. It also has a major solid waste collection and disposal problem, which has caused much embarrassment to the city's municipal administration three separate times during the 1990s. Furthermore, Chiang Mai has no public transportation system, because of mismanagement and eventual bankruptcy of the existing municipal system in the early 1990s. The collapse of that system brought about the appearance of a fleet of unregulated and uncoordinated motorcycle-engine-vehicles plowing the streets of the city and creating horrendous traffic jams during peak-traffic periods, in addition to serious air pollution from the cars' exhausts (For more on this problem, see Romanos et. al. Chapter 8 of this book, and Schultz, Chapter 10 of this book).

However, the major regional transportation infrastructure connecting the city and the province to the rest of the country is well planned and more than adequate. It consists of a good airport and terminal facilities, good rail service, and more than adequate road network. The regional road network in the province is still not completely developed, but has been given priority by the central government, because of the rising interest in traveling between Chiang Mai and the northern boarder with Laos, as well as the other attractions of the Golden Triangle region, the Burma highlands, and the Mekong River valley.

At the same time, its tourist superstructure, in terms of hotels, resorts, and general accommodations, travel agencies, and commercial establishments catering to tourists, is adequate and improving continuously. Chiang Mai is famous for its eating establishments, and already has some international class resort hotels. Its tourist enterprises offer services in a variety of classes, price ranges, and sophistication. Its attractions and tourist destinations, both within the city and in the broad area of the farmland and the hills and mountains of North-East Thailand are many, diverse, and of major interest. And the city can accommodate all types and budgets of visitors. But all this comes at a cost. Construction of hotels and commercial establishments catering to tourists is a constant presence in the city, causing disruptions of urban life and contributing to deteriorating environmental conditions. The land speculation associated with the expansion of tourism facilities removes large swaths of land from agricultural uses. And, keen competition for customers among a growing number of businesses has caused lower prices and has triggered noticeable declines in the quality of services and products available locally.

This brings us to the issue of the quality of services provided to visitors in Chiang Mai. This is one of the problems that, unless corrected, may in the not-too-distant future contribute to a significant decline of visits to the area. Outside the major hotels, resorts, and tourist establishments and businesses catering to the wealthier groups of visitors, the service available in most other tourist establishments and businesses is marginal. Hotel personnel and tourist operators are mostly untrained and uninformed workers at low pay. They contribute little to the performance of the sector locally, while helping perpetuate a spreading reputation of incompetence and laziness. This extends to the quality of excursions organized around the region and to the hill tribe areas, the conditions of the buses and other vehicles used, the knowledge of the guides leading the tours, the language proficiency of tourist industry workers, and the degree of sophistication of the programs offered, including specialized programs like nature trails and mountain trekking.

One of the reasons for this inadequacy of the quality of tourist services is the weak role, absence of training and educational programs, lack of enforcement of regulations and lack of coordination of government agencies responsible for tourism policy formulation, the establishment of rules and regulations for efficient sector operations, and the planning, management, coordination, and monitoring of sector conditions and operations. The Tourism Authority of Thailand, the sole central government agency in charge of the sector, has done much to market the country and promote its tourism industry overseas. It has organized national and regional programs, and has collected and published data on the conditions and statistics of the sector. But it has done all of this in a centralized way, in which all plans and decisions are made in Bangkok, often in the total absence of participation by local governments and businesses. Its office in Chiang Mai did not have essential information about visitor statistics, had no access to any of the studies and plans prepared for the city's tourism development, and was unable to even provide information on the numbers and types of local tourism establishments. The municipality of Chiang Mai is even more unprepared to deal with the issues of concern to this important sector. Its Municipal Tourist Office, Tourist Information Center, and Tourist Police, which are supposed to provide many types of visitor assistance and information, were unable to supply much specific information. Questions were routinely referred to TAT's Bangkok offices, even though TAT also maintains offices in Chiang Mai. The city had no copies of the tourism development plan prepared by TAT, and had no police power to enforce any of the regulations regarding the operation and conditions in tourism businesses. In short, the role of the public sector in the operations of the tourism industry in Chiang Mai is practically non-existent.

4.4. Environmental Impacts of Tourism on Chiang Mai

A great deal of research has been published on the impacts of tourism on the natural environment and on approaches to mitigating these impacts (Gunn, 1997; Jenner and Smith, 1992; Kuss et. al., 1990; Pearce, 1989; Murphy, 1985). Much of this literature is directly applicable to Thailand. Thailand's environment is considered among the most polluted in Asia, and the country is notorious for its traffic congestion and air pollution, especially in Bangkok (Phiu-nual, 1996; Shenon, 1993; Klinsong, 1992; Tunsarawuth, 1991; Moreau, 1991). Tourism cannot be blamed for all of these problems, especially in Bangkok and in the few other major urban centers of the country, where rapid urbanization, high levels of construction activity, expansion of roads and highways, and intensive industrialization, have all delegated environmental concerns to secondary priority levels. But tourism is recognized as a sector responsible for several serious impacts on the country's environment (Chon and Singh, 1994). As early as 1993, TAT issued an unusually self-critical report assessing the many and diverse impacts of tourism sector growth on the country's environment (TAT, 1993). Among the most significant of these impacts were:

- Deterioration of visitor sites and pollution. The report identifies sites as degraded by garbage and dirt, surface and ground water contaminated, sea shores polluted, solid waste not collected, odors and stench overwhelming sites, and heavy, polluting, and dangerous traffic. These problems are described as being harmful to the health of visitors, causing visitor distress, and resulting in such negative impressions that tourists did not intend to return for a second visit.
- Encroachment of public land. The report identifies private commercial developments all over the country being built on public lands – mountain areas, forests, islands, beaches – without legal titles or proper building permits and impact studies.
- Intrusion into natural settings by tourist installations and building construction. This problem occurs when scenic landscapes and views are being compromised by unsightly buildings for the accommodation of tourism activities. Mostly this phenomenon occurs along beach areas, but farmland, mountain vistas, and forest settings are all prone to the same kind of environmental damage and visual intrusion.
- Invasive infrastructure and public works. These include disturbances of the natural settings from soil excavation, damage to rivers and waterways, injured mountain and coastal landscapes from road projects, flooding of lowland areas because of inadequate drainage, and drying of mangrove coastal areas.

Since the publication of this report, the Eighth Five-Year Plan, covering the period 1997 to 2001, has proposed a new vision of sustainable development, protection of natural resources and the environment, and stronger administrative services to meet the needs of the people (See Romanos et. al., Chapter 8, in this book). But implementation of these goals has been slow, as the Plans are not accompanied by decisions for budgetary allocations on their priority areas. Thus, to date the country has failed to provide adequate piped water, sewerage, and solid waste disposal facilities to its urban communities, and that has affected the overall quality of the urban environment in all the cities of Thailand. Chiang Mai stands out among them. The city has experienced a series of solid waste crises throughout the 1990s, which chased away tourism for periods of time, and were the focus of attention by national and international press for weeks at a time (Romanos, 2000). The surface waters of the Ping River are heavily polluted, and the Mae Kah canal, which runs through the center of the city, carries thick, dark, unhealthy water in it. In addition, construction of high-rises, mostly hotels but also other commercial structures, has been a constant source of dust, traffic, and annoyance, making the city unpleasant and discouraging tourists from longer stays. This is ironic, given that some of the main factors attracting visitors to Chiang Mai are its beautiful landscape, its rice fields, and its serene environment. Damaging that environment seems an imprudent thing to do, because it would reduce the appeal of the area to visitors. And yet, there appears to be a general lack of concern among the public and the tourism industry investors, as well as the public officials.

But the issue of environmental protection is even more general. In their race for profits, Thai entrepreneurs in Chiang Mai and elsewhere will go out of their way to bypass or avoid compliance with environmental regulations and impact assessments. The management of natural resources – in these cases land and water resources – a key component of sustainable development, is given lip service as new buildings are being raised, often in violation of building codes or the height restrictions associated with proximity to temples and historic sites. And the weak position of the municipal administration regarding police powers and code enforcement exacerbates these violations of the law and the principles of sustainable community development.

Outside the Old City, a major environmental intrusion is the conversion of land in the urban periphery from agricultural to urban and tourism uses. There is a trend to build resorts in the countryside, because such locations offer visitors the illusion of staying in a rural setting, while also they are captives of the resorts for everything from meals and souvenir purchases to the arrangements for tours and site-seeing. This trend has been causing major sprawl in the greater Chiang Mai region, as more and further outlying land becomes idle in anticipation of being sold to hotel enterprises. Accelerated land speculation, in turn, results in large scale farmer displacement and migration to the city in search of alternative urban employment, and the gradual transformation of the functions and appearance of the rural areas surrounding Chiang Mai. Rational economic thought would surmise that there is nothing wrong with this phenomenon, because rationally thinking farmers calculate that selling their land to a hotel interest will yield more economic rent at perpetuity than farming that land. But from the point of view of externalities, disappearance of agricultural activity, increase of the region's rice dependence on imports, and immigration of farmers to the city as unskilled labor, are all undesirable effects with high social costs.

4.5. Public Health Impacts of Tourism

Thailand has been known as the "sex tourism capital" of the world. It is debatable where this trend originated. One explanation has been that it is a leftover of the Vietnam war, during which time soldiers on R&R in Thailand encouraged the expansion of prostitution. Another explanation is that this is just the continuation of a custom by Thai men to visit prostitutes even after they get married. Either way, tourism probably cannot be blamed as the sole factor contributing to the rapid spread of AIDS in Thailand, but it has played, and continues to play a major role in it. AIDS is an epidemic in Thailand, and more than half of the AIDS cases in Thailand are located in the upper Northern region (Ritchie, 1996). AIDS has an especially high prevalence among the hill-tribes people of the Northern region, whose risk factors include poverty, lack of formal education, pre-marital sex with multiple partners, involvement in the commercial sex trade, and IV drug use. It has been reported that over 80 percent of the tribal women working in Chiang Mai as prostitutes are HIV-positive (CDC, 1994). The country, belatedly, has recognized the threat, and to its credit has confronted the issue directly and openly, though initially much of this focused on ethnic Thais and ignored the ethnic minorities where the problems were most prevalent. Hospitals specialize in the treatment of AIDS, and orphanages have been established in some parts of the country to care for, and monitor the babies of AIDS patients until an age when they can safely be adopted. The incidence of AIDS has had an adverse effect on the number of "family" tourists (the most desirable category of visitors) coming into the country, because they are reluctant to bring their families to destinations with the reputation of "sex tourism" and related health risks.

The poor public health issue is more general than just the incidence of AIDS. The uncollected garbage, the contaminated water which is dangerous to drink from the tap of even the most luxurious hotels, the unclean and polluted seashores and beaches, the odors and stench of central cities and slums, and the overwhelming air pollution caused by traffic, which forces urban residents to wear breathing masks in the streets, are all clear health hazards for visitors, and cause visitor distress and lasting negative impressions discouraging them from returning for second visits (TAT, 1993).

The tourism sector itself must take some of the responsibility for creating these negative impressions. It was reported during our interviews that while national environmental regulations require large hotels to provide on-site sewage treatment, in Chiang Mai many of these facilities were poorly maintained and ineffective, and substantial quantities of untreated sewage was being dumped from the hotels into the city's canals and the Ping River. Also, previous study found that nearly all (98%) of the tour buses in Chiang Mai discharged their on-board toilets untreated into the environment, either into storm drains, or onto road surfaces while traveling (Karnchanawong, et. al., 1992).

Much of the focus on the relationship between public health and tourism considers only the health of the tourists, but it is important to consider, in terms of the city's sustainable development, the impact of tourism on the health of Chiang Mai's residents as well. As the discussion above suggests, there are a number of practices by tourist businesses that put contamination into the environment and create significant health risks. However, in terms of community health, one of the strongest risk factors is poverty. Any effect of tourism on increasing or decreasing urban poverty is likely to have a significant impact on the general health of the city. In this respect, a positive development has been that because the tourists themselves may require health care while visiting the city, the growth of tourist traffic has served to provide marginal improvements in the city's health infrastructure.

4.6. Social and Cultural Impacts of Tourism on Chiang Mai

I consider the issues associated with cultural preservation critical to my interpretation and model of sustainable development in rapidly changing regions like Chiang Mai. I see cultural values and the tangible and intangible heritage they express themselves through as fundamental linkages connecting the past with the present of a community, or more generally of a human-natural system. Without this continuity, what happens today has no relation to what happened in the past. New values and decisions are not founded on history. Thus, the community of the present, while it may be able to survive in the future, cannot be considered a continuation of its past. It is, and will be, a new, different form of that system, and therefore the community of the past cannot be considered successful in sustaining itself in the future. This view of cultural heritage makes it imperative that such continuity and linkages connecting the past of a community to its sustainable future be preserved and maintained at strength.

In Thailand, and even more so in Chiang Mai, cultural identity has been one of the most powerful manifestations of the country's civilization, and has been treasured for its ability to provide this historical continuity. Anywhere in the country, its citizens address current events and problems in terms of their past. History is present not only in the monuments and the temples, but in everyday life systems of reference of the common people. Government leaders frequently refer to that past to explain or justify decisions, and all plans and programs for the future development of the country start by pointing the significance of that historical continuity[11]. It is therefore important that the cultural integrity and preservation of the country, and our city, be addressed with regard to the possibilities and the threats represented by tourism activities.

In this respect, the 1993 assessment report of the Tourism Authority of Thailand mentioned earlier addresses, in addition to environmental and health impacts, a number of concerns about the future of threatened cultural resources and social values. It identifies six main dimensions of that problem:

- Buildings and structures. This refers to the illegal construction of buildings, mainly hotels, restaurants, bars, and other tourist establishments in restricted locations such as beaches, scenic landscapes, or in proximity to monuments and archaeological sites. These buildings are illegal because they are located in places where construction is prohibited by the Department of Fine Arts or by local ordinances. They are also illegal because they are built without a building permit, and because they violate code requirements regarding height, set-back, open space, and architectural design standards. Chiang Mai is specifically cited in the report as an example of intrusion of stalls and other buildings into the protected periphery of temples and archaeological sites.

- Commercial sex. The report acknowledges the point mentioned in the Public Health section above, regarding the long tradition of Thai men to visit prostitutes. But it also emphasizes the increase of prostitution as a result of sex tourism, and the by-products of it, especially child prostitution, drug use, and diseases, both sexually transmitted (STD) and AIDS.
- Crime and hostility. Crime is recognized as being on the rise in the urban areas and resort communities of Thailand. Some of it is the by-product of prostitution and drug use. But also crime against tourists is identified as a growing concern. Much of this is incidental to a broader concern about the interaction between tourists and local people. Much reported, for example, is the exploitation of tourists through extortion or imposition of unreasonably high prices. Another aspect of the same issue is the hostility occasionally displayed by local people towards tourists because they feel offended by the intrusion in their societies. This issue is elaborated in the following paragraphs.
- Creation of inequalities caused by development. Places with tourism attraction have been receiving more attention by the government and more public and private funds for the improvement of their infrastructure (roads, electricity, piped water, garbage collection) and services (bank branches, police, municipal services, entertainment) than the rest of the country. This inequality of treatment and availability of resources, even though it is not deliberate, is causing jealousy among communities and regions, especially between the extreme south and northeast of the country on the one hand, and the rest of the provinces on the other. It also is one of the factors triggering in-migration into tourism rich areas by people from other regions of Thailand.
- Manipulation of traditional culture. In Thailand, culture is inextricably tied to religion and religious ceremonies. The majority of the significant monuments and tourist attractions are temples, statues of Buddha and monasteries. (For more on Buddhism, See Matz, Chapter 2, in this book.) Most of the cultural events attracting tourists are religious ceremonies. Thus, the commercialization of culture for tourism consumption affects the local people expressing their faith. The intensive marketing of the country by TAT and by national and international operators has intensified this problem. Other aspects of this issue have to do with the general commercialization of cultural elements. For example, dances and tribal ceremonies have specific meanings and purposes for a community. But transforming them into displays for tourists takes that cultural meaning away from them. This is exacerbated when these ceremonies are simply staged for the benefit of tourists, as for example in the staged swing

performances of the Akha tribal people in the region between Chiang Mai and Chiang Rai (Ruamsap and Piriyasurawong, 1998; Toyota, 1996)[12].

- Decline in the quality of handicrafts. While tourism has helped revive many ancient arts in Southeast Asia, and has even been responsible for the renaissance of artistic expression in places like Bali (Picard, 1996), it is also responsible for the commercialization of arts and crafts and the gradual deterioration of their quality. In Chiang Mai, some of the handicrafts produced in the region are still maintaining their high quality – silver and gold jewelry, ceramics and celadon, textiles, wood sculpture – while others are loosing it rapidly, as wholesale production is attempting to meet the growing demand for cheap artifacts from the orient. Thai paper, brass, and wood handicrafts especially are now flooding the international market and can be found in tourist shops in the United States, all over Europe, and in many Asian tourist destinations.

In addition to the issues identified by the TAT report, some other concerns need to be addressed. One of them is the issue of conservation and preservation of the built environment. As Rincon and Rahmi discuss in Chapter 15 of this book, Thailand is doing little to protect and enhance its buildings, archeological sites, and significant monuments, but it does even less to preserve these components of everyday culture – the simple Lanna houses made of teakwood, or the brick Chinese shop houses lining up most of the old commercial roads of its cities – which are responsible for providing the every day connections between the old and the new, between tradition and modernization, and which define the path of a community's future cultural evolution. Individuals and businesses are currently shouldering the responsibility of preservation in Chiang Mai, but the results have been mixed, because they do so without government supervision, quality controls, or trained assistance and consultation. Thus, the significant cultural heritage of the city is being threatened by rapid development and the pressures of replacing the old structures with modern and more convenient and practical ones. And yet, the presence of the traditional built environment is one of the most important attractors of tourism to the city. If that heritage disappears, tourism to the Chiang Mai region is also bound to suffer considerably.

This all brings us to the end point I want to make here, that of values and traditions. The issue I mentioned above, of hostility towards visitors in some conservative communities, is a manifestation of the concern of local people about these threatened values. For the older generations, their traditional values are the bedrock of their lives. Intrusion of alien values and mores threatens these foundations, not because they can be challenging to the old people but because they can tempt and corrupt the young. What the old timers see is the danger of creating discontinuities in their cultural heritage, and the breakage of the links that hold together their past with the present and their community's future. Traditional societies see this as a challenge. Part of it is a class difference. Rural people mistrust urbanites even if they are not threatened by them, because urbanites represent a different culture, which some people may consider superior. Similarly, traditional people who believe in the customs of their ancestors may be threatened by the sophisticated outsiders who may mock their beliefs. And, this can happen if the two worlds have the time and the opportunity to explore, and perhaps even approach each other. But with tourism, individual behaviors of entertainment come into play. Tourists are visiting a place to escape their own norms and mores, to enjoy themselves, to draw pleasure from what they see and experience and, often, to behave in much more free and uninhibited ways. Locals do not understand this urge for leisure, but they see the casual, disrespectful, even dismissing behavior. And they react to it with negativism and hostility. Their values do not dictate that behavior, but the threat to them does.

5. WHAT DO THESE ISSUES SUGGEST? AN AGENDA OF RECOMMENDATIONS FOR ACTION

This chapter has reviewed issues of tourism impacts or potential impacts on the community of Chiang Mai, but has done so without taking positions. It purpose was not to assess what is positive and what is negative about today's tourism industry in Thailand and its future, but rather to identify the dimensions of the conflicts between the demand for tourism and the supply of culture, natural beauty, and societal values in the location of consumption. Many of the dimensions of the issues identified here did not particularly inspire optimism. There are many indications that tourism may become a threat to Chiang Mai's sustainable development in the future, if it continues to intrude on the city's carrying capacity, helps adulterate its culture, and exhausts it natural, cultural, and capital resources. But tourism is also the fastest growing sector of the national economy, generates the largest share of foreign receipts, and has encouraged the private sector to attempt actions – from the preservation of old structures to the construction of infrastructure to the training of personnel – which the government alone could not have accomplished, given its many obligations, its inefficient structure, and its political instability during the last fifteen years. So, should Thailand continue to encourage the expansion of the tourism sector, and if so, under what conditions?

One of the points hinted at above is the need for an informed state apparatus making policies which correctly address issues, solve problems, and prepare the ground for a better future. Tourism is a powerful sector, both because it brings wealth into the system and because it has the potential to impact the system significantly. The state should, therefore, explore the issues we discussed above carefully, measure their magnitude, assess their possible outcomes, and then use this knowledge to develop sound policies that would guide the sector into a path supportive of the city's, and the nation's, sustainable development

A second point identified is the interaction between tourists and the place and people receiving them. Both groups need to become educated about each other. This may sound like a triviality, but if it is logical to expect people visiting a place to know what is there to see, where the hotels are, what is the local currency, what language the locals speak, why then should it not be expected that visitors should also learn about the customs, norms, cultural taboos, and fears of the receiving populations, so that they become more sensitive to the people they visit?

A third point is the encouragement and, simultaneously, the regulation of the private tourism sector. To date, business has bailed out the government with regard to tourism sector needs by providing infrastructure and services at a quantity that helped the sector expand in Thailand. But this happened at the expense of decline in quality of service, commercialization of culture, intrusion into sacred places and sites, and exploitation of the cultural and natural resources. If this trend continues, much of the lure and attractiveness of the country to visitors may deteriorate, lose its appeal, and motivate visitors to seek leisure and entertainment elsewhere. Before that happens, the state and the private sector should come to an understanding. Reasonable regulations should be imposed by the state, and they should be enforced systematically and objectively. And, the private sector should stop trying to bypass every possible rule and to infringe on places and codes designed to protect the public and its heritage.

Fourth, the state, with the support of the private sector, should develop norms and standards for the improvement of the quality of services provided by the tourism operators and agencies, and should develop educational programs for the training of personnel. Local human resources can take advantage of these programs to learn not only how to perform a specific service in support of tourism, but also a range of useful skills, such as accounting, management, computer software, graphics, and related skills, which they could then use in other professions as well. This would result in a better educated and trained labor force flexible in terms of its employment capabilities. A skilled labor force, both in the tourism sector and in the rest of the economy, would be less subject to the whims of employers, more competitive, and more likely to be able to negotiate higher wages.

Lastly, local and state governments need to realize the volatility of tourism and the fragility of the global traveling and leisure system, and begin to build capacity in alternative fields of development. Tourism may continue to be a tremendous source of revenue in the future, but that future is uncertain. During the last ten years, tourism in Thailand and all other countries of Southeast Asia suffered major fluctuations as a result of the Gulf War, the Asian fiscal crisis, the Japanese economic slump, and more recently the unrest in the Muslim world. We do not know what the next crisis will be or when it may come. But diversification of a national and regional economy is an essential protection against sudden and unexpected political and economic tremors[13].

NOTES

[1] Several authors, as well as local governments, believe that tourism may have the highest potential to distribute the benefits of development to peripheral areas in the provinces of Thailand, and to help diversify the rural economies of the areas (Parnwell and Khamanarong, 1996; Hitchcock, King and Parnwell, 1993)

[2] The Plan emphasized the role of tourism, but did little to move from the abstract goals to the practical issues of their implementation. According to Schmidt (1996, 74): "Being broad, lacking operational content, devoid of statutory enforcement, and unbinding of budget allocation, it provides much lee-way to the line ministries to interpret and exploit the plan as a useful reference only when they find it useful to do so".

[3] Wuttithep et. al., (1989) however estimate that tourism is an expensive enterprise for Thailand, as his team calculated that the cost of the imports needed to support the tourism sector represented more than 50% of the total revenue of the sector in 1987.

[4] The tourism component of the Joint Project was conducted via a series of 35 structured interviews in Bangkok (6) and Chiang Mai (29), supplemented with information from published and unpublished sources. Of these, 8 interviews were conducted with central and local government agencies, 15 interviews were with hotels, tour operators and travel agencies, and 12 were with museums, universities and trade associations.

[5] 123,135 million baht in 1992 (TDRI, 1994).

[6] A study conducted in the island of Koh Samui, 600 km south of Bangkok in the Gulf of Thailand, illustrates the dynamics of local-outside labor mix in the tourism sector. The study shows that locally-owned establishments have been more likely to employ local workers, but even so, more than half of their labor comes from outside the island, and the trend is changing in favor of hiring more non-local workers. In the rest of the resort establishments, there is a clear preference to hire non-local workers, and about one-third of the resorts had an established policy against hiring local labor. The reasons cited included unreliability, tendency to take time off to attend family and community activities, and cost: migrant workers are more willing to work long hours without extra compensation, and less likely to walk off the job if they are being abused (Williamson and Hirsh, 1996, 193-4).

[7] Our survey of tourism handicraft vendors on Changklan Road indicated that 25% of them were Chiang Mai natives, another 15% came from the rest of the Chiang Mai province, and fully 60% came from other provinces.

[8] For example, TAT has set up a staged village near Chiang Rai where Akha tribal people sell their handicrafts, and private operators have set up another Akha village for the demonstration of the swing custom to busloads of tourists (Ruamsap and Piriyasurawong, 1998; Toyota, 1996). See also endnote 12 below.

[9] Cars in Chiang Mai have in recent years increased by an average of 5,000 per month (Associated Press, 1999).

[10] It should be noted here that the growth of tourism in Thailand in the early and mid-1990s was caused by larger numbers of tourists visiting the country, and by longer stays of these visitors, rather than from higher daily per capita expenditures. In other words, more tourism demand did not translate into "higher quality" tourism, if quality were interpreted as wealthier visitors.

[11] It is no coincidence that the Thais continuously emphasize the fact that Thailand has always been Thai (Siamese) and has never been occupied or colonized by foreign powers. Even though that claim is to a certain extent questionable – the Lanna and the Burmese occupied large swaths of what today is Thailand for centuries – the point is significant because it implies that no western, culturally different, system of values and administration was ever imposed on the country.

[12] The Akha is a tribal group living in several villages in the hills between Chiang Mai and Chiang Rai. They originate in Burma, where about 180,000 of them still live today. In late August or early September, they perform a sacred annual ceremony celebrating the coming of the harvest season. During the ceremony, the men of the village go into the forest, cut down straight tree saplings, and use them to

construct a towering swing, which is then used by all the members of the village to swing high for several days. This is a popular show with tourists, who see it devoid of all the multi-day ceremonial activities accompanying it. But tourist operators and the government have capitalized on it. An Akha village has appeared on the road leading to Mae Salong Hill, where tourists by the busloads come to see a few Akha villagers swing back and forth, and TAT has built an artificial village along the track to the Huay Mae Sai waterfall, near Chiang Rai, for the same purpose. The locals deride the performances and call them a farce (Ruamsap and Piriyasurawong, 1998).

[13] I would like to thank Hugo Rincon and Dwita Rahmi for their participation in the Tourism Team of the Joint Study and for their assistance with the surveys and structured interviews.

REFERENCES

Associate Press. (1999). Thais fight to preserve Chiang Mai. *APnews*, February 3.

Ballinger, Rhoda. (1996). Recreation and tourism management in an area of high conservation value: The Gower Peninsula, South Wales. In Gerda K. Priestley, J. Arwel Edwards and Harry Coccossis (Eds.), *Sustainable Tourism? European Experiences* (pp. 35-53).Wallington, UK: CAB International.

Bangkok Post Economic Review. (1991). Dec. 31.

Bell, Peter F. (1996). Development or maldevelopment? The contradictions of Thailand's economic growth. In, Michael J. G. Parnwell (Ed.),*Uneven Development in Thailand*, (pp. 49-62). Aldershot: Avebury.

Bosselman, Fred P., Craig A. Peterson, and Claire McCarthy. (1999). *Managing Tourism Growth: Issues and Applications*. Washington, D.C.: Island Press.

Butler, Cherly, Harold Goodwin and Julia Leslie. (1999). *Opinion Survey, Chiang Mai*. Canterbury: University of Kent, Ecotourism Department, November '98-January '99.

CASE. (1998). *Report on Chiang Mai Slums*. Chiang Mai: Community Architects for Shelter and Environment.

Centers for Disease Control. (1994). Test of Tribal People and HIV in Chiang Mai. Cited in Ritchie, Mark. (1996). *Project Evaluation Report - The Health Project for Tribal People: A Model of Successful HIV/AIDS Education in Upland Minority Communities in Northern Thailand.* Chiang Mai: The Health Project for Tribal People.

Chon, K. S., and A. Singh. (1994). Environmental challenges and influences on tourism: The case of Thailand's tourism industry. In C. P. Cooper and A. Lockwood (Eds.), *Progress in Tourism, Recreation and Hospitality Management*, Vol. 6 (pp. 80-91). Chichester: John Wiley and Sons.

de Kadt, Emanuel Jehuda (Ed.). (1979). *Tourism: Passport to Development? Perspectives on the Social and Cultural Effects of Tourism in Developing Countries*. New York: Oxford University Press (Published for the World Bank and Unesco).

Dearden, Philip. (1996). Trekking in northern Thailand: Impact distribution and evolution over time. In, Michael J. G. Parnwell (Ed.),*Uneven Development in Thailand* (pp. 204-225). Aldershot: Avebury.

Dixon, Chris. (1996). Thailand's rapid economic growth: Causes, sustainability and lessons. In, Michael J. G. Parnwell, (Ed.), *Uneven Development in Thailand* (pp. 28-48). Aldershot: Avebury.

Elliot, D. (1978). *Thailand: The Origins of Military Rule*. London: Zed.

Gunn, Clare A. (1997). *Vacationscape: Developing Tourist Areas*. Washington, D.C.: Taylor and Francis.

Hewison, Kevin. (1989). *Power and politics in Thailand: Essays in Political Economy*. Manila: Journal of Contemporary Asia Publishers.

Hitchcock, Michael, Victor T. King and Michael J.G. Parnwell (Eds.), (1993). *Tourism in South-East Asia*. London: Routledge.

Jenner, Paul, and Christine Smith. (1992). *The Tourism Industry and the Environment*. London: *The Economist* Intelligence Unit.

Karnchanawong, Seni, Somjai Karnchanawong, Vililuck Kijjanapanich, and Hiroyuki Araki. (1992). Proceedings of the Fourth National Symposium on Environmental Technology and Management. Chiang Mai: Chiang Mai University, Department of Environmental Engineering.

Klinsong, Patima. (1992). Traffic-clogged Bangkok turns to canals for help. *The Jakarta Post*, February 12.

Kuss, Fred, Alan R. Graefe and Jerry J. Vaske (1990). *Visitor Impact Management: A Review of the Research*. Volume 1. Washington, D.C.: National Parks and Conservation Association.

McLaren, Deborah. (1998). *Rethinking Tourism and Ecotravel: The Paving of Paradise and What You Can Do to Stop It*. West Hartford: Kumarian Press.

Moreau, Ron (1991). Life in the slow lane. *Newsweek*, September 30, 16-17.

Murphy, Peter. (1985). *Tourism: A Community Approach*. New York and London: Methuen.

Muscat, R. J. (1994). *The Fifth Tiger: A Study of Thai Development Policy*. Tokyo: United Nations University Press.

Parnwell, Michael J. G., and Suranart Khamanarong. (1996). Rural industrialization in Thailand: Village industries as a potential basis for rural development in the North-East. In, Michael J. G. Parnwell (Ed.), *Uneven Development in Thailand* (pp. 161-185). Aldershot: Avebury.

Pearce, Douglas. (1989). *Tourism Development*. Second edition. New York: Longman.

Phiu-nual, Kunchit. (1996). Congestion and pollution in a rapidly expanding city of South-East Asia: The case of Bangkok. In, Yoshitsugu Hayashi and John Roy, (Eds.), *Transport, Land-Use and the Environment* (pp. 27-45). Dordrecht: Kluwer Academic Publishers.

Picard, Michel. (1996). *Bali: Cultural Tourism and Touristic Culture*. Singapore: Archipelago Press.

Putsatee, A.M. et. al. (1992). *Impact of Trekking Tourism on the Environment and Local People*. Bangkok: Mahidol University, Faculty of Social Sciences and Humanities.

Ritchie, Mark. (1996). *Project Evaluation Report - The Health Project for Tribal People: A Model of Successful HIV/AIDS Education in Upland Minority Communities in Northern Thailand*. Chiang Mai: The Health Project for Tribal People.

Romanos, Michael. (2000). Need for structural reform in intermediate size cities: Environmental management in Chiang Mai, Thailand. In, Carla Chifos and Ruth Yabes, (Eds.), *Southeast Asian Urban Environments: Structured and Spontaneous* (pp.275-326). Tempe: Arizona State University, Program for Southeast Asian Studies,

Ruamsap, Pongsak, and Prathai Piriyasurawong. (1998). Akha highs....and lows: TAT's tourist shows may kill the thrill. *Bangkok Post*, July.

Santikarn Kaosa-ard, Minsarn. (1994). A fast track to growth – and disillusion? *Bangkok Post*, September 12.

Schmidt, Johannes Drasbaek. (1996). Paternalism and planning in Thailand: Facilitating growth without social benefits. In, Michael J. G. Parnwell (Ed.),*Uneven Development in Thailand* (pp. 63-81). Aldershot: Avebury.

Shenon, Philip. (1993). What's doing in Bangkok. *The New York Times*, January 31, 10.

Thailand Development Research Institute. (1994). *Quarterly Review*. Bangkok: TDRI, September.

Thailand Provincial Investment Gateway. (2002). http//www.investmentthailand.com.

Tourism Authority of Thailand. (1993). *An Outline of Problems and Remedies of Tourism Impact on the Environment*. Bangkok: TAT, Environmental Conservation Section.

Tourism Authority of Thailand. (1996). Annual Report 1996. Bangkok: TAT.

Tourism Authority of Thailand. (1996a). *Statistical Report 1996*. Bangkok: TAT.

Tourism Authority of Thailand. (1998). *Quarterly Statistics, Chiang Mai*. Bangkok: TAT, September.

Toyota, Mika. (1996). The effects of tourism development on an Akha community: A Chiang Mai village case study. In, Michael J. G. Parnwell (Ed.),*Uneven Development in Thailand* (pp. 226-240). Aldershot: Avebury.

Tunsarawuth, Sinfah. (1991). Thailand starts major efforts to cure Bangkok's traffic woes. *The Jakarta Post*, November 4.

United Natons ESCAP. (2000). www.unescap.org/stat/statdata/thailand.htm

Williamson, Peter, and Philip Hirsh. (1996). Tourism development and social differentiation in Koh Samui. In, Michael J. G. Parnwell (Ed.), *Uneven Development in Thailand* (pp. 186-203). Aldershot: Avebury.

Wuttithep, Indhapanya, et. al. (1989). *Distribution and leakages of tourism revenues in business, hotels, and tour agencies*. Bangkok: National Institute of Development Administration.

PART III

ENVIRONMENTAL MANAGEMENT ISSUES IN DEVELOPMENT

TAXI
2748
2245

CHAPTER 8

MICHAEL ROMANOS
THOMAS E. WAGNER
SUWATTANA THADANITI

ENVIRONMENTAL MANAGEMENT IN THAI CITIES: LOCAL CHALLENGES, STATE RESPONSES

Abstract This chapter examines the organizational, institutional, legal, and fiscal setting within which urban planning, development, and environmental management are conducted in Thailand; and more specifically, how that setting affects these functions in intermediate size cities. It also identifies components of the urban environment that are in need of improved management in Chiang Mai, an intermediate size city in northern Thailand which we studied intermittently over a period of five years, and the management problems associated with these environmental components. The chapter finds that local governments in Thailand are impeded in their development and management efforts by: (a) the high degree of centralization and policy dominance of the central government; (b) the lack of clear lines of authority between central agencies and local administrations; (c) a shortage of financial resources for efficient program implementation; (d) the inability of secondary cities to take initiatives to solve their own problems; and, (e) a neglect of the secondary cities by the central government and a lack of recognition of their potential as agents of growth and decentralized development.

1. INTRODUCTION

Southeast Asia's cities are centers of national and regional economic growth. But new manufacturing facilities, construction of physical infrastructure, expanding transportation systems, and heavy use of natural and energy resources place a significant strain on the natural environment and culture of these cities and their urban regions (Kingsley, 1991; Lea and Courtney, 1985). Pressures are also caused by uncontrolled population growth, traffic congestion, and slum housing as well as the generation of solid wastes, hazardous wastes, and water and air pollution. Together these forces result in declining quality of life in both major urban centers and intermediate size cities (Racelis, 2000; Meier and Rauch, 2000; Hardoy and Satterthwaite, 1981, 191-5). In the absence of sound planning and management, these urban problems hinder national economic development and contribute to the inefficient allocation and use of human, capital, and natural resources. As the impacts of modernization and economic and demographic growth become more evident, many planners and policy makers are recognizing that continued development success is dependent on sustaining natural resources and improving urban environmental management.

M. Romanos and C. Auffrey (eds.), Managing Intermediate Size Cities, 155—184.

An underlying premise of this book and the Joint Study[1] it reports on is precisely that urban environmental management is a viable tool which can significantly contribute to the long-term sustainable development of cities of all sizes. In this context, environmental sustainability, and sustainable development, are defined as providing the physical and institutional infrastructure for growth and improved quality of life, which at the same time would protect natural resources, cultural resources, and environmental quality. A more thorough discussion of sustainable development and its premises for the Joint Study is presented in Chapter 1 of this book (Romanos and Auffrey).

The main purpose of this chapter is to examine the organizational, institutional, legal, and fiscal setting within which urban planning, development, and environmental management is conducted in Southeast Asia, and Thailand in particular; and more specifically, how that setting affects these functions in intermediate size cities. A second purpose is to identify components of the urban environment that are in need of improved management in Chiang Mai, an intermediate size city in northern Thailand which we studied intermittently over a period of five years, from 1994 to 1998, and the management problems associated with these environmental components[2]. The first of these is addressed in Sections 2 and 3 below. The second is discussed in Sections 4 and 5. Section 6 reviews the findings of this component of the Joint Study.

2. URBAN ENVIRONMENTAL MANAGEMENT

Modern management of the public sector, in terms of flexibility of organization and outcomes, leadership models, use of networks, use of market mechanisms, competition, coping with change, and employment of modern tools and techniques, had been transferred from the private sector in earnest, or at least began to accelerate in the early 1970s (Thompson and Ingraham, 1996, 286-7; Ranson and Stewart, 1994, 6). Yet, it makes more sense to apply tools and systems of effective management to the public sector, because of the chronically lean resources available there and the legal obligation for accountability of public sector operations and units to higher levels of government and to the public. Urban management is the efficient and effective production and distribution of public goods and services in support of economic growth, economic development, increased government revenues, and improved quality of life. Urban development management, and urban environmental management in particular, are public sector operations in support of planning and development implementation programs, the improvement of environmental infrastructure and environmental services, and the enhancement of the social, as well as the built and natural environment in cities (Romanos, 2000; Margerum, 1997; Ranson and Stewart, 1994; Foster, 1992). Such management of capital and human resources becomes critical as the resources available to municipal governments for carrying out routine and developmental operations dwindle while demand for them expands (Romanos, 2000; Leitmann, 1995; Atkinson and Vorratnchaiphan, 1994; Gordon, 1993).

The concept of urban environment embraces two interrelated notions, that of the built environment where urban residents live, work, and perform other social functions, and the natural support system, which provides the natural resources necessary for the city and its people to survive. As the built environment expands, it infringes on the natural support system. However, as the expanded built environment brings in more capital, technological, and human skill resources, it can contribute to the enhancement of the natural support system and the increase of its capacity. The two components of the urban environment are thus closely interlinked. The study of the urban environment and its needs for resource management focuses mainly on the degradation of the quality and quantity of resources and the decline in the quality of environmental infrastructure networks and service provision. The urban environment is usually broken down into: solid waste collection and disposal, hazardous wastes, water supply and sanitation, flooding, air quality, and traffic congestion.

Urban growth is accompanied by several environmental issues. First, there are the urban production processes, foremost those used in industries which use raw materials and produce waste and pollution which negatively affect local communities (Foster, 1989). Next, there is the problem of high population concentrations in urban areas, specifically actual numbers of people, human settlement densities, and more upper income households, which all generate more solid and liquid wastes. The increased need for urban services is another issue. Many urban residents use services whose production, i.e. electricity, transportation, consumes resources and creates pollution. The inadequate supply of these services, i.e. lack of sewers and piped water, causes environmental degradation, contamination of resources, and often diseases. The growth of urban areas also sees an exchange of non-biodegradable materials, metals and plastics, for biodegradable ones. Finally, the delicate nature of the urban landscape and its inadaptability for residential use are issues which need resolution (Foster, 1989, 3).

All levels of government are to be held accountable for these environmental issues which stem from urban growth. However, local administrations bear the greatest responsibility for ensuring that all rules and regulations issued by various levels of government are adhered to in order to improve environmental quality and preserve environmental resources. These local governments must also supply urban residents with multiple environmental services which are part of an area's environmental infrastructure network and its daily management operations. To accomplish this, they must employ and manage substantial human and capital resources.

Urban environmental management can be seen as the whole set of operations public and private sector partners must comply with to ensure the following:

- Delivery of municipal environmental services;
- Support and expansion of environmental infrastructure and facilities;
- Maintenance and monitoring of environmental quality standards;
- Enforcement of environmental rules and regulations;
- Mitigation of the environmental impacts of other routine and development activities, and;

- Assurance that future actions by the public and private sectors do not diminish present environmental conditions in the city (Romanos, 1999; 2000).

3. URBAN PLANNING AND MANAGEMENT IN THAILAND

3.1. Structure of Thai Governance and Lines of Planning Authority

Thailand is administratively organized into a three-tier system - central government, provincial government, and local jurisdictions - highly centralized in its top-down approach to legislating, planning, funding, and implementing programs. While the system is run by a trained civil servant force, including many well qualified bureaucrats, its rigid structure impedes the ability of lower level decision-making and reduces the effectiveness of public sector operations. Major decisions, by-and-large, are made in the central government agencies in Bangkok. While this may be justified for planning and development programs because of the need for high-level coordination and long term funding commitment, it suffers from the absence of consultation between central, provincial and local tiers, and thus it may or may not - and often does not - promote the best interest of the provincial and local administrations and their constituents.

At the central government level, Thailand is a parliamentary democracy with the King as the head of state. The running of the country is the responsibility of a Prime Minister elected by the country at large, and a parliament of Deputies. The national administration consists of ministries and other agencies. The Ministry of the Interior is in charge of local administration, and two of its units have the decisive role in the running and planning of the country. These are the Department of Local Administration (DOLA), which is responsible for all local government affairs, personnel appointments and payrolls, program and project action approvals, (Ministry of Interior, 1992); and the Department of Town and Country Planning, which is in charge of the preparation and approvals of all plans and development programs for Thai cities and all other local jurisdictions (Romanos, 2000; Ministry of Interior, 1994).

At the second level, that of the 73 Thai provinces, provincial governors are appointed by the Ministry of the Interior - rather than being elected by the residents of the province - are mostly from another part of the country so as to not have personal interests in the region, and are only allowed to serve one five-year term in office. In their majority they are civil servants, and at least some of them come from the military. Governors are in charge of the provincial government, and mainly serve to carry out to the provincial level the orders, decisions, and regulations of the central government agencies. Provinces are administered by Provincial Authorities, bodies consisting of an executive branch headed by the governor and a legislative body called the Provincial Council (*sapa changwad*). Provincial Councils act as a form of provincial parliament, in the sense that they can pass and enact provincial laws, question the governor on matters of administration, and review other matters affecting the province. Even though, in theory, this creates a province-level layer of checks and balances, in reality, these Councils rarely go beyond the routine ratification of central government decisions and guidelines (Romanos, 2000).

While at the top two levels of administration both the structure and the responsibilities of public bodies are relatively clear and unambiguous, at the local level a number of jurisdictions exist side by side. There, administrative designations vary depending on whether an entity is an urban or a non-urban jurisdiction.

- *Rural jurisdictions* are under the direct control of the provincial governor. The largest of these jurisdictions are called districts (*amphur*) and their administrative officers, appointed by the state Department of Local Administration, report to the provincial governor. They represent all the smaller villages in their region to the provincial and the central administration. Administrators of smaller villages are locally elected and have a limited range of authority concentrating strictly on local issues. In 1991, there were 678 districts, 6,946 communes, and 63,368 villages in Thailand (Romanos, 2000; Dolan, 1993).
- *Urban entities*, on the other hand, are self-governed municipalities (*tesaban*). They are of three kinds: *nakorn*, with a minimum population of 50,000 and a population density of at least 3,000 residents per square kilometer, subdivided into districts (*amphur*); *muang*, with populations of at least 10,000 and a density of 3,000 residents per square kilometer; and *tambon*, designated irrespective of size or density by the Ministry of the Interior (mostly low population, fast growth areas expected to grow into *nakorn* or *muang* in the near future). In 1990, there were two *nakorn* (Chiang Mai and Hat Yai), 86 *muang*, and 44 *tambon* in Thailand (Dolan, 1993, 3)[3]. Municipalities have a legislative and an executive body, like provinces. The elected legislative body, or assembly, is responsible for approving the municipal budget, enacting municipal ordinances, and overseeing the operations of the executive branch, while the executive branch, consisting of the head administrators of four or more administrative units – usually administration, finance, public works, and health -- is appointed by the mayor, who usually is the elected leader of the majority party in the municipal council. Thus, the structure for democratic representation and for bottom-up procedures and decision making appear to be in place. But where the system breaks down is in the interventions of the central government agencies in this process: the elected mayor and his administrators must be approved by the appointed provincial governor, while local assembly-passed regulations must be approved by the Ministry of the Interior. In addition, all local government personnel hiring is under the direct control of the Ministry of the Interior through its Municipal Service Commission. The extent of this central government intervention is a major point of contention between the local (elected) and the provincial (appointed) administrations (Romanos, 2000)[4].
- *Sanitary districts*. These are local jurisdictions that resemble *tambon* municipalities in terms of their administration. Usually, they designate areas which, even though they may be very urbanized, do not yet meet the

criteria to become municipalities. Their purpose is primarily to provide sanitary services to the urban populations they serve but, in recent years, they have evolved to provide other functions, such as electricity, education, and fire protection as well.

3.2. Planning and Public Management Functions

Planning and management authority and responsibilities in Thailand are distributed among several ministries and departments of its central government. Under the Office of the Prime Minister, the National Economic and Social Development Board (NESDB) and the National Environment Board are responsible for developing policy at the highest level for social and economic development and for the establishment of environmental standards respectively. NESDB is the generator of the country's Five Year Development Plans. Environmental policy and planning are the responsibility of the Ministry of Science, Technology and Environment, the Office of the Prime Minister, the Ministry of the Interior, and the Ministry of Industry, all of which have jurisdiction over different aspects of environmental service provision and management as well. Land use planning and housing are the domain of the Ministry of the Interior. All policies, plans and programs for the nation's municipalities, and the funding for them are decided, prepared, and managed by the agencies of these central ministries from Bangkok. Table 1 illustrates the distribution of responsibility for planning and plan implementation among the various central ministries. That distribution, as well as the unclear lines of authority among the three levels of government in Thailand, have severely impeded the ability of local governments to manage their resources effectively and to provide their services to the public efficiently. Both planning and urban environmental management functions suffer as a result of these conditions.

As mentioned previously, NESDB is responsible for the production of Thailand's series of Five Year Development Plans. These are national social and economic development plans that have helped organize all development programs and priorities under one medium term programmatic agenda. NESDB was originally established in the 1950s to consider and elaborate a fiscal program for economic investments and to cooperate with foreign specialists in planning the country's economic development. With assistance from the World Bank, NESDB created a fiscal plan as a baseline for development in the country. It completed the First Five Year Development Plan in 1961, and released its Eighth Plan in 1996.

Each plan covers a five-year period and is intended to "develop" the country in ways that improve the quality of life for the citizens (Table 2). The development concepts in the national plans have changed over the years. Usually, the changes reflect recent perspectives on development and respond to issues and concerns that have arisen since the last plan was formulated (Mekvichai, 1996; *Eighth National Economic and Social Development Plan (1997-2001)*, 1996); *Thai Development Newsletter*, 1995; Vorathanyakit, interview, 1998). In many respects, these national development plans have been very successful in guiding Thailand's economic growth over the past 35 years.

Table 1. Thai Government Agencies Responsible for Development & Environmental Management

Agency	*Reporting Line*	*Responsibilities and Duties*
National Economic and Social Development Board (NESDB)	Office of the Prime Minister	Responsible for preparing a national plan every five years to guide the Kingdom's social and economic development. Also responsible for reviewing major projects administered by other ministries and state agencies.
National Environment Board	Chaired by the Prime Minister	The highest authority on the environment. Consisting of seven ministers, the Board has the power to establish environmental standards.
Ministry of Science, Technology, and Environment (MOSTE)	National Environment Board	The administrative arm of the National Environment Board.
Office of Environmental Policy and Planning (OEEP)	Ministry of Science, Technology, and Environment	Prepares and coordinates environmental policy and plans, advises provincial and local authorities on environmental matters, and assists in preparing and reviewing provincial environmental plans.
Pollution Control Department (PCD)	Ministry of Science, Technology, and Environment	Prepares and recommends policy, plans, and measures for pollution control and prevention.
Department of Environmental Quality Promotion (DEQP)	Ministry of Science, Technology, and Environment	Responsible for providing information on environmental quality protection to other government agencies, businesses, and the public.
Ministry of the Interior	Office of the Prime Minister	Includes several national and local agencies responsible for administration and services and the environmental infrastructure in provinces and municipalities.
Public Works Department	Ministry of the Interior	Responsible for advising and assisting local governments in matters related to water supply and sanitary and other engineering works, including the design, construction, and administration of waste treatment plants.
National Housing Authority	Ministry of the Interior	Responsible for low-income housing, services, and infrastructure improvements. Also administers a semi-autonomous credit program for low-income communities through the Urban Community Development Office.
Department of Industrial Works	Ministry of Industry	Regulates utilities, services, and standards for most major industries.
Industrial Real Estate Authority of Thailand	Ministry of Industry	Services and monitors industries located on the Industrial Estates.

Sources: Suphapaodok and Chueyprasit, 1994; Krüger, 1996.

Table 2. Thailand's Five-Year Plans

Plan Years	*Plan Emphasis*
First and Second Plans: 1961-1971	The first two plans emphasized economic growth in the agriculture and industrial sectors. The nation's natural resources were considered important and development focused on building infrastructure to improve access to rural areas.
Third Plan: 1972-1976	The third plan continued to encourage economic growth but focus shifted toward urban development with an emphasis on the development of the Bangkok area.
Fourth Plan: 1977-1981	Although economic growth continued to be the focus, the fourth national plan emphasized resource protection and rehabilitation and laid the groundwork for planning and management of natural resources.
Fifth and Sixth Plans: 1981-1991	These plans addressed the unevenness in growth and the vast differences in development between the country's various regions and between urban and rural areas. The basic strategy was to create regional urban "growth poles" in the belief that this would spread growth more evenly throughout the nation. The plans emphasized control of land use, improving the infrastructure and municipal services, and protection and rehabilitation of natural resources. The concept of "sustainable development was a major element of the sixth plan.
Seventh Plan: 1992-96	The seventh plan stressed sustainable development through the objectives of (1) sustaining the country's economic growth at a level compatible with stability; (2) promoting more equitable income distribution and rural development; and (3) developing human resources, improving the quality of life and enhancing the quality of the environment by setting goals and standards. The plan is also distinguished by its attention to citizen participation in the planning process.
Eighth Plan: 1996-2001	The eighth plan proposes a vision of sustainable development and continued national growth, protection of natural resources and the environment, balancing the quality of life between Bangkok and the provinces and rural areas, and stronger administrative services that serve the "real needs of the people." The plan also calls for greater citizen involvement in planning and decision-making and greater collaboration between the levels of government responsible for implementing the plan.

Source: Vorathanyakit, interview, 1998.

Each plan has served as the general framework for development throughout the country. Added to the national plans are individual development plans for administrative ministries, regions, provinces, and municipal areas. The ministerial, regional, and local plans provide details for specific planning areas but are consistent with the goals, objectives, and targets of the national plan. Both national and local levels of government are responsible for implementing and managing the programs based on the national plan. The national government sets policy, formulates plans for projects, and provides financing.The provincial and local levels are responsible for overseeing and operating the development projects and programs. These provincial and local governments have responsibility for public safety, health, education, and building and maintaining the local infrastructure. Central government agencies, however, manage the police, health, education, highway, and other departments. Thus, in many respects, the provincial and local governments are in charge of these programs in name only. This situation, of course, creates many problems in implementing local plans; and has led to interagency conflicts and program delays in the past.

4. CHIANG MAI: A MODEL CITY FOR THE STUDY OF ENVIRONMENTAL MANAGEMENT IN INTERMEDIATE SIZE CITIES

Chiang Mai is an intermediate size city located in Chiang Mai Province in Thailand's northern region, and is the second, third or forth largest city in Thailand, depending on which enumeration is accepted[5]. Significantly smaller than Bangkok, Chiang Mai in many respects is the second most important city in the nation[6]. It was designated a regional growth center in the Fifth National Development Plan; and more recently, it was identified as the center of the "Development Square Region" in which Thailand, China, Myanmar, and Laos are cooperating for trade and development.

4.1. City Profile

Chiang Mai has historically been one of Thailand's major economic and cultural centers[7], and today it is also a prime tourism destination. The city has experienced sustained economic growth and social change since the 1970s. Most of the growth has occurred in tourism, followed by agriculture; these sectors, when combined, account for an estimated 50% of the gross provincial product. Universities and other educational institutions grew substantially through the 1980s and are considered the third largest contributor to the local economy. Some industrial and manufacturing development occurred in the 1980s, but well below that found in other regions of the country. The construction industry boomed in the 1980s but declined in the 1990s because of the completion of several major projects, new restrictions on land use, a decrease in speculators due to an oversupply of hotel and condominium space, and the recession caused by the monetary problems in Southeast Asia in 1997 and 1998.

An important feature of Chiang Mai's economy is the small scale of its businesses. In 1990, 8,435 of the private homes, or nearly 20%, were used for business. This figure reflects the presence of artisans and small shops catering to tourists. As for the city's labor force, the 1990 census shows that 19.5% were employed as craftsmen, production workers, and laborers; 25.8% in sales; 17.9% as professional, technical, and related workers; and 13.8% as service workers (*Pocket Thailand in Figures 1994*, 1994).

It is difficult to assess the impact of the "informal sector" on Chiang Mai's economy. Many members of the city's labor force work as peddlers, vendors, and unmetered taxi drivers, and in some home craft industries such as lacquer ware and woodcarving. Many of these informal-sector workers are women and children earning supplemental family income; although they may not appear in the statistics, they certainly affect the economy. In 1990, the average monthly per capita income for people living in Thailand's northern region was 1,791 baht (about U.S. $72), with an average monthly household income of 6,267 baht (about U.S. $ 250). This is somewhat below the average for the nation as a whole and falls well below the average for the central region of Thailand, where Bangkok is located (*Key Statistics of Thailand 1996,* undated).

The housing stock has grown phenomenally in Chiang Mai in the last two decades, from 26,289 housing units in 1982 to 35,591 in 1989 to 42,571 in 1990. 58.7% of the 1990 units were owner-occupied. Over 98% of the dwellings have electrical lighting and sanitary flush toilets. These figures do not reflect the condominiums and new residential areas constructed in areas outside the city (*Pocket Thailand in Figures 1994*, 1994).

One negative aspect of Chiang Mai's growth and development is the flight of true Chiang Mai natives. Many of those who lived in the city have moved to the countryside to escape the noise and chaos. In the words of one long-time resident, "Real Chiang Mai people do not live in the city now. They like peace, don't like crowds, and are very religious. Any restoration or new development around the temples upsets them, so most have moved out into the country" (Harmon, 1997).

The urbanized land area of Chiang Mai has been expanding rapidly in the last fifteen years. Domestic and international tourism, economic growth, and building speculation have been sited as the main driving factors behind this expansion, the last factor mainly because the area has become a desired second home location for wealthy Bangkok residents. Little is being done to contain urban sprawl, and conversion of farmland to urban uses is being reported by our respondents as one of the major growth problems of the urban region. Within the city itself, construction is subject to few controls. The Chiang Mai master plan contains some loose zoning recommendations, but they are not being enforced, while its building regulations are inadequate and ineffective in dealing with uncontrolled construction. For example, no provisions for building height controls existed until 1990 when the construction of a high-rise by the east bank of the Ping River triggered public outrage and mobilized the community and local NGOs into vocal protests, the result of which was the adoption of a more restrictive code regulating building heights. However, the height restrictions were limited to land along the riverbanks and within a 100-meter radius around the *wat* (temples) dotting the city. Thus, in the last eight years several high-rise hotels and condominiums have been completed or are under construction within the municipal boundaries of Chiang Mai.

4.2. The Municipal Administration

Chiang Mai is a municipality administered by a mayor and a 24-member Municipal Council, both elected to four-year terms. The mayor carries the title of "Lord Mayor", is the chief officer of the municipality, and manages the city with the assistance of the Executive Board, the executive branch of the local government, the heads of the departments which are appointed by the mayor from among the members of the Council[8]. The Council is the body responsible for the development of local policies and regulations.

At the provincial level, the governor of the province of Chiang Mai is appointed to the post by the central Ministry of the Interior, as are all other provincial administrators. Despite his appointed status, the governor has major powers over the municipal administration, including the right to remove members of the elected Municipal Council and the mayor-appointed Executive Board. The governor can also act at will, hiring or firing members of the municipal administration, and thus directly interfering with, and undermining the authority of the mayor and the municipal bodies to run the city effectively. The ability of central government agencies to interfere at the local level so powerfully through the office of the governor, and their power to annul the authority of the locally elected bodies, has been a cause of constant conflict among the political elites of Chiang Mai, and has resulted in local government ineptness in resolving even its most critical public sector management problems. The conflict is further exacerbated by the opposing goals of the people in office at the local and provincial levels. Because of their weak position to manage local resources without interference by the governor, mayors have traditionally not been able to pass any kind of even slightly unpopular legislation regarding the more efficient operation of the city departments, including the imposition and collection of service fees. At the same time, governors are using their time in office in the province to position themselves for higher office with the central government, and thus are not likely to involve themselves in the solution of long-term provincial problems. The conflict, not unique to Chiang Mai, but a serious impediment to the city's ability to streamline its municipal and regional operations, has been recognized by the country's five-year plans as a priority objective, but has not been addressed to any satisfactory degree to date.

5. ENVIRONMENTAL MANAGEMENT CONCERNS IN CHIANG MAI

Because of its rapid economic and physical growth, the Chiang Mai Municipality is facing some serious environmental issues. Its rural areas are facing deforestation, loss of biological diversity, soil erosion, flooding, and groundwater shortages, while its urban neighborhoods suffer from overcrowding, poor housing conditions, solid and hazardous waste disposal, low water quality, air pollution, soil contamination, and traffic congestion. In this respect, Chiang Mai has been experiencing most of the serious environmental problems of the large and intermediate size cities of Thailand, in some cases closely resembling Bangkok in magnitude and severity. Respondents to our surveys[9] have indicated that among the worst threats to the city's sustainable development are:

- air quality and traffic congestion;
- solid and hazardous wastes and soil contamination;
- water availability, quality and pollution;
- slums and slum improvement;

- lack of open space, and;
- seasonal flooding and drainage[10].

No attempt is made here to rank these problems in terms of severity of impacts or costs of mitigation. Rather, the focus is on the government structures responsible for the management of these areas and for the solution of their problems[11]. Thus, in the following sections, we provide summary information on the nature of these problems, and on the city's efforts to deal with them.

5.1. Air Quality and Traffic Congestion

Air quality in the cities of Thailand is affected by (a) the exhausts of ground transportation vehicles, especially diesel-powered buses, trucks and older vehicles producing "black smoke"; (b) industrial production using fossil-fuels and generating toxic gases and bad odors; (c) dust generated from quarrying, mining and factory production (like cement production) and construction activity; and, (d) the burning of garbage and other agricultural and urban wastes (Urban Environmental Guidelines Project, 1994). While air pollution is a serious problem in Bangkok, caused primarily by high concentrations of vehicles, high volumes of traffic, defective automobile exhaust systems, and low quality gasoline (Xoomsai, 1988, 12-14), Chiang Mai, whose traffic congestion is limited to certain parts of town, and only for short periods of time during the day, is not seen as consistently having air pollution problems according to our survey's respondents. But this perception is not supported by the Krüger (1996) study, which found that the city does have an air pollution problem, manifested mainly during peak-hour-traffic periods, and consisting of increased levels of carbon monoxide and dust. These air pollution indicators are observed in major concentrations in the city, especially during the dry season. Furthermore, transportation is not the only cause of air quality decline in the city. Both agricultural practices, such as the burning of the rice stems after the harvest, and building construction are major contributors to the city's air quality problem, especially during the dry season, when rainfall is not present to help filter the air (Romanos, 2000; Krüger, 1996).

To address air pollution problems in its cities, Thailand has established a number of legal devices, including the Industrial Factories Act 2535 (1992) to regulate factory generated air pollution, the Car Act 2522 (1979) requiring cars and transport vehicles to undergo environmental checks before registration can be renewed, Public Health Act 2535 (1992) regulating other nuisances such as odors, and the Enhancement and Conservation of National Environmental Quality Act 2535 (1992) prescribing air pollution treatment systems and regulating vehicle exhaust. Further, a notification published in the *Government Gazette*, Volume 100, Section 38, dated 31 March 2536 (1993) empowers the mayors of Thai cities to issue written orders to suppress causes of nuisance arising in public or private places, and to promulgate municipal regulations to ensure air and noise pollution controls. Further, several central government agencies have formal responsibilities for the enforcement of these laws and the protection of air quality in the nation's cities, including the Department of Pollution Control, the Office of Environmental Policy and Planning, the Department of Environmental Quality Promotion, and the Department of Scientific Services, all under the Ministry of Science, Technology and Environment. Other agencies with air pollution protection responsibilities include the Department of Industrial Works of the Ministry of Industry, the Department of Land Transport of the Ministry of Transport and Communications, and the Department of Health in the Ministry of Public Health. Unfortunately, all these Departments are regulatory, but have no police power for the enforcement of the laws and regulations designed to protect the public from air pollution problems. Periodically, combined efforts by regulatory and enforcement agencies to solve this problem have met with mixed results. For example, the Royal Thai Police cooperated with the Department of Land Transport to initiate on-the-road vehicle emissions testing, in Bangkok. But there was a public outcry when it was found that the police were not measuring exhausts correctly and that they were soliciting bribes from negligent drivers. This necessitated the additional involvement of officers from the Department of Pollution Control, the Department of Land Transport, and volunteer organizations, to ensure transparency of operations. As a result, the experiment became too big and complicated, and had a short life.

Chiang Mai itself has no regulatory framework or management program to address its air pollution problems directly. It has repeatedly focused on the problem of its public transportation, in the hope that solving that problem will also reduce the incidence of air pollution, but to date, no solution has been found. Public transportation in effect does not exist in Chiang Mai. The municipal bus system in service until 1997 went bankrupt that year because of mismanagement, low quality of service, and lack of patronage. Once the system seized operations, it was replaced by privately operated vehicles such as motorcycles, *tuk-tuk* (small motorized vehicles which can carry up to two passengers) and *song tau* (converted pick-up trucks providing flexible door-to-door service throughout the city, with a capacity of 8-10 passengers). These alternatives are all highly polluting, and are in addition the cause of many traffic accidents. But they provide forms of transportation efficient and inexpensive enough to make it difficult for the municipality to come up with a competitive alternative[12].

5.2. Solid and Hazardous Wastes, Soil Contamination

Waste collection and disposal, including the location and siting of urban landfills, was considered by our survey respondents to be the most important environmental issue and concern that should be addressed by Thai city governments. Practically all other urban environmental issues, they felt, were secondary until such time as this critical problem had been solved. The collection and disposal of solid waste is indeed seen as an increasingly serious problem in modern societies, and one that requires a lot of organization by municipal administrations and major campaigns of public information to make communities aware of the magnitude and the seriousness of the issues involved. In Thailand, solid waste management is one of the most urgent environmental service areas in need of attention, because of the large amounts of garbage generated daily, and the inadequate provision of collection, disposal, and recycling available in its urban areas. In Bangkok, residents generate approximately one kilogram of solid waste per day, and other urban areas do not fall far behind. Cities which attract a lot of tourism have even larger rates of solid waste generation. As urban societies modernize, the composition of that solid waste is also changing. Today, larger percentages of it are metal (about 10%), plastic (about 11%), wood (about 11%) and paper (about 19%) in addition to general waste (Urban Environmental Guidelines Project, 1994, "Solid Waste Management", 2).

The responsibility for the collection and disposal of solid waste is clearly vested with the country's municipalities through the 1953 Municipal Act 2496 which stipulates that the municipality shall be responsible for disposing of waste within the municipal area, and Section 77 of the Enhancement and Conservation of National Environmental Quality Act 2535 (1992) which stipulates that municipalities shall install a combined waste treatment system within the municipality, sometimes with the assistance of a private contractor, and charge a fee to the public for this service. Further details about the provision of waste collection containers in public and private places, and the determination of methods of collecting and disposing of municipal waste are included in the Public Health Act 2535 (1992). But again, where the system has failed so far is not for lack of regulations, but rather for the ability of local governments to implement them effectively.

The municipality of Chiang Mai generates about 174 metric tons of solid waste per day (1994 figure). For solid waste collection and management purposes, the city and its surrounding area have been organized into five sanitation districts. Four of these districts are the four quadrangles of the urban area within the boundaries of the municipality. The fifth is a sanitary district designated to include all the urbanized region around Chiang Mai but outside the incorporated city limits. Collection takes place on the average twice a week, and is done by hired private contractors. Collected waste is transported and dumped in the municipal landfill (Romanos, 2000; Saowapon, 1997). The residential and commercial areas outside the municipal boundaries, but within the sanitary district also use the city landfill, because the sanitary district has no disposal site of its own[13]. The landfill used until 1989 was located in Mae Hia, approximately 10 km away from Chiang Mai and outside the boundary of the municipality. When landfills were located, the city made an agreement with the leaders of the villages surrounding the site, for the proper sanitary disposal of the waste in the dump. This commitment was never carried out by the municipality, thus resulting in serious environmental hazards for the villages, including leachate and water seepage that caused severe groundwater contamination. Thus, when the dump site was completely filled in 1989 and was forced to close down, the city found itself without an alternative site and without any good faith with which to negotiate with the surrounding communities. For a few years, the city was forced to dispose of its solid wastes at a number of temporary sites, including cemeteries. In each case, local residents forced the site to shut down within a short time. The situation reached crisis proportions in September of 1994 when the city ran out of dumping options, and left its residential and commercial garbage uncollected for several days. Within a week the city was overwhelmed by the odors and the vermin roaming among the piles of garbage. Then torrential rains in the mountain areas caused the Ping River to flood, inundating the streets of the downtown, and carrying with it the garbage. By October, several blocks along the river were covered with up to one foot of dried mud within which the city's garbage was trapped (*The Sunday Post*, Sept. 4, 1994, 20-21; October 9, 1994, 24). In

September 1998, the story repeated itself as for over a week garbage was again not collected from the city streets because there was no place for it to be taken. Odors overwhelmed the neighborhoods and the tourist-frequented downtown. Garbage trucks attempted to haul garbage to graveyards and nearby forests at night, but were met by groups of local residents who prevented them from approaching. By the end of the month, the mayor resigned stating that she was unable to cope with the politics of the situation (*Bangkok Post,* 1998). Currently the problem has been solved by designating a new landfill much more distant than the previous one; but the solution is not a long term one, and the issues associated with the transportation of that waste are still unresolved.

One of these issues is the fee for the collection and disposal service. Presently, that fee is inconsequential for residential customers, but substantial (around 1000 baht, or US $25 per month) for business customers. The fee levels are determined by the central government, not by the municipalities. Collection is done by sending municipal collectors door-to-door. Because households are often not at home when the collectors come, or refuse to pay, actual collection of fees is miniscule as compared to its potential to generate city revenue. It is estimated that Chiang Mai presently collects about 10 percent of the amount that it could collect under this system. Since Chiang Mai has limited manufacturing activity, hazardous wastes are not considered a major environmental threat to the community at present. But there is concern that industrial wastes from the Lamphun Industrial Estates, 30 kilometers to the south, could eventually be dumped in Chiang Mai's disposal site if a proposed specially designated and equipped hazardous waste disposal facility is not built in Bangkok, as planned (Supporn, 1998). The degree of soil contamination in Thailand is largely unknown, but certainly it exists, as indicated by the impact of the closed landfills. In addition to the contamination of the soil from leachate leakage, chemical and hazardous waste disposal, and commercial activities such as gas stations, auto repair shops, photo shops, and dry cleaners all are causes of soil contamination. No mechanisms for the reduction of the poisonous repercussions of such pollution are currently in place in Thailand.

5.3. Water Quality and Quantity

Clean water, a basic human need, becomes contaminated after use, and therefore unsuitable for immediate reuse. Polluted water is useless and dangerous to human health and aquatic life, but its treatment carries a cost. Water pollution is caused by: (a) organic matter, which takes up dissolved oxygen and can result in the death of aquatic life; (b) nutrients, when sewage is discharged into rivers or lakes, causing excessive fertilization and growth of weeds; (c) pathogens (disease germs) present in sewage; (d) toxic substances, such as industrial and chemical wastes, pesticides, and some household wastes, which are poisons to animals and plants; and, (e) siltation of rivers and canals caused by careless deforestation, mining, construction and agricultural practices. Water quality deterioration becomes evident in the dirty and smelly waterways of a country or city. Thailand's waterways have been showing such deterioration for years, and Chiang Mai has not been an exception to this. The Krüger report rates the city's water quality "poor" (Krüger, 1996). Sections of the Ping River near Chiang Mai suffer from severe water pollution, including a stretch just south of the city that is suitable only for transportation and industrial uses[14]. Most of that water pollution is due to inadequate sewers and uncontrolled dumping, directly into the waterways, of household and community wastes, and as leach into soils or groundwater (Chaithawat, undated). As a result, many of the *klong* (canals) in the city are eutrophic; their water has become black, and they resemble foul-smelling open sewerage systems that detract from the quality of the city's environment.

In terms of water availability to the residents and business of the city, Chiang Mai's water supply is considered adequate, largely because of the city's physical location in the mountains, where rainfall, groundwater, and streams are plentiful. The city's water supply is treated in three facilities which are considered adequate in capacity and efficient in treatment. At least 65% of Chiang Mai's households are connected to the city system, while the rest of the urban households depend on shallow or deep water wells. A new water treatment plant, completed by the end of 1998, is expected to provide the additional capacity needed to meet the city's anticipated demand for water (Chaithawat, undated).

The municipality recently completed a wastewater collection system on the west side of the Ping River. The system is based on state-of-the-art technology and is designed for a 20-year city growth period. However, no funds are available to extend the system to the east side of the river, so half the city will not be served by a treatment facility. In addition, there have been concerns about the municipal staff's training and technical ability to operate and maintain the system (Nithiwat, 1997).

The city's water and sanitation system is lacking a fair and comprehensive system of fees and charges to the residents. As a result, no fees are currently being collected that could be used for the maintenance and servicing of the system.

5.4. Slums and Slum Improvement

Slums are low-income squatter settlements that lack basic infrastructure and services, or legal housing areas that have fallen into bad times, are overcrowded, and have deteriorated housing and services. Often these settlements occupy unhealthy sites unsuitable for other development because of flooding or proximity to industrial sites. Their inhabitants are mostly rural people who have migrated to cities because of crop failure or lack of land security, and are in search of employment and a way out of poverty. Their low skills confine them into near-poverty jobs, but they provide all sorts of useful and inexpensive services for urban populations in Asian cities, and constitute the backbone of the "informal sector" of their economies. Yet, little, if anything, is done by municipal administrations to alleviate their poor living conditions and to provide even the absolutely essential services for their survival. Slums typically lack toilet facilities, and discharge effluents directly into rivers and canals; thus, exacerbating the unhealthiness of the local environment. Open-discharge sewerage also contaminates the groundwater, which is often the only source of drinking water for the slum communities, thus further spreading the incidence of disease. Waste collection is also inadequate, since municipal administrations usually have limited resources to service slum communities. And since most slum housing is illegally built, electrical connections are also mostly absent, thus forcing the residents to make illegal and dangerous electrical system connections.

Approximately 10% of Chiang Mai's population lives in 19 slums or squatter settlements in and on the periphery of the city (Krüger, 1996). Even though this is considered a large number of people for a prosperous city like Chiang Mai, other studies bring that number much higher. For example, a more recent study conducted by Chiang Mai University identified a total of 58 separate slum communities within the boundaries of the municipality. Most of these communities are on public or government land such as old temple sites or canal right-of-ways. Many are hidden from view behind buildings or walls built by adjacent landowners (Usdanka, 1997; Patama and Damiano, 1997).

Slum improvement is one of the most difficult planning and development management interventions in a city, because of the high densities of the settlements, the usually low quality land and location, the fact that land is illegally occupied, and the high cost of installing new infrastructure into an area already built up. In addition, the slum dwellers are often not even local residents, do not vote in the local elections, and thus have no political power or leverage. But slum relocation is even more difficult, because it will require large amounts of land to build housing and the needed community facilities, that land must be accessible to public transport systems so that residents can get to their employment places, and standards must be met because of the high visibility of these programs and the accountability required by foreign aid and government programs funding them.

No specific legislation exists in Thailand at present to deal with the need to upgrade slums. Four related acts provide a framework for slum improvement projects, but are not being used to their full extent. They are the Urban Planning Act 2518 (1975) and Act 2535 (1992) which empower municipalities to develop their own comprehensive land use plans, including the designation of land for community rehabilitation projects and the establishment of measures to prevent the emergence of slums; Public Health Act 2535 (1992) which directs municipalities to maintain health and hygiene in their communities; and the Municipal Act 2496 (1953) which requires municipalities to keep slums clean and tidy and to improve their conditions.

A slum improvement program which has been successful in Chiang Mai is the "Community Visits" initiative. It consists of regular visits to the slum communities by the mayor, municipal councilors, the City Clerk, and chief officers of the municipal departments. The city has been divided into zones, and a program of visits by zones is planned annually, with advance notice given to the slums to be visited. During these visits the city officials and their technical personnel prepare inventories of problems, the officials have meetings with the local community residents, and discussions take place on how best to address problems. Simple problems are solved immediately, more involved problems are assigned to specific city departments, and commitments are made for their solutions. The visits are also opportunities to pass on information on hygiene, child rearing, drug use, and other environmental issues such as the use of clean water. Both slum dwellers and city officials consider this one of the most effective ways to deal with pressing city problems. (For more on informal settlements, see Auffrey, Chapter 14, in this volume.)

5.5. Lack of Open Space

The transition of Thai cities from smaller, quaint environments into bustling larger urban centers, is taking a toll on their open space resources. In the past, smaller towns and cities had little traffic, low building volumes, and plenty of land devoted to temples, monasteries, schools and squares that could be used for recreation and children's play. But today's Thai cities, both large and intermediate size ones, are built very densely. Chiang Mai, as most other cities in Thailand, has a shortage of open space. The Krüger report rates the city as "poor", stating that, at 7.82 square meters of open space per capita, Chiang Mai falls way below the international standard of 15 square meters per capita (Krüger, 1996). Still, this may be more a matter of definition than a real problem. It is true that the Chiang Mai Municipality has little designated park space, since it contains only one public park within its boundaries, and there is only one other park outside of the city. Open space, however, is abundant in Chiang Mai. There are substantial open, park-like areas around the old city moat and walls and along the central riverfront, which are attractive and easily accessible to the public. Further, there are 39 temples (*wat)* within the walls of the old city and many more in the rest of the city. These temples have accessible open space that is used by the city's residents. Many of the *wat* also have schools adjoining them, and they include playing fields and courts that are used by young people.

In addition, a common architectural style outside the old city is the use of open courtyards and of green space around many buildings. West of the central business district, toward the Doi Suthep Mountain, the spaces include Chiang Mai University's large open campus, the zoo, and several other educational institutions. As a result, a great deal of open space is accessible to the public. If properly planned, controlled, and maintained, this space can be a significant asset to the city's residents and visitors. This however, does not discount the need for more city greening. The city does not currently have a policy of green development, not only with regard to the establishment of new parks, but even associated with the cleaning and greening of water bodies, the riverfront, and the landscaping of streets, thoroughfares and major transportation hubs or corridors. In this respect, both slum clearance and the improvement of waterways will add to the greening of the city, by providing opportunities for the reclamation and planting of open public spaces.

5.6. Seasonal Flooding and Drainage

Seasonal monsoon rains bring enormous quantities of water down in very short time spans, and invariably cause flooding. This flooding disproportionately affects the poorer segments of urban populations, because they are the ones mostly living on marginal lands, flood plains, and slums by waterways. Yet drainage is a municipal service that is only addressed during the monsoon season, and goes away as soon as the dry season sets in. Flooding problems in areas like Chiang Mai are bound to get worse, as deforestation of the hillsides and poor agricultural practices cause more land erosion and water run-offs; these, in turn, cause more river siltation, and that in turn is one of the main causes of seasonal flooding. This is mostly a problem of water management and waterway maintenance. Clear waterways allow rapid passage of large quantities of water, and drainage can accomplish that easily. Thai cities are given the legal means to act so as to prevent flooding in their jurisdictions, while the central government is making more resources available for flood control programs. A major initiative among them is the Regional Cities Development Programme, financed by the Department of Local Administration. The program allows the preparation of studies and plans for urban drainage, coordination among communities, cities and provinces for the comprehensive combating of floods, the development and implementation of a sanitation strategy, and coordination of the plan with the overall comprehensive plan of the city. Funding is made available for the implementation of all these provisions (Urban Environmental Guidelines Project, 1994, "Drainage", 4).

The view of the survey respondents, confirmed by the available data, is that Chiang Mai does not have a history of heavy flooding. The city appears to fall into a 10-year cycle of moderate floods. In recent years, it had a flood in 1984, and another one in September of 1994, which attracted a lot of national publicity because it coincided with the city's landfill-closing problem. Future flooding problems may have been prevented by the recent damming of the Ping River, which has resulted in a more regulated flow of water. But at the same time, land use and ecological change in the region may be causes for more severe flooding in the future. In the Chiang Mai region, these include illegal logging in the hillsides and mountains upstream of the city and large scale conversion of rice paddies into housing estates. This environmental change appears to be taking place already. Survey respondents mentioned that now the flood rains come more often, stay longer, and cause more damage to crops and urban property.

6. INSIGHTS INTO THE ROLE AND ABILITIES OF LOCAL GOVERNMENTS

This brief review of some key environmental problems and their management needs is enough to illustrate the inadequacies of the Thai administrative structures and mechanisms to deal efficiently with them. Of concern to this chapter is mainly the ability of local, municipal and related government agencies, such as sanitary districts, to cope with the many responsibilities given them by the laws of the state, and have resources made available to them. A number of issues emerge from this expose:

A. *Policy dominance of the central government, and lack of clarity regarding rules, regulations, and laws.* A large number of laws guiding the protection of the environment and quality of life exist in Thailand. They are the domain of different central government agencies and are used to determine the turf of these agencies. Data generated as a result of the passage of these laws are not made public, and are used as power instruments by the responsible agencies. Laws are passed to promote the objectives of these agencies, rather than to address effectively families of problems and the solution of social ills. As a result, a multiplicity of laws exists, with considerable overlaps.

The development of environmental, developmental, and planning policy is concentrated at the central government level. This is not only limited to the passage of general enabling legislation, but it expands into the domain of local government policies and programs as well. Thus, comprehensive and land use plans for cities are prepared in Bangkok, mostly in the absence of the cities concerned. These cities do not even have copies of the relevant studies and maps, but have to go through elaborate and time-consuming procedures to access information related to the planning of their own cities.

B. Lack of clear lines of authority and empowerment. Even though municipalities are by law empowered to make decisions in a number of areas, in reality their powers have been effectively usurped by administrators at or related to the central government agencies. Given the democratic procedures involved in the election of the municipal councils and the mayors themselves, such policies would be expected to be, on the one hand consistent with central government policy, and on the other hand, in line with the programmatic statements of the local candidates, and the wishes of the majority of the local electorate. Nevertheless, the law gives absolute rights to the provincial governors and their administration, to overrule and reject policies passed by local administrations.

The role of the Ministry of the Interior is extremely powerful in this respect. By having complete control of the municipal manpower, and having the authority to approve all hires and dismissals of municipal personnel, in effect it introduces the possibility of corrupt practices, creates a labor force of civil servants lacking loyalty to its local government supervisors, and prevents the development of local and regional professional expertise. In addition, it renders the municipal administrations powerless to manage their own resources, to select their own people and hone their own skills, and to build a personnel body focused on the issues of the municipality with a stake in their solution.

Thus, the overlap of legal jurisdictions, the superseding of elected by appointed bodies, the suppression of local priorities in favor of central government ones, and the power games played at the central government levels among central agencies, all contribute to an ineffective administration allowing little progress towards meeting local development goals.

C. Shortage of financial resources for program implementation and enforcement. While a sound national legal framework is in place in most instances, and indeed the nation's legislators have thought carefully about issues of development, planning, environment and quality of life, and have developed sensible guidelines to achieve social and economic progress, few resources have been made systematically available by the central government to finance the implementation of the laws' mandates. Thus, central government laws become only frameworks, sets of official wishes and goals, with little power of enforcement, and thus little potential to bring about change. Even in the provisions of the Five-Year plans, which are supposed to be the foundations of the annual budgets for development planning, the resources allocated are inadequate, do not reflect the priorities of the Plans, and in many instances are soon diverted to pay for other emergencies or changing priorities, thus limiting the ability of the Ministries to carry out the Plans and making the Plans impotent instruments for development purposes. At the provincial level, implementation of programs is limited to those priorities determined by the central Ministries and allocated as part of larger scale regional/provincial plans. Provincial administrations, appointed by the Ministry of the Interior, function primarily as enforcers at the regional level of the programs of the central government agencies. Thus, local jurisdictions, i.e. city, town and *tambon* municipalities, *tambon* administrations and sanitary districts, which contain most of the population of the country and thus carry most of the burden for the provision of public services, are left to cope with inadequate resource transfers from the central government, and with limited leverage to raise their own funds through fees and local taxation. Even when they are tempted to do so through the passage of local policies, procedures and regulations, often they are overturned or their decisions are disapproved by the appointed representatives of the central government in their region. Thus, local governments in Thailand to date are in a weak position to provide the services and manage human, capital, and natural resources in their jurisdictions as demanded by their populations, as made necessary by the needs of increasing populations, and as made imperative if emphasis remains on urban decentralization and the enhancement of the role of intermediate size cities*D. Disproportionate severity of urban environmental problems in secondary cities.* Even though Bangkok is experiencing very severe environmental problems, other cities of Thailand, such as Chiang Mai, are not too far behind in terms of magnitude of the problems. Considering the large population difference between the 7.5 million people of Bangkok (1999 figures) and the 250,000 people of Chiang Mai, one would expect the degree of deterioration of the environment in Bangkok to be considerably higher than that of Chiang Mai. Yet the Krüger study places Chiang Mai very close to Bangkok in terms of most environmental performance indicators. One simple explanation for this phenomenon is that, while intermediate size cities are also experiencing rapid growth and major population, economic growth, and resource use pressures, they have been paid little attention by the government, proportionately speaking, and thus have been falling behind in their efforts to stay at the same or higher quality of life levels as Bangkok.

E. Lack of recognition of the potential role of intermediate size cities. Even though the policies set forth by the Five-year Plans recognize the contribution of the nation's system of intermediate size cities to the solution of the growth and development problems of the country, in reality only selected cities have been given attention as potential development centers besides Bangkok. Still today, the capital of the country is a giant metropolitan region which absorbs most of the national resources, gets the most educated and productive labor force, is able to attract most domestic and international investments, and has the most sophisticated administrative machinery in place. Some other major secondary cities, such as Pattaya, Chiang Mai, Phuket, and Songkla, also receive special attention from the ministries of the central government, but they do so because of their special role as major tourism destinations, not as part of a plan to enhance the secondary cities of the periphery. And yet, the population of the country clearly indicates preferences for living in the smaller urban areas of Thailand. Chiang Mai is a huge real estate market of condominiums purchased by residents of Bangkok who use them as vacation homes and are planning to retire on them, but now live in Bangkok because that is where the jobs are.

In summary, the central government of Thailand has recognized the need for good planning and effective management of its local resources. It already has in place a strong, modern legal framework enabling development to proceed. It also provides continuity of its social and economic development programs and policies via its Five-Year Development Plans which are used as guides for the determination of development priorities and for the implementation of projects in all the ministries and agencies of the government at all levels. Thus, Thailand has a complete legal and administrative enabling framework in place to proceed with its growth and development programs. Where it is still in need of improvement is the system of mechanisms through which the country will decentralize its decision making, especially for regional and local level development priorities. It must empower in fact, rather than just on paper, the local governments to pass their own development policies and adopt their own resource management procedures, develop an objective and consistent system of resource allocation and distribution from the center to the periphery, eliminate the over-concentration of resources from fees and taxes at the center, and enable local governments to exercise true local autonomy in the pursuit of their local development and resource management goals. If these changes are accomplished, secondary cities in Thailand will be able to adopt and employ modern management procedures that will enable them to manage their human, capital, and natural resources efficiently and effectively in providing superior quality service to the public. They will be free to determine their own priorities for the solution of social, economic and environmental problems, will feel justified to facilitate the development of stronger participatory mechanisms at the local level, because it will demonstrate to its citizens that decisions can be made locally through a political process, and will be compelled to define their own path and pace towards long-term sustainable development.

NOTES

[1] The research and field studies on which this book reports (from here on referred to as "the Joint Study"), were conducted mostly in the summer of 1998 by a joint team of researchers from four universities and one municipality. The academic institutions represented were Chulalongkorn University in Bangkok, Thailand, Chiang Mai University in Chiang Mai, Thailand, Gadjah Mada University in Yogyakarta, Indonesia, and the University of Cincinnati, in Cincinnati, Ohio, U.S.A. The municipality of Chiang Mai was the host and collaborator of the research team, and the entity to which the results of the study were reported. Funding for the study was provided by the United States Information Agency (now U.S. Department of State) College and University Affiliations Program, the University of Cincinnati Faculty Development Council and its Institute for Global Studies and Affairs, Chulalongkorn University, and the Municipality of Chiang Mai. The authors and editors of this book are grateful to these institutions for the support they have received towards the completion of the Joint Study and this book.

[2] The research for this chapter was launched in 1994, while Romanos was an Asia Foundation Fellow at the Asian Institute of Technology, in Bangkok, Thailand. It was completed in 1997-98 when Wagner and Romanos conducted research in Bangkok and Chiang Mai in collaboration with Thadaniti of Chulalongkorn University, under a grant from the United States Information Agency (now Department of State) College and University Affiliations Program. The authors are grateful to the three institutions for their financial and other support during the Joint Study.

[3] Two other cities, Bangkok and Pattaya, are designated as special self-governing municipalities.

[4] An additional disadvantage of the Commission's practices is that municipal personnel are not considered part of a specific local administration, but rather part of the national municipal labor force and, therefore, are transferable from municipality to municipality. This creates uncertainty of employment and inefficiencies in productivity, because transferred employees take a long time to learn the new city and its issues and problems, and to adjust to new living and working conditions (Romanos, 2000; Kaothien et. al., 1994).

[5] The level of urbanization in Thailand, defined as the percentage of total population living in urban communities, is low in comparison with other countries. In 1970, only 13.24% of the population lived in municipal areas. By 1990, this figure had risen to only 18.73%; just over 50% of the labor force works in agriculture. In 1980, however, nearly 45% of all city dwellers in Thailand lived in the Bangkok

metropolitan area. Although this proportion declined to about 40% in 1990, Bangkok was still 22 times larger than Nonthaburi, now considered the country's second largest city (*Pocket Thailand in Figures 1994*, 1994).

[6] Although statistical records rank Chiang Mai fourth among the cities of Thailand, with a population of 166,889 (Ministry of the Interior, Department of Local Affairs, 1990), a 1995 study conducted to project solid waste levels estimated that 172,714 people lived in the city and 74,126 around the city's edge. An additional "disguised population" of 88,591 unregistered residents (mostly students and squatters), tourists, and commuters who come to the city each day must be considered as well (Suree, 1997).

[7] The city was established in 1296, when King Mengrai of the Anachak Lan Na Tai built a stone fortress on the plains between Doi Suthep Mountain and the Mae Ping River. It became a major commercial center engaging in trade with the Shan State of Burma and kingdoms in China, Laos, and Siam. Historically important because it was the capital of the Lan Na Tai Kingdom (country of the "million rice fields"), Chiang Mai was occupied several times by Burmese kings and was a vassal state of Bangkok. The Kingdom of Lan Na Tai continued under the rule of the "Seven Chiefs Family" until it was annexed in 1892 by King Chulalongkorn, who married the princess and heir of Chiang Mai (*Chiang Mai,* undated; Koyano, 1985; Narumit-Rekagarn and Pranom,1988).

[8] Under the supervision of the Municipal Executive Board are the following administrative offices:
The Bureau of Public Works, which consists of the Division of Construction Control, the Division of Civil Works, the Division of Building and City Planning, and the Division of Sanitary Works; The Division of Sanitation and Environment; The Division of Medical Services; The Division of Education; The Division of Technical Services and Planning; The Division of Social Welfare; The Division of Finance; and The Office of the Municipal Clerk (Chiang Mai Municipality, 1997).

[9] About fifty-five structured interviews were conducted for the preparation of this chapter, some of them in 1995, the rest of them in 1998. The structured interviews followed a standard questionnaire format of open and closed questions, and were conducted by the authors themselves. Each interview lasted approximately two hours. Respondents included local government officials, municipal technical personnel, central government agency officials in Chiang Mai, university faculty, public and private sector engineers and environmentalists, NGOs, business people, and citizen activists.

[10] Other problem areas directly related to sustainable development and indirectly related to environmental management, include:

- lack of public transportation,
- rampant real estate development for investment speculation,
- unmanageable tourism,
- loss of cultural identity,
- loss of quality of life as its old character as an agricultural market and cultural center disappears.

[11] The environmental problems of Chiang Mai and other cities in Thailand are assessed in a 1996 report on the urban environment prepared by Krüger Consult and funded jointly by the Thai Office of the National Economic and Social Development Board and the Danish Cooperation for Environment and Development. The report, titled *Urban Environmental Management in Thailand: A Strategic Planning Process* (Krüger, 1996), outlines environmental concerns in eight categories--air quality, sewerage or water quality, water supply, solid wastes, ground pollution, traffic, slums, and open space--and ranks them from 1 (good) to 5 (very poor) on an environmental profile scale. Chiang Mai Municipality is rated poor in five of the categories: air quality, traffic, open space, slums, and sewerage; borderline in solid waste and ground pollution; and acceptable in just one category, water supply.

[12] An ambitious, and to many, unrealistic, proposal is presently being considered by the city, which calls for a rapid transit system of two lines crossing the city north to south and east to west. These lines would

pass by the historic center of Chiang Mai, and their stations would create serious damage to the old town and the archaeological sites. No one seems to be concerned that a rapid transit system may be too drastic a solution for a city of the size of Chiang Mai. No decision has been made as of September 1998.

[13] A further burden to the city garbage disposal system occurs seasonally, in the month of April, when two local fruits, the *lam-yai* (logan) and the *lin-jie* (litchi) are being harvested. The workers employed cut the branches of the trees in order to pick the fruits, and these branches end up in the municipal landfill as well, increasing exponentially the amount of garbage generated daily during the season.

[14] Even so, the river is in better condition now than in past years, largely because of new dams north of the city which have allowed for continuous water flow during the dry season (Supporn, 1997).

REFERENCES

Atkinson, Adrian and Chamniern Paul Vorratnchaiphan. (1994). Urban Environmental Management in a Changing Development Context. *Third World Planning Review, 16:2,* October, 147-169.

Bangkok Post, (1998). "Firing of Chiang Mai Council Members by the Provincial Governor," June.

Bartone, Carl, Janis Bernstein, Josef Leitmann, and Jochen Eigen. (1994). *Toward Environmental Strategies for Cities: Policy Considerations for Urban Environmental Management in Developing Countries.* Washington, D.C.: World Bank, Urban Management Programme, Policy Paper No. 18.

Blair, Thomas L.(Ed.). (1985). *Strengthening Urban Management: International Perspectives and Issues.* New York: Plenum Press.

Chiang Mai Municipality. (1997). *Local Government,* Chiang Mai: Chiang Mai.

Cohen, Saul B. (Ed.). (1998). *The Columbia Gazeteer of the World,* New York: Columbia University Press.

"Development and the Environment". (1998). *The Economist,* March 21, 3-16.

Devas, Nick and Carole Rakodi. (Eds.). (1993). *Managing Fast Growing Cities: New Approaches to Urban Planning and Management in the Developing World.* Essex, United Kingdom: Longman Scientific and Technical.

Dolan, Lawrence W. (1993). Local Administration and the Provision of Environmental Infrastructure in Thailand. Bangkok: U.S. Agency for International Development, Regional Housing and Urban Development Office, January (mimeo).

The Eighth National Economic and Social Development Plan (1997-2001). (1996). (Summary) National Economic and Social Development Board, Office of The Prime Minister: Bangkok, Thailand.

Foster, David. (1989). Urban Environmental Concept Paper. Washington, D.C.: U.S. Agency for International Development, Office of Housing and Urban Programs, (mimeo).

_____ (1992). The Role of the City in Environmental Management. Bangkok: U.S. Agency for International Development, Regional Housing and Urban Development Office, January (mimeo).

The Fourth National Symposium on Environmental Technology and Management: Appropriate Technology for Human Wastes Treatment. (1992). Department of Environmental Engineering, Chiang Mai University & Japan International Cooperation Agency: Chiang Mai, Thailand, June 8-9.

GTZ. (1992). *Urban Environmental Management Guidelines for Thailand: Summary.* Bangkok: Office of Urban Development, Urban Environmental Management Guidelines Project.

Gordon, Gerald L. (1993). *Strategic Planning for Local Government.* Washington, D.C.: International City/County Management Association.

Hardoy, Jorge E., Diana Mitlin, and David Satterthwaite. (1992). *Environmental Problems in Third World Cities.* London: Earthscan Publications.

Hardoy, Jorge E., and David Satterthwaite. (1981). *Shelter: Need and Response.* New York: Wiley & Sons.

Hart, Stuart L. (1997). Beyond Greening: Strategies for a Sustainable World. *Harvard Business Review, January-February,* 66-76.

Hewison, Kevin. (1993). Nongovernmental Organizations and the Cultural Development Perspective in Thailand: A Comment on Rigg (1991). *World Development, Vol. 21, No. 10,* 1699-1708.

Huysman, Marijk. (1994). Waste-picking as a Survival Strategy for Women in Indian Cities. *Environment and Urbanization, 6:2,* 157-174.

Joint Study Project on Development of the Appropriate Technology as a Primary Health Care for Human Wastes Treatment Disposal in Northern Thailand, Final Report. (1992). Department of Environmental Engineering, Faculty of Engineering, Chiang Mai University, Thailand, The Institute of Public Health, Ministry of Health and Welfare, Japan, and Japan International Cooperation Agency: Chiang Mai, Thailand, February.

Kaothien, Utis. (1998). *East Asia: The Unfinished Agenda,* Speech presented to the Asia Development Forum, Manila, Philippines, March 9-13.

Kaothien, Utis, Douglas Webster, Jaroonsri Chyehard, Ksemsan Suwarnarat, and Voravit Vorathanyakit. (1994). *Local Government Strengthening in the Context of Decentralization: The Thailand Case.* Paper prepared for the Regional Workshop on National Urbanization Strategies and Local Government Development in the Context of Decentralization, September 26-29.

Key Statistics of Thailand 1996. (undated). National Statistical Office, Office of the Prime Minister: Bangkok, Thailand.

Kingsley, G. Thomas. (1991). Land, infrastructure and the urban poor: Changing roles for different levels of government. In, *The Urban Poor and Basic Infrastructure Services in Asia and the Pacific*, Volume III. Manila: Asian Development Bank, and Washington, D.C.: World Bank, Economic Development Institute.

Komai, Hiroshi, (1985). Does Strengthening of Intermediate City Lead to the Alternative Development?: Chiangmai as the Case. In Shogo Koyano (Ed.), *Sociological Study of Urbanization in Southeast Asia*, International Joint Team For Overseas Scientific Surveys, Japan, 46-54.

Korkeatkachorn, Wipaphan and Suntaree Kiatiprajak (Eds.). (1997). *Directory of Non-Governmental Organizations, 1997,* Thai Development Support Committee (TDSC), Bangkok, Thailand.

Kottatep, Suporn. (1996). Integrated Solid Waste Disposal System: The Solution of Chiang Mai Municipality. In George G. Van der Meulen and Peter A. Erkelens (Eds.), *Urban Habitat: The Environment of Tomorrow*. Faculty of Architecture, Eindhoven University of Technology.

Krüger Consult. (1996). *Urban Environmental Management in Thailand: A Strategic Planning Process*. Office of The National Economic and Social Development Board and Danish Cooperation For Environment and Development: Bangkok, Thailand.

Lea, John P. and John M. Courtney. (Eds.). (1985). *Cities in Conflict: Studies in the Planning and Management of Asian Cities*. Washington, D.C.: The World Bank.

Local Government Chiangmai Municipality. (undated). Public Relations Section Chiangmai Municipality.

Lubeigt, Guy. (1994). Traditional and Recent Aspects of the Urban Development of Chiang Mai, Thailand. In Marc Askew and William S. Logan (Eds.), *Cultural Identity and Urban Change in Southeast Asia: Interpretive Essays*. Deakin University Press: Victoria, Australia.

Matsushita, Hidetsuru, Ching-Tang Kuo, Monthip S. Tabucanon, and Suporn Koottatep. (1987). Comparative Study on Carcinogenic Polycyclic Aromatic Hydrocarbons in Airborne Particulates in Bangkok, Chiang Mai and Tokyo. A paper presented at the Third Joint Conference of Air Pollution Studies in Asian Areas. November 30-December 2, 1987, Tokyo, Japan.

Meier, Gerald M., and James E. Rauch. (2000). *Leading Issues in Economic Development*. New York: Oxford University Press.

Mekvichai, Banasopit. (1994). "Patterns of Urban Development, Planning, and Government in Thailand". A paper presented at the Workshop on Urban Governance and Poverty Alleviation in Southeast Asia: A Comparative Framework, Manila, Phillippines, November, 1994.

Narumit-Rekagarn, Sirichai and Pranom Tansukanan. (1988). Special Report of Thailand. Problems and Proposals on Conservation: The Case of Chiang Mai, Thailand. A paper presented at the UNESCO/SPAFA Workshop on Principles and Methods of Preservation Applicable to the Ancient Cities of Asia, Chiang Mai, Thailand, November 6-12.

National Urban Development Policy Framework: Final Report, Volume 2. (undated). National Economic and Social Development Board: Bangkok, Thailand.

Nernhad, Anu. (undated). *Bike: An Important Solution to City Traffic Problems*. Faculty of Architecture, Chiang Mai University, Chiang Mai, Thailand.

The New Era of Development in Thailand: NESDB and Its Role. (undated). National Economic and Social Development Board, Office of the Prime Minister.

O'Meara, Molly. (1997). Riding the Dragon: Can APEC Be Held to a Sustainable Course? *World Watch, Vol. 10., March/April,* 8-18.
Pocket Thailand in Figures, 1994. (1994). Alpha Research Co. Ltd. Bangkok, Thailand.
Porananond, Usdanka. (1997). Housing for the Poor in Chiang Mai, Thailand. In Helmut Schneider and Karl Vorlaufer (Eds.), *Employment and Housing: Central Aspects of Urbanization In Secondary Cities In Cross-Cultural Perspective.* Ashgate Publishing Ltd.: Hants, England.
Racelis, Mary. (2000). Community empowerment, people's organizations, and the urban poor: Struggling for shelter, infrastructure, services, and dignity in the Philippines. In, Carla Chifos and Ruth Yabes (Eds.), *Southeast Asian Urban Environments: Structured and Spontaneous* (pp.3-27*)*. Tempe, Arizona: Arizona State University Press.
Ranson, Stewart, and John Stewart. (1994). *Management for the Public Domain: Enabling the Learning Society*. New York: St. Martin's Press.
Ratanawaraha, Apiwat. (1995). *Physical Development of Chiang Mai, Thailand*, University of Tokyo, Japan, Unpublished Paper.
Rigg, Jonathan. (1991). Grass-Roots Development in Rural Thailand: A Lost Cause? *World Development, Vol. 19, No. 23*, 199-211.
_____ (1993). A Reply to Kevin Hewison. *World Development, Vol. 21, No. 10*, 1709-1713.
Romanos, Michael. (2000). Need for structural reform in intermediate size cities: Environmental management in Chiang Mai, Thailand. In Carla Chifos and Ruth Yabes (Eds.), *Southeast Asian Urban Environments: Structured and Spontaneous* (pp. 275-326). Tempe, Arizona: Arizona State University Press.
______, and Carla Chifos. (1996). Contributions of the Urban Informal Sector to Environmental Management. *Regional Development Dialogue, Vol. 17, No.1, Spring.*
Roonrakwit, Patama and Giuseppe Damiano. (1997). Project Proposal: Community Mobilisation Programme for Shelter in Chiang Mai. Proposal submitted to Asia Fund, Bangkok, Thailand.
Saowapon, Chaitawat. (undated). "A Perspective on the Water Related Issues in the City of Chiang Mai," Department of Civil Engineering, Faculty of Engineering, Chiang Mai University, Thailand, Unpublished Paper.
Summary of People's Development Plan: Its Concept and Direction. (1995). *Thai Development Newsletter, No. 29*, 30-34.
Suphapaodok Arthorn and Chartree Chueyprasit, (1994). Environmental and Urban Management in Thailand. In Asman Awang, Mahbob Salim, and John F. Halldane (Eds.), *Environmental and Urban Management in Southeast Asia.* Institute Sultan Iskandar: Johor, Malaysia.
Tapananond, Nopanant, et.al. (undated). Methodology for the Implementation of Urban Planning in Thailand: The Revitalization of the Inner Area of Bangkok. Unpublished paper. Department of Urban and Regional Planning, Chulalongkorn University and Department of Engineering, University of Tokyo.
Theerapappisit, Polladach.. (1995). The Visual Impact of Building Constructions on Historic City: A Case Study in Chiang Mai, Northern Thailand. Unpublished Special Study, Asian Institute of Technology.
Thompson, James R., and Patricia W. Ingraham. (1996). Organizational Redesign in the Public Sector. In Donald F. Kettl and H. Brinton Milward (Eds.), *The State of Public Management.* Baltimore: The Johns Hopkins University Press, 286-306.
Tourism Authority of Thailand, *Chiang Mai.* (undated).
Urban Environmental Guidelines Project. (1994). *Urban Environmental Management Guidelines – Thailand.* Bangkok, Thailand: Department of Local Administration, Local Government Development Division, and Deutsche Gesellschaft fur Technische Zusammenarbeit.
Urban Population, Employment Distribution and Settlement Patterns, National Urban Development Policy Framework, Draft Final Report, Area 2. (undated). National Economic and Social Development Board, Thailand Development Research Institute Foundation: Bangkok, Thailand.
van der Meulen, George G. (1996). Solid Waste Management Systems for Chiang Mai. In George G. Van der Meulen and Peter A. Erkelens (Eds.), *Urban Habitat: The Environment of Tomorrow*, Faculty of Architecture, Eindhoven University of Technology.
Waste Disposal Management Project Feasibility Study. (1996). Department of Environmental Engineering and Social Research Institute: Chiang Mai University, Chiang Mai, Thailand. (Unpublished).

World Development Report 1994: Infrastructure for Development. (1994). World Bank, Oxford University Press: New York, NY.
Yamklinfung, Prasert,. (1985). Chiangmai: Urbanization of an Intermediate City. In Shogo Koyano (Ed.), *Sociological Study of Urbanization in Southeast Asia*, International Joint Team For Overseas Scientific Surveys, Japan , 30-43.
1990 Population and Housing Census: Chanwat Chiang Mai. (undated). National Statistical Office, Office of the Prime Minister, Bangkok, Thailand.

INTERVIEWS AND PRESENTATIONS

Aquino, Karan. Northern Development Foundation, Interview, Chiang Mai, Thailand, July 7 and 8, 1998.
Boonyanupong, Kirksak. Chiang Mai Social Research Institute, Interview, Chiang Mai, Thailand, November 11, 1997.
Boonyanupong, Suree. Chiang Mai Social Research Institute, Interview, Chiang Mai, Thailand, November 11, 1997.
Burarak, Krittiyab. E-mail from Krittiyab to Roger Barry, September 10, 1998.
Charoenmuang, Duongchan Apavatjrut. Chiang Mai Social Research Institute, Interview, Chiang Mai, Thailand, November 11, 1997.
Chuntaketta, Prateeb. Ex-Governor, Industrial State of Thailand, Presentation and Interview, Bangkok, Thailand, June 23, 1998.
Damiano, Giuseppe. Community Architects for Shelter and Environment (CASE), Interview, Chiang Mai, Thailand, November 11, 1997.
Harmon, Jonathan. Faculty of Architecture, Chiang Mai University, Interview, Chiang Mai, Thailand, November 12, 1997.
Hupertz , Robert Dieter. Faculty of Architecture, Chiang Mai University, Interview, Chiang Mai, Thailand, November 12, 1997.
Jumroonrat, Nithiwat. Department of Sanitary Engineering, Chiang Mai Municipality, Interview, Chiang Mai, Thailand, November 12, 1997.
Koottatep, Dr. Suporn. Department of Environmental Engineering, Chiang Mai University, Interview, Chiang Mai, Thailand, November 11, 1997, and July 6, 1998.
Manmontri ,Anandh. President, Walk For A Better Environment, Interview, Chiang Mai, Thailand, November 12, 1997.
Narintarakul Na Ayudhaya, Kingkorn. Institute of Sustainable Development Education Program, The Northern Development Foundation, Interview, Chiang Mai, Thailand, July 8, 1998.
Narumit-Rekagarn, Sirichai. Founding Member, For Chiang Mai Group, Presentation and Interview, Chiang Mai, Thailand, July 8, 1998.
Norden, Regine. Environmental Education Advisor, The YMCA for Northern Development Foundation, Interview, Chiang Mai, Thailand, July 13, 1998.
Patcharin, Aviphan. Director, The YMCA for Northern Development Foundation, Interview, Chiang Mai, Thailand, July 13, 1998.
Porananond, Usdanka. Department of Geography, Chiang Mai University, Interview, Chiang Mai, Thailand, November 11, 1997.
Roonrakwit, Patama. Community Architects for Shelter and Environment (CASE), Interview, Chiang Mai, Thailand, November 11, 1997.
Supik, Jurgen. Managing Director, Dorsch Consult, Presentation and Interview, Bangkok, Thailand, June 23, 1998.
Tantayanusorn, Niwat. Former Deputy Mayor for Education, Chiang Mai Municipality, Interview, Chiang Mai, Thailand, July 13, 1998.
Vorathanyakit, Voravit. Presentation by Urban Development Coordination Division Staff of National Economic and Social Development Board, National Economic and Social Development Board Office, Bangkok, Thailand, June 25, 1998.
Yanasarn, Dr. Sanay. Head, Department of Geography, Chiang Mai University, Interview, Chiang Mai, Thailand, November 11, 1997.

CHAPTER 9

XINHAO WANG
LIN LIU

WATER RESOURCES FOR A SUSTAINABLE CHIANG MAI

Abstract The ultimate goal of the study was to assess the human disturbance in an urban system and to provide valid information for sustainable urban planning. This chapter identifies current major issues in Chiang Mai, Thailand, related to water supply and demand, wastewater management, and water quality. Existing institutional and government structures regarding environmentally sustainable development are analyzed to identify players in urban sustainable planning and development. Data obtained from the field show that the quality of surface water has been negatively affected by the urbanization of Chiang Mai. The chapter explores the nature of urban stream degradation and the relationships between this degradation and specific urban activities or "human factors". Currently, the water supply only reacts to the demand from land development. There is not an overall plan to guide development. The authors recommend that the Chiang Mai municipality establish a water quality monitoring system, integrate water supply planning and water quality management with land use planning, and use GIS to share data and to support decision making.

1. INTRODUCTION

While Chiang Mai strives to sustain its role as the socio-cultural, economic, educational, and tourist center of the northern region of Thailand, securing resources and protecting the environment must be integral components of the city's development policy. In particular, water is the most essential resource for human activities and the engine of economic growth (Abu-Zeid, 1998). A vital urban area must have sufficient water of good quality for consumption, manufacturing, food production, transportation, and other forms of life-support activities. The challenge to governmental officials is that most people do not realize the importance of water until there is a crisis (Priscoli, 1998). In this aspect, the importance of water management, including planning and investment, cannot be over-emphasized. The study presented in this chapter describes the current condition of water resources in Chiang Mai using data collected during the summer of 1998. Water quality samples collected from different parts of the city clearly demonstrate the impact of human activities on water quality. Based on the analysis, the authors provide strategies and recommendations for properly managing water resources to support sustainable development in Chiang Mai.

M. Romanos and C. Auffrey (eds.), Managing Intermediate Size Cities, 187—206.

2. EXISTING CONDITIONS

2.1. Water Supply

Water supply in Thailand is a state enterprise. Agencies with authority on water resources include national, provincial, and municipal agencies. New development must be approved by the central government to have access to water. The Provincial Waterworks Authority (PWA) regulates the water uses and its Regional Water District controls the water supply for the city. According to the director of the Chiang Mai Municipal Waterworks, the water service area covers 100% of the area within the city moat, a canal surrounding the old city, and 50% of the areas in the rest of the municipality (total of 92.5 km^2). (Interview with Chit Noorit, Director, Chiang Mai Water Works.) There are three water intake plants pumping water from the Mae Ping River above the municipality and the Mae Taeng Irrigation Canal (Saowapon, 1997). Three purification plants treat water at rates of 10,000 m3/day, 25,000 m3/day and 23,000 m3/day, respectively. It exceeds the estimated present demand of 58,000 m3/day. A seventh plant is under construction at the Mae Kuang Dam, approximately 25 kilometers northeast of the city. By the end of the first phase, the new plant will be able to produce water at 48,000 m3/day and the capacity will be 70,000 m3/day by the end of the second phase. There is enough water to accommodate the city's future water needs. It is expected that meeting the water supply demand in the city will not be a problem with the new plant operating at full capacity.

The waterworks have 367 km of pipes, and plan to add 62 km of pipes to expand the service. The pipe extension follows the request from development plans and responds to complaints. The density of developed area often guides where new pipes go. To achieve the goal of serving the most customers within the water work's limited resource, service extensions add low elevation area users first because that is where the water pressure is higher. Until recently, new housing projects demanding water supply were numerous and the number of users increased at a rate of 200 users per month. Since the economic crash in 1997, the rate has decreased to 100 new users per month. There are seven telephones to answer reports about where pipe repairs are necessary to increase the service capacity of existing plants (Noorit, 1998).

Although water from the treatment plants is fit for human consumption, many people still do not drink the tap water. Bottled water is regarded as the cleanest drinking water and the price is low (five baht per liter). People are used to drinking bottled water and many fear they will get sick from drinking the tap water. Several reasons contribute to the situation. Some of the pipes are old and may be broken. As a result, water at the end of the pipe may be unclean. A survey in the early 1990s reported that 28.8% people drank bottled water, 29.8% drank tap water, and 34.4% drank well water (Araki, 1992). It is not uncommon for a household to use well water for non-drinking purposes. For example, the director of the provincial town and country planning development had a 20-meter-deep well in his home, which is just outside of the municipality border. In general, people living in municipal areas have access to much cleaner drinking water than that in sanitary districts and villages.

2.2. Water Quality

In 1985, the Ministry of Science, Technology and Environment published the surface water quality classification standards. Surface water in Thailand is grouped into five classes based on the classification standards for 20 different water quality parameters (Araki et. al., 1992). Class 1 represents the water that is in its natural state without contamination from human activities. It is used for conservation and requires only ordinary process for pathogenic destruction. In addition, to be consumed after ordinary water treatment process, water in Class 2 can be used for aquatic organism conservation, fishery assistance, and recreation purposes. The Class 3 water can be consumed after an ordinary treatment process, while the Class 4 water requires special water treatment process before consumption. Surface water in Class 5 is contaminated to the level that only navigational use is permitted. Concentrations of selected parameters for the surface water quality standards are listed in Table 1.

Table 1. Water quality standards and classification of surface water

		Standards (mg/l)		
Class	*Condition*	*BOD5*	*DO*	*NO3-N*
1	Extra clean	Naturally	Naturally	Naturally
2	Very clean	≤ 1.5	> 6	≤ 5.0
3	Medium clean	≤ 2.0	> 4.0	≤ 5.0
4	Fairly clean	≤ 4.0	> 2.0	≤ 5.0
5	All other	> 4.0	≤ 2.0	> 5.0

Source: Based on Araki, 1992, 28, and Koottatep, 1997, 4.

The Mae Ping River is a major river flowing through Chiang Mai. The water quality in this river has improved in recent years. After a dam was built upstream from the city a few years ago, the Mae Ping River dry season flow has been increasing to dilute contaminants. (Interview with Suporn Koottatep, Professor, Environmental Engineering, Chiang Mai University.) Maps 1 through 4 display the water quality measured at different places in Chiang Mai based on data collected from May 1994 to December 1996 (Firoz, 1996; Napattalung, 1997; Peetragorn, 1995; Yang, 1995). Sites C, D, G and I are on the Mae Ping River. The waterworks draw water from the river for public water supplies for the city. Site C is on the river near the Chiang Mai super highway. Site D is near the center of the city. Sites G and I are above and below the confluence of the Mae Kha Canal, respectively. Sampling sites A and H are located on the Huay Kaew Stream above the city. The values from these two sites are used as background concentration for the comparison.

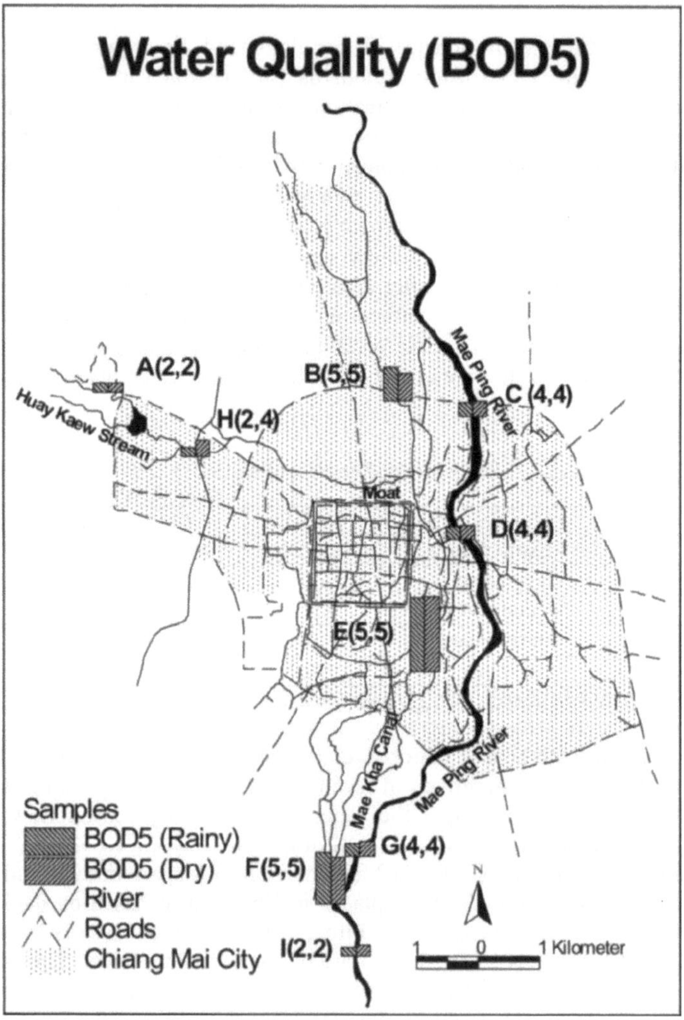

Map 1: Biochemical Oxygen Demand (BOD5) in Surface Waters in Chiang Mai

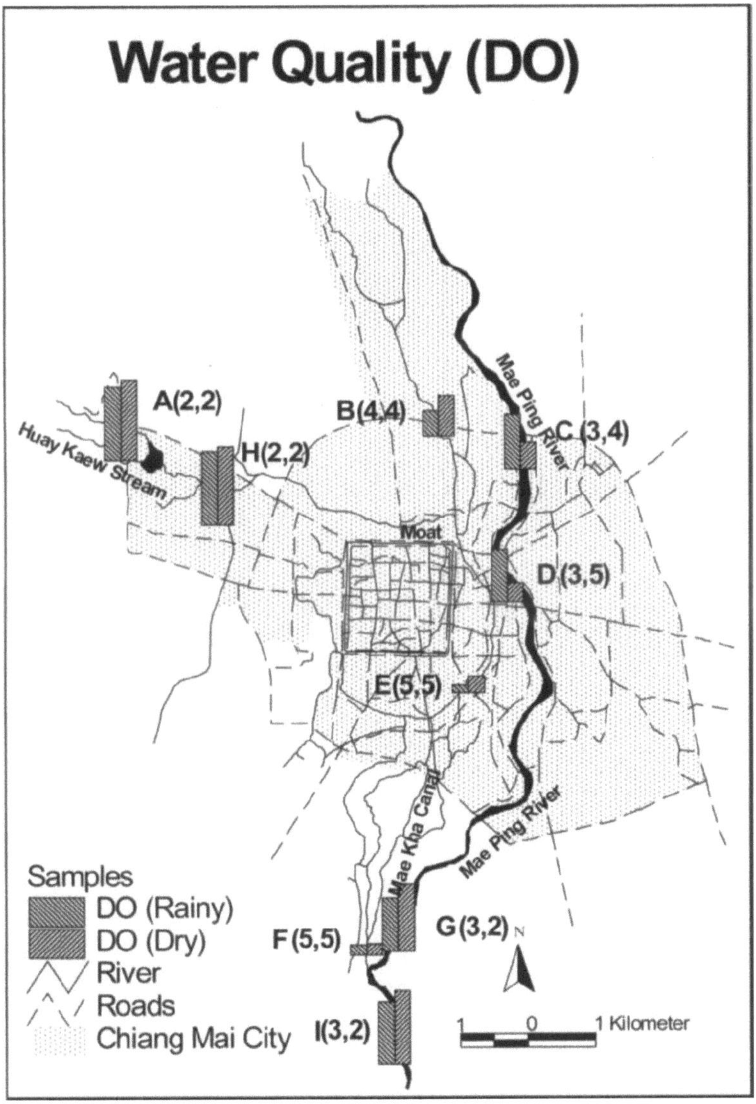

Map 2: Dissolved Oxygen (DO) in Surface Waters in Chiang Mai

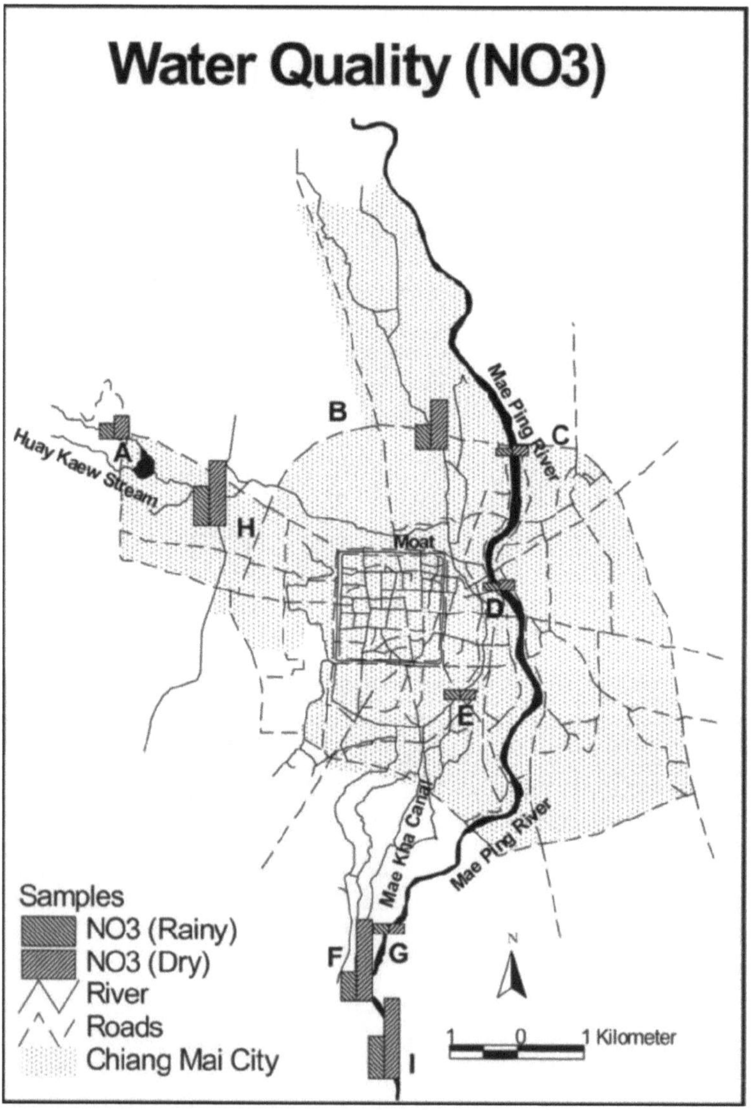

Map 3: Nitrate-Nitrogen (NO^3N) in Surface Waters in Chiang Mai

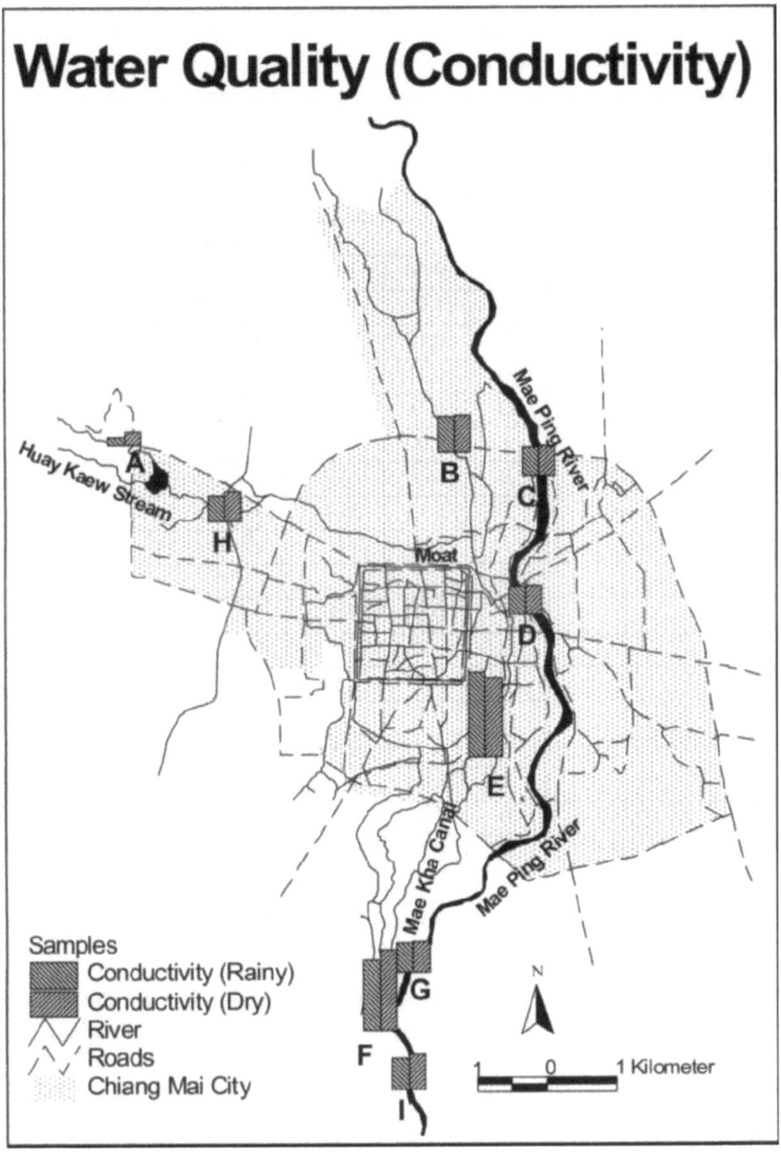

Map 4: Conductivity in Surface Waters in Chiang Mai

The Mae Kha Canal was originally constructed as a water supply and drainage system for the Chiang Mai municipality (Firoz,1996). It travels through the municipality between the east side of the city's ancient wall and the Mae Ping River. It drains into the Mae Ping River about three kilometers down stream from the city boundary. It has been used as an open sewer for its portion within the city and is severely contaminated by urban waste. Site B is on the Mae Kha Canal near the Chiang Mai super highway. Site E is on the canal near the center of the city. Site F is located at the outlet of the Mae Kha Canal as it enters the Mae Ping River.

Table 2. Surface water quality in Chiang Mai

	SITE	*NO3*	*BOD5*	*DO*	*Conductivity*
Rainy Season					
Huay Kaew Stream	A	0.48	1.00 (2)	6.83 (2)	45.10
	H	1.79	0.65 (2)	6.76 (2)	155.14
Mae Ping River	C	0.10	2.48 (4)	5.06 (3)	169.63
	D	0.11	2.42 (4)	4.66 (3)	164.43
	G	0.18	2.18 (4)	4.94 (3)	175.38
	I	2.00	0.40 (2)	5.80 (3)	188.63
Mae Kha Canal	B	1.00	10.43 (5)	2.15 (4)	214.97
	E	0.13	25.42 (5)	0.54 (5)	520.86
	F	1.24	17.10(5)	0.68 (5)	430.92
Dry Season					
Huay Kaew Stream	A	0.88	1.03 (2)	7.52 (2)	75.24
	H	3.28	3.67 (4)	7.30 (2)	177.73
Mae Ping River	C	0.23	2.95 (4)	2.34 (4)	180.00
	D	0.25	2.85 (4)	1.45 (5)	174.40
	G	0.20	2.85 (4)	6.28 (2)	185.80
	I	4.17	1.38 (3)	6.80 (2)	208.57
Mae Kha Canal	B	2.43	8.34 (5)	3.58 (4)	221.20
	E	0.14	25.90 (5)	1.34 (4)	482.17
	F	4.21	15.35 (5)	0.79 (4)	496.56

Source: Calculated based on Firoz, 1996; Napattalung, 1997; Peetragorn, 1995; and, Yang, 1995.

Table 2 is the calculated mean values of the monitoring data for the dry season (November to May) and the rainy season (June to October). Numbers in the parentheses indicate the classes according to the Thailand surface water classification standards. The sampling results clearly demonstrate the impact of the Chiang Mai municipality on the surface water. Five-day biochemical oxygen demand (BOD5) is a conventional water quality parameter that represents the amount of oxygen consumed while biodegradable matters are oxidized in water (Arms, 1994). The higher the BOD5 value, the more biodegradable matter in the water, and therefore, the poorer the water quality. The BOD5 values in raw sewage are between 40 and 150 mg/l. According to the Thai surface water classification standards, any surface water with BOD5 concentration greater than 4 mg/l is Class 5, suitable for navigation use only. The concentration values from the sites on the Mae Kha Canal are significantly higher than the sites on the Huay Kaew Stream and the Mae Ping River. The concentrations in the Mae Kha Canal are all much higher than 4 mg/l year round. Concentrations at the site near the center of the city are higher than 25 mg/l.

Although the water quality in the Mae Ping River is better than that in the Mae Kha Canal, it also reflects the impact of the city. The concentration from the section near the city, 2.42 mg/l during rainy season and 2.85 mg/l during the dry season, places the river in Class 4, the second worst classification.

The same pattern can be seen from the level of dissolved oxygen (DO), another parameter commonly used in measuring surface water quality. When a water body is contaminated by oxygen-consuming organic material, the dissolved oxygen is often low; the heavier the contamination, the lower the dissolved oxygen level. The DO level in the Huay Kaew Stream is quite close to the saturated oxygen level, which indicates that the river is at its natural condition. The Mae Ping River maintains a medium clean level most of the time. The Mae Kha Canal has very low DO levels. The standard for Class 5 river is 2 mg/l and the DO levels from two sampling sites in the Mae Kha Canal are lower than 1. Nitrate nitrogen (NO3-N) concentration represents the nutrient level in surface water. The NO3-N concentrations in both the Mae Kha Canal and the Mae Ping River are not excessive, indicating that the N concentration is not a problem in surface water bodies in Chiang Mai.

The water quality research literature indicates that conductivity is one of the most widely monitored water quality parameters. This may be because conductivity measurements are usually included in automated multi-parameter monitoring systems monitoring changes in the quality of surface water (Allen & Mancy,1972). Conductivity reflects the strength of major ions in water (Maasdam and Smith,1994; Prygiel and Coste,1993). Although conductivity has been primarily used to study the extent of salinization or salt water intrusion (Gondwe,1991; Kumaraswamy and Sivagnanam,1991; Beke et al.,1993; Reichman and Trooien,1993), it is also a good estimation of total dissolved solid (TDS). Conductivity is also linearly related to the sum of cations (McCutcheon et al.,1993; Hounslow,1995). In addition, conductivity measurements provide information about the total concentration of ionic species in a water sample (Tyson,1988; Allen and Mancy,1972). Since many pollutants contain soluble salts, it is conceivable that the conductivity may reflect the concentration of these materials (Rikard and Kunkle,1990; Howard and Haynes,1993; Gurnell et al.,1994). In addition, conductivity has been related to dissolved oxygen or oxygen demand (Rajagopal et al.,1993) and nutrient levels in water (Johnston et al.,1990; Edwards and Helvey,1991; Hakamata et al.,1992), which in turn may influence aquatic plant and fish populations (Jensen,1990; Conlon et al.,1992). After comparing the conductivity upstream and downstream from a sewage effluent discharge point on the Mukuvisi River, Mathuthu et al. (1995) reported that conductivity level increases after the sewage discharge point. Similar results were found by Chesterikoff et al. (1991). The monitoring data at the three surface water bodies demonstrate that conductivity concentration in the Mae Kha Canal was the highest of the surface water in Chiang Mai. The Mae Ping River is in the middle and the concentration in the Huay Kaew Stream is the lowest. This is more evidence that the Mae Kha Canal has been heavily contaminated by human activities in the Chiang Mai.

During the summer of 1998, the authors observed portions of the surface water in Chiang Mai. The observation helped to link the water quality measured with water chemistry parameters containing pollution sources. Figures 1 – 5 are photos taken along the Mae Kha Canal. The section of Mae Kha Canal in the middle of the city was heavily polluted. The impact of urban dwellers on surface water can be seen from many places.

Figure 1: Mae Kha Canal

The color of the canal water was black and smelled like a sewerage pipe. Garbage was dumped near or into the canal. Some people built their houses on top of the canal. It was clear that some residents were actually using the canal as their open sewer. The color of the Ping River reflected high concentration of suspended solid. In some extent, erosion from the deforestation is a bigger problem than chemical pollution. The moat looked less contaminated. There was a water fountain in the middle of the moat, which was used for aesthetic purposes. Meanwhile,it helped to increase the dissolved oxygen in the moat (Figure 6). During the dry season, water from the irrigation canal is used to flush the city moat. Since the city does not have an adequate wastewater collection system, only 26% of households in Chiang Mai were connected to the public drainage system (Araki,1992; Saowapon,1997). Further, there is evidence that some people may find it convenient to dump garbage into a canal nearby. Improving the management of domestic waste will be necessary for the improvement of water quality.

Figure 2: Housing at the edge of the canal

Figure 3: Waste dumped by the canal

Figure 4: Residential areas by the canal

Figure 5: The Chiang Mai City Moat

3. MANAGEMENT AND PROTECTION OF WATER RESOURCES FOR A SUSTAINABLE CHIANG MAI

Sustainable development in Chiang Mai can be achieved by integrating activities and functions of different organizations within a spatial context. Specifically, we recommend that the Chiang Mai municipality: 1) tie water resource management to land use management; 2) coordinate decisions for water supply and development; 3) solicit collaboration of governmental agencies, universities, and non-governmental organizations to increase awareness of environmental status; and, 4) develop a comprehensive database to support informed decisions.

Currently, there are two types of urban land use plans being implemented in Thailand. The first is a general comprehensive plan and the second is a specific plan. To develop a general plan, an advisory committee is organized. There are up to twenty-five land use categories in a general comprehensive plan. Each category includes permitted major and minor uses and prohibited uses. At least two public participation meetings are required during the process of developing a general plan. In the end, the general plan has to be passed by the National Town and Country Planning Board. Traditionally, the general land use plan focuses on compatibility among different land uses and the connection between land use and water quality is often omitted. In recent years, many efforts have been seen in promoting the watershed protection approach. There are studies on the impact of wastewater discharged from point pollution sources on surface water quality and estimates of pollutant loadings from nonpoint pollution sources. In an effort to establish the relationship between the land use and water quality, this study has demonstrated that river water quality in Chiang Mai is very likely related to urban land use. This calls for a pollution prevention approach that is built upon the understanding of the impacts of land use on stream water quality. Once the water quality component is included in the general land use plan, the city will have an enforceable tool to protect water resources in its jurisdiction. For example, under the Building Control Act, a development proposal must comply with the general plan for a municipality to issue a building permit. (Interview with Wichai Kajonpredanon, Provincial Town and Country Planning.)

Distribution of water supply is another critical factor affecting the development of Chiang Mai. The Regional Water District of the Provincial Waterworks Authority (PWA) has the sole control of water supply for the city. Under the current system, decisions about extending water supply pipes follow the demand of development. The PWA had developed a water plan according to the general land use plan. Unfortunately, this could not be confirmed with the director of the PWA Regional Office. Also, it is unclear if the National Department of Town and County Planning consults with the waterworks while developing future land use plans. (Interview with Chit Noorit, Director, Chiang Mai Water Works.) This obvious disconnection between water resources and development planning is probably the reason why water suppliers are simply trying to follow the development. The Chiang Mai municipality should carefully analyze the natural conditions of the area and develop a water service plan that is consistent with the general plan and can provide sufficient water to the city for a long time.

In an effort to introduce environmental dimensions to development planning, the environmental status should be clearly reported and specific guidelines should be formulated to require environmental consideration in the development policy making process. Protecting water resources requires the participation of many parties. Currently, environmental assessment is under the jurisdiction of the national and provincial environmental agencies, while the authority for development projects is held by national development planners. The Ministry of Science and Technology (MST) is responsible for developing environmental regulations. Several MST departments, such as the Department of Planning and Policy, the Department of Pollution Control, and the Department of Standards, and MST's regional office, the Northern Environmental Planning Center, oversee environmental issues in Chiang Mai. Other players related to environment and development include local politicians, non-governmental organizations (NGOs), and universities.

Making the general public aware of the water quality problem and deriving solutions should also be the priority of an environmental protection effort. A survey conducted in 1991 demonstrates an interesting spatial variation of the water quality issues among residents in Chiang Mai's four sub-districts (LBI-CMU 1991). Map 5 shows the results from part of the survey related to the surface water in Chiang Mai. People in Chiang Mai were aware of the level of water pollution in surface waters in the city. More people living in the three sub-districts through which the Mae Kha Canal flows felt that the water quality of the canal was very poor than those living in Sriwichai, the sub-district that was far away from the canal. The public awareness demonstrates the urgency of improving the water quality in the city. Planners and government officials should guide this awareness into actions to improve the environmental quality in the city.

To better manage the water resources, Chiang Mai needs to develop an effective and comprehensive database that can tie different components together. Currently, different agencies keep their own data, most of them in paper format. Although some are in digital format, sharing data is a major obstacle due to technical and human reasons. Basic census data by sub-district, such as household income, number of people below poverty level, health condition, and number of people completing elementary or high school are not available in any form from Chiang Mai University, nor the Chiang Mai municipality. The municipal finance department does have income data at the sub-district level. The education department does not have any data on the number of people at different education attainment levels. There is evidence, though, that some sub-district vital statistics are collected.

Currently, there are two levels of databases in Thailand maintained with Geographic Information Systems (GIS). The first level is at the provincial level, (1:50,000 scale, paper) and data include topographic maps being prepared by the Royal Thailand Survey Department. The Office of Policy and Planning, Ministry of Science, Technology and Environment, has prepared data for about thirty provinces (not including Chiang Mai), which includes ten data layers, such as political boundary, land use, and tourist information. The Thai Environmental Institute (a private organization) has 1:250,000 scale data for Thailand, including geology, agriculture, energy, automobile, etc. A second level is available for some urban areas (1:12,000 scale, paper). The Land Development in Bangkok is also working on land use data. The National Research Council of Thailand maintains satellite imagery. The Northern Center, located in the Chiang Mai University Remote Sensing and GIS lab, transfers data between the lab and the center in Bangkok via tapes (Interview with Pong-In Rakariyatham, Dean, Faculty of Social Sciences, Chiang Mai University).

The Department of Town and County Planning has 1:4000 data (252 sheets to cover the planning area that covers a larger area than the municipality). The data contain road networks, contours, building footprints, and natural features. Other layers are sanitary districts and sub-district boundaries. AutoCAD, MapInfo, and Idrisi were used to create the data from spot images (1:30,000 scale, black and white) three years ago and checked with field survey. We were told that a consulting firm has prepared detailed data for the city inside the wall. The Area Traffic Control Center (ATC) of the provincial police station maintains a new automatic traffic control system developed by a Japanese company. The system contains data on street networks, number of lanes, and real time traffic flow.

GIS was initially used in Chiang Mai about 1991. Users included Chiang Mai University and the Chiang Mai municipality. One problem in GIS application in Thailand is the collaboration among different agencies that are operated at different levels, such as the central and provincial governments, rural departments, and municipalities. Each agency has different priorities and interests. (Interview with Krairop Luanguthal, Sales and Marketing Director, Environmental Systems Research Institute (Thailand) Co., LTD.) Consequently, there are difficulties in data collection and data sharing. Developing and maintaining a digital database will significantly help municipal officials make informed decisions. GIS is a system that processes data which has spatial significance, including spatial data, non-spatial data, and topological data. The operation includes data collection, data management, data manipulation and analysis, and data output. The system adds meaningful value to data representing the environment, to explore the relationships between environmental components, and to facilitate planning, management, and decision-making.

Just as important as the technological advances it offers, GIS development initiates a higher-order systematization of geographical thinking, which is crucial to sustainable development for a community. Geographic information systems can help sustainable development in Chiang Mai at least in two aspects. First, growth management requires extensive data about existing land uses, utility supplies, and traffic circulation. GIS, with its data input, storage and manipulating capability, can help to collect, manipulate, and provide pertinent data. Second, management of a sustainable Chiang Mai needs a comprehensive consideration of many factors, such as population, development trends, availability of public services, and environmental constraints. GIS, with its data retrieval, analysis, and reporting ability, can facilitate the analysis and support decision-making.

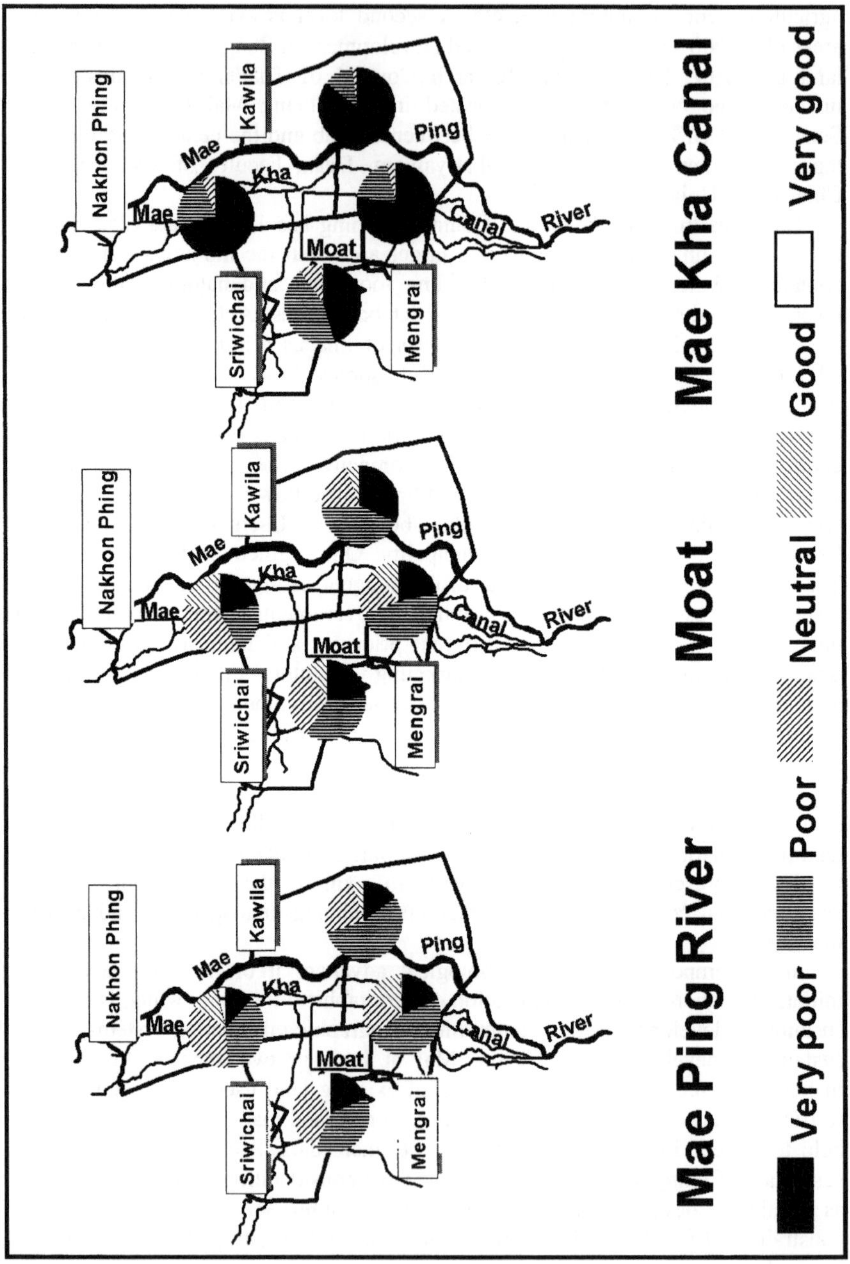

Map 5: Resident Perception of Water Quality by Sub-District, Chaing Mai Municipality

A GIS accepts data from a variety of sources, such as maps, tables, and aerial photographs, and stores data as different data layers. With GIS, a municipality can explore relationships between many independent entities simply by comparing their locations. GIS helps to perform spatial analysis efficiently and effectively. Therefore, in a short time, planners or decision makers can examine several alternative scenarios that weigh physical, economic, and political constraints and demands differently. Because a GIS usually includes a set of tools for the manipulation of data and the display of these manipulations, it can be used as a visualization tool, helping planners, decision makers, and the public realize the nature of problems.

Appropriate growth management needs detailed spatial data and spatial analysis in various respects, such as information about the distribution of land and resources, the routes of roads, water, gas, electricity, telephone lines, and sewerage systems. Development of digital databases can substantially decrease the costs for collecting data and ensure data consistency. Furthermore, digital databases can be constantly updated. Using the most recent data adds much value to any kind of analysis. A more detailed discussion of GIS can be found in Chapter 11 (Liu and Wang).

4. CONCLUSIONS AND RECOMMENDATIONS

Rapid growth and its accompanying development in Chiang Mai require municipal officials to consider growth management in order to secure the city's quality of life. Many consequences of uncontrolled growth are readily and frequently observed in the city. This study demonstrates that the recent rapid development in the Chiang Mai municipality has significantly affected the water quality in the area, especially the Mae Kha Canal that flows through the city. If this problem of water pollution is not solved, it will be a limiting factor for the development of Chiang Mai. This problem can, and should be, resolved with integrated planning and management. Although the discharge standards were set up by the Office of the National Environmental Board (ONEB) in 1985, they were not widely implemented (Araki,1992). With the completion of a new wastewater treatment plant, much of the sewage from the city should be treated. However, in many parts where septic systems are still used, the systems are connected with the storm sewer. Previous studies have concluded that about 70% of the surface water pollution in Chiang Mai is from household waste (Interview with Wasant Pongsapich, Director, Environmental Research Institute, Chulalongkorn University). To control this problem, a water quality monitoring system should be developed to monitor the water quality in different parts of the city.

Although the drinking water supply is not a problem in Chiang Mai, the distribution of drinking water can be improved with a more proactive approach, instead of only following the lead of real estate developers. The connections to the water supply should be integrated with land use plans that will direct the area of new development and protect environmentally sensitive areas. In fact, problems with the water supply, wastewater treatment, and surface water quality are largely the result of inadequate or absent land use controls.

A healthy urban development must consider both economics and environment together. The term environment here has both physical and social meanings. Although it is unrealistic to stop all land development in order to protect a city from various environmental problems, it is possible and reasonable to control the spatial distribution of the type and the intensity of land development to prevent those problems. Appropriate growth management helps the city to achieve its growth goals. It avoids a situation where a current growth opportunity becomes a growth problem, and at the same time, helps to convert a growth problem to a growth opportunity. No community needs to grow so fast that it is adversely affected environmentally, fiscally, or socially. Scheduled development provides access to previously undeveloped areas in both a steady and calculable manner; thus, enabling a city to plan and service growth at a rate that assures a favorable economic balance. Applying this concept, we recommend a phased development process starting with planning and design of utilities and major intersections and interior streets. It is necessary that the infrastructure and services, including water supply providing access to the designated development areas, be made available before the construction of structures. The experience of urban development in the United States demonstrates that it is important to manage development at an appropriate rate.

The development of a city-wide database incorporating land use and other data layers will assist the use of data for many purposes, such as creating land use and street maps, and maintaining streets, drainage facilities and utilities. The development of geographic information systems brings together many techniques into a single system framework that provides powerful new tools for spatial analysis. Promoting data sharing is another advantage of developing digital databases. Many of the local departments, such as planning, transportation, recreation, health, and public works, may take responsibility for collecting certain data particularly useful to their programs and make that data accessible to every other cooperating program as well. This has the potential to reduce the cost and improve the performance of various programs. It also decreases the occurrence of conflicting results because of inconsistent data used by different programs.

The Chiang Mai Municipality has the responsibility to endorse appropriate growth management to ensure sustainable development, whereby there are increased revenues and improved environmental characteristics. This approach will attract more investment from all over the world. The city government should regulate the timing and location of subdivisions in order to maintain high level services and facilities, to control the development character, and to maintain a desirable balance of land uses. To complement these strategies, the city government should increase public awareness on the water issues in the city. This includes using the mass media and primary and secondary schools on the one hand, and making timely information related to water issues available to the general public on the other. Ultimately, preserving water resources for a sustainable Chiang Mai must be the product of a public and private partnership. Always considering water resources in development decisions will assist in making Chiang Mai a viable city well into the future.

REFERENCES

Abu-Zeid, Mahmoud A. (1998). Water and sustainable development: the vision for world water, life and the environment. *Water Policy 1(1)*, 9-20.

Allen, H. E. & Mancy, K.H. (1972). Design of measurement systems for water analysis. In: Ciaccio, L. L., (Ed.), *Water and Water Pollution Handbook*. New York, NY: Marcel Dekker, Inc.

Araki, Hiroyuki, et al. (1992). Study Report on development of the appropriate technology as a primary health care for human wastes treatment and disposal in Northern Thailand. Chiang Mai University, Ministry of Health and Welfare, Japan and Japan International Cooperation Agency.

Beke, G.J., Entz T. & Graham, D.P. (1993). Long-term quality of shallow ground water at irrigated sites. *Journal of Irrigation Drainage Engineering*. ASCE 119: 116-128.

Chesterikoff, A., Garban, B. & Ollivon, D. (1991). Daily rhythms in the River Seine: relative impacts of natural and anthropogenic factors. *Water Research, 25,* 1523-1528.

Conlon, M., Gunn, J.M. & Morris, J.R. (1992). Prediction of lake trout (Salvelinus namaycush) presence in low-alkalinity lakes near Sudbury, Ontario. *Canada Journal of Fishery Aquatic Science*. 49 (Suppl. 1), 95-101.

Edwards, P.J. & Helvey, J.D. (1991). Long-term ionic increases from a central Appalachian forested watershed. *Journal of Environmental Quality*. 20:,250-225.

Firoz, Abu Saleh Mohammad. (1996). *Water Quality Assessment of Huay Kaew Stream, Mae Kha Canal,and Mae Ping River*. Master's thesis. Chiang Mai: Chiang Mai University.

Gondwe, E. (1991). Saline water intrusion in southeast Tanzania. *Geoexploration* 27:,25-34.

Gurnell, A.M.,Brown, G.H. & Tranter, M. (1994). Sampling strategy to describe the temporal hydrochemical characteristics of an alpine proglacial stream.*Hydrological Processes*, 8, 1-25.

Hakamata, T., Hirata, T., & Muraoka, K. (1992). Evaluation of land use and river water quality of the Tsukuba Mountains ecosystem. *Japan Catena, 19*, 427-439.

Hounslow, A. W. (1995). *Water Quality Data*. Boca Raton, FL: Lewis Publishers.

Howard, K.W.F. & Haynes, J. (1993). Groundwater contamination due to road de-icing chemicals - salt balance implications. *Geoscience Canada, 21*, 1-8.

Jensen, J.W. (1990). Diversity of Ephemeroptera and Plecoptera in Norway relative to size and qualities of catchment area. *Fauna Norvegica* 37, 67-82.

Johnston, C.A.,Detenbeck, N.E., & Niemi, G.J. (1990). The cumulative effect of wetlands on stream water quality and quantity, a landscape approach. *Biogeochemistry 10,* 105-141.

Koottatep, Suporn. (1997). *Water pollution problems in Thailand: Past, present and future*, in The Second Shizuoka Asia-Pacific Forum. November 26.

Kumaraswamy, K. & Sivagnanam, N. (1991). Diagnosing salt water intrusion in Vaippar Estuary of the Cholamandal Coast, India. *Transactions of the Institue of Indian Geographers, 13*, 43-51.

LBI-CMU. (1991). Chiang Mai Planning Project, survey of the attitudes in the Chiang Mai Municipal area, a supplement to consultant's final report, Chiang Mai policy-based action for historic and environmental preservation. Chiang Mai: Louis Berger International, Inc. and the Faculty of Engineering, Chiang Mai University.

Maasdam, R.; & Smith D. G. (1994). New Zealand's national river water quality network. 2. Relationships between physico-chemical data and environmental factors. *New Zealand Journal of Marine and Freshwater Research, 28,* 37-54.

Mathuthu, A.S., Chibanda, B., Tafirenyika, S., & Zaranyika, M.F. (1995). Impact assessment of sewage effluent discharges on the quality of the receiving lower Mukuvisi river waters in Harare, Zimbabwe. *Journal of Environmental Sciences and Health. Part A: Environmental Science. Engineering, 30,* 281-297.

McCutcheon, S.C., Martin, J., & Barnwell, T.O., Jr. (1993). Water Quality, In D. Maidment, (Ed.), *Handbook of Hydrology*. New York, NY: McGraw-Hill, Inc.

Napattalung, Manasa. (1997). *Quality of Running Water in Some Areas of Chiang Mai Based on Chemical Criteria*. Master's thesis. Chiang Mai: Chiang Mai University.

Peetragorn, Yupin. (1995). *Assessment of Water Quality by Means of Phenolic Compounds around Chiang Mai City*. Master's thesis. Chiang Mai: Chiang Mai University.

Priscoli, Jerome D. (1998). Perspectives on the problematique of water on the inauguration of water policy, *Water Policy 1(1),* 1-8.

Prygiel, J.; Coste, M. (1993). The assessment of water quality in the Artois-Picardie water (France) by the use of diatom indices. *Hydrobiologia,* 269-270, 343-349.

Rajagopal, S., Lalithaparameswari, N., Padmalatha, G., Geetha, S., & Balasubramanian, S. (1993). Water quality predictions in the River Bhavani (south India). *Environment and Ecology. 11,* 534-539.

Reichman, G.A. & Trooien, T.P. (1993). Corn yield and salinity response to irrigation on slowly permeable subsoils. *Journal of Soil Science Society of America.* 57, 1549-1554.

Rikard, M. & Kunkle, S. (1990). Sulfate and conductivity as field indicators for detecting coal-mining pollution. *Environmental Monitoring and. Assessment, 15,* 49-58.

Saowapon, Chaithawat. (1997). *A Perspective on the Water Related Issues in the City of Chiang Mai.* Unpublished paper.

Tyson, J. (1988). *Analysis: What Analytical Chemists Do.* London: The Royal Society of Chemistry.

Yang, Zhenbo. (1995). *Uses of Several Species of Floating Aquatic Plants as Biomontors in Assessing Heavy Metal Accumulation from Aquatic Ecosystems in Chiang Mai.* Master's thesis. Chiang Mai: Chiang Mai University.

CHAPTER 10

JUDITH A. MEYER SCHULTZ

A STUDY OF CHANGING TRADITIONAL URBAN SOLID WASTE MANAGEMENT APPROACHES TO SUSTAINABLE PRACTICES

Abstract As a component of the 1998 Joint Study in Chiang Mai, Thailand, this chapter examines solid waste management approaches in Chiang Mai in terms of human impact as a function of three major criteria: population size and growth rate, per-capita consumption, and environmental damage caused by the technology used to produce what is consumed and inability to absorb the resultant waste. Survey analysis, problem evaluation, and strategy formulation were used to focus on the feasibility of shifting management practices from traditional waste handling and disposal to an increasing prioritization of sustainable practices such as waste reduction, recycling, and reuse. In the midst of the current Asian economic crisis and faced with probable major growth in the near future, Chiang Mai has had a unique opportunity to assess its ecological capital using municipal solid waste generation as an indicator, and to subsequently rapidly develop the mechanisms to shift to sustainable solid waste management practices. This paper begins by reviewing the debate over sustainable development and considers the question of how and whether an urban center in a developing nation could shift from traditional waste management approaches to environmentally sustainable practices. In Section 2, the solid waste study site is assessed. Sections 3 and 4 reveal research results and describe specific proposals for sustainable solid waste management as applicable to unique conditions in Chiang Mai.

1. BACKGROUND

What are the realities of shifting environmental management practices in cities from traditional to sustainable? Blassingame (1998, 1) argues that the words "sustainable cities" are an oxymoron since sustainability means that a unit will use no more than it can replace and argues that city residents eat food they don't grow, etc. Daly (1990) similarly argues that sustainable growth is contradictory; that economy, the human subsystem of the earth's ecosystem, must eventually stop growing, but can continue to develop. Drakakis-Smith (1995, 4) argues that economists often confuse sustainable with "sustained development". Berg (1990) claims there are no sustainable cities today, but, whether or not they seem utopian, they are inevitable if humankind is to survive on the planet with projected population growth. Hawken (1993, xiii) amplifies this point in indicating that the real problem is the climactic explosion in the number of human species for the foreseeable future, since a majority of the six billion people on earth, and the billions yet to come, will live in large, central, unsustainable cities. Environmentalists are highly critical of the sprawl that emerges from such population centers. However, Blassingame (1998, 1-3) suggests that sprawl provides opportunity to test for the values and planning

M. Romanos and C. Auffrey (eds.), Managing Intermediate Size Cities, 209-220.

necessary for achieving sustainable cities. The challenge is to ask what is the best form for human habitation in such numbers and can it be sustainably designed. Reducing pollution and increasing efficiency are perhaps the easy part. David Orr (1992, 17) asks if we will not also have to be "good enough" in both a moral and ecological sense, embracing a basic systematic change in thinking to perhaps pedestrian societies. Dominski (1992) proposed a three-stage model of transition over a century to sustainable cities where the first stage is recycling, and the second stage is reuse, as roadways become rights-of-ways for electric trains and bicycles, and roofs become gardens or energy collectors. The third state is reduction of energy use and virgin materials. Brown, et. al. (1992) suggests that for the most of the twentieth century, economic debates focused on whether capitalism or socialism was the best way to organize a modern industrial economy, but this has evolved into the more fundamental question of how can we design a vibrant economy which does not destroy the resource base and environmental systems upon which it depends.

An attempt to identify quantitative indicators for sustainability was proposed by William Rees (1992) to measure the "ecological footprints" of cities, by comparing the resource base and environmental impact of a city, which could be 20 times greater than the land that it occupies. Solid waste disposal management is one appropriate indicator of the ecological footprint of cities. Historically, as economic growth progresses in unsustainable cities, the amount of waste generated per capita increases exponentially. While Whitney (1992) indicates that in the past almost all wastes were recycled between urban and rural areas, because of the present unprecedented scale of solid waste production which is increasing many times faster than urban populations, most solid waste of third world nations is dumped at open sites, or incinerated with little concern for control of air emissions (Yeung, 1992). As a result, land, air, and water pollution, which are all interconnected to solid waste disposal, are intensified (Drakakis-Smith, 1995, 10). In most third world cities, there are also many examples of recycling of waste material in the city by individual households or by scavengers, termed "the waste economy" by Whitney (1992).

Freeman (1979) suggests that in order to provide measurable solid waste management disposal criteria, surveys can also give a good indication of accuracy of willingness to pay for solid waste disposal and illustration of behavior that would be revealed if the public good could be offered in a market where exclusion was possible. Van Diner (1995) suggests that ecological capital is a complex category that provides resources for production and absorption of wastes. Nations and even cities need to assess not only willingness to pay for public good, but also for ecological capital. In addition, Van Diner suggests that where these wastes add to or improve the stock of ecological capital, they can be considered as investment in such capital; but more often, however, they destroy, or pollute, and are then agents of negative investment or capital consumption.

2. ASSESSMENT OF STUDY SITE

2.1. Profile of Chiang Mai

The location of this sustainable development research study was Chiang Mai, the third largest city in Thailand, which occupies 40 square km. and has a (1998) population of approximately 185,000. It is located in the province also called Chiang Mai, with a population of one and one-half million. Population density is approximately 6,600 /sq. km. compared to 85 in the surrounding province. Land to the north of the city is mountainous and in the surrounding provincial region is agricultural flood plain used primarily for growing rice and other vegetable crops. Thailand's growth rate is approximately 1% per year, with a doubling time of 60 years due to the current large age cohort of the population in their reproductive years. Thus, while Thailand attempts to cope with major existing environmental management problems, the doubling of its population in just sixty years will require new infrastructure, and excellent planning in the short and long terms.

In 1990, projections by the Thai government for growth of the population of the City of Chiang Mai were 197,481 by 2005 and 229,184 by 2015, but these figures appear to have been conservative estimates. The current predicted annual rate of population growth for the entire province is 1.18% and for the Greater Chiang Mai Area (GCMA, often referred to as the Muang District) 2.16%. Outside the actual City limits, growth is only 0.64%. The population of the GCMA in 1995 was 247,767; for the years 2005 and 2015, the population of the GCMA is expected to reach 306,798 and 379,893 respectively. This makes the City of Chiang Mai one of the fastest growing cities in Thailand (Saowaan, 1996).

Chiang Mai is significant as a regional agricultural production center, and has been designated by the central government as an emerging industrial and population growth center, as well as one of the top tourist destinations in Thailand for Hilltribe trekking, elephant performance centers, and entrance into Burma and the Golden Triangle region. It is a dynamic urban area currently facing several development issues, including maintenance of the quality of urban environment, preservation of historic and cultural heritage, improvement in the delivery of urban services, and achieving balance between tourism and other domestic economic activities. The establishment of conditions for social harmony among diverse economic and ethnic groups of the region, while improving the environmental quality, has been perhaps the most challenging aspects of the city's recent evolutionary dilemmas.

Changes in Chiang Mai over the past decade have raised issues for solid waste management at present and for the future. The Greater Chiang Mai region has experienced a number of demographic changes, with growth contributed by the influx of people from rural areas, the purchase of second homes by Bangkok's upper class, the increase in the number of domestic visitors, immigration, and the large expansion of the numbers of international tourists. Each of these growth components has potential for contributing to the growing complexities of solid waste management both in volume and in type of solid waste. Continuing current trends towards acceptance of western influence, as part of globalization, in terms of a consumption-based society with greater demands for processed and manufactured goods will presumably produce increased packaging and waste.

Currently, Chiang Mai's 8.10% annual economic expansion yields the highest gross provincial product in the North with a GPP equivalent to 19% of the GPP of the North, or 35.4% of the upper north. The average income is ranked first in the North and 21st in the country. Service and industrial sectors are expanding at a rapid rate, causing a continuing labor shift from agricultural sector to the other two sectors (PCMP, 1998). Population growth and economic expansion have resulted in increases in urban waste generation and erosion of air and water quality, noise pollution, traffic congestion, and slums. Waste disposal planning must be formulated in accordance with the master plan of solid waste management and urban management as a whole. Short and long-term plans, of necessity, must take into account population growth and waste generation rate, their future forecasts, economic growth rate, and the local or city development plans (Ikeguchi, 1992).

2.2. Historical Considerations

In 1958, the Mae Hin Solid Waste disposal site opened to service the GCMA which processes waste at a rate of 15 tons/day with use of soil cover, but with no leachate collection system. From 1973 to 1983, wastes were disposed of by both sanitary and open dumping methods. By 1983, because the municipal population had increased so significantly, open dumping (at a rate of 150 tons/day) was reported as the sole means of waste disposal. This was continued until the landfill closed in 1989 amidst protests that the site was contaminating shallow well water supplies (Interview, 1998).

Urban Thais in 1993 disposed of .66-.84 kilos of waste per capita per day, usually collected by municipal employees using trucks. In 1994, 1974 tons/day were collected. Households produce 90% of total waste, but contribute only 50% of total charges. Various sites were temporarily used for dumping, including a local cemetery, until a major garbage crisis ensued in 1995. In 1994, the City could not site a new dumpsite due to the NIMBY syndrome, and garbage was left uncollected for several days. The crisis led several governmental agencies to discuss the concept of a solid waste disposal system, i.e. a sanitary landfill or an incineration plant (Saowapun, 1996).

2.3. Research Approaches: Macro and Micro

In order to assess the base-line status of solid waste management practices in Chiang Mai, both macro and micro levels of research were conducted. Macro level reviews of existing applied solid waste studies and papers were conducted. In addition, numerous meetings were arranged with representatives of the central, provincial, and municipal governments that have various jurisdictions over planning components of solid waste management. Visitations were made to the municipal landfill, and to several recycling businesses. At the micro research level, a random survey of households and businesses was made to determine the accuracy of the macro data. The areas sampled were neighborhoods randomly selected by walking or driving for a stated period of time in varied directions.

The problems specific to solid waste management in the municipal area can be mainly categorized into: on-site management, collection system, disposal method, and infectious waste management. Currently, the municipality collects 80% of the solid waste generated and the generators, such as Chiang Mai University and the military restricted area, collect the rest. In all cases, it is then transported to the municipal waste disposal site.

2.4. Current Solid Waste Management Crisis

The Chiang Mai solid waste crisis initiated a political thrust for selecting a new disposal site. A temporary landfill was finally sited, but with a life span of only two years. This landfill was constructed with a leachate collection system, clay liner, and HDP1.5 mm plastic back-up liner. Leachate is transported to the local sewage treatment plant. Seventy trucks dump garbage daily from residences and businesses that is compacted, spread, and covered. This temporary landfill was slated to close in early August of 1998. A future site was selected in the suburbs, but after considerable geographical and geological evaluation, citizens vehemently opposed it. A new site was proposed in July of 1998, and funds appropriated. Debate ensued as to whether a permanent or temporary landfill should be built on the site. A permanent landfill was seen to discourage extraction of recyclables and reusable, and required time-delaying approval, funding, and control of the planning by the central and provincial governments, eliminating local control. As a result, at the time this study was in progress, the Municipal government favored the new temporary landfill.

At the time of this study, the existing landfill, which was slated for closing, was reaching within two weeks of capacity. An alternative site had been selected outside of the city in the suburbs, but the NIMBY syndrome had prevented final approval of the new landfill site. Within one month after the completion of the study, the older landfill had closed with no alternative new site approved, and solid waste was temporarily placed in the Christian cemetery. (In a primarily Buddhist society, Christian believers constitute about 1% of the population.)

3. RESEARCH RESULTS

3.1. Economic Considerations

Currently, the Municipality of Chiang Mai operates solid waste management by deficit spending of [10x] revenues collected. There are two reasons for this. One is that many people avoid the collection by taking their waste to a public container. The second is that the officials interviewed indicate that the fees are set too low. The maximum allowed to be charged by law is 40 baht ($1U.S.) for residences and 2,000 baht per month for hotels and businesses (Schultz, 1998). Fees are collected in person, which may allow for negotiations. More cost effective management could be achieved by using a better billing system connected to an accounting system, a revised tariff schedule, review of contract terms, improved budgeting, and a use of cost centers. A transfer station would make transportation more efficient by transferring waste to larger trucks with better capacity (Joint Report, 1996).

3.2. Reuse, Recycling, Reduction

There is no formal procedure in place for reuse, recycling, or reduction of waste in Chiang Mai. As such, it is difficult to obtain a complete picture of reuse in Chiang Mai. However, the author visited numerous waste-reuse businesses and interviewed business owners and homeowners. The micro survey indicated that an informal process works effectively in which many individuals place recyclables and reusables outside their doors, in which case, others, who take them to the reuse businesses, pick them up. Individuals (scavengers) were also observed at the Chiang Mai landfill, picking reuseables in the midst of the operation of large machinery. (This is a typical process throughout Asia.) Numerous reuse businesses in Chiang Mai operate to collect glass, plastic, metal, corrugated, paper, numerous other reusable items and were interviewed as part of the study. All metal and plastic items are taken to Bangkok; paper is sent to a plant in a nearby city (Interview #2).

3.3. Characterization of Solid Waste in Chiang Mai

Table 1 indicates that the solid waste stream in Thailand and in Chiang Mai is high in organic and vegetable matter, in some instances up to 70%. The waste is also 50% wet in content (Saowapon, 1996). Neighborhood fresh markets are a major source of vegetable matter, much of which becomes garbage. Thus, a recent proposal to build a municipal incinerator for co-generation was not pursued for the reason that the garbage is simply too wet. The monsoon season further exacerbates this situation.

Table 2, was calculated by using annual figures and an assumption of 50% based upon percentage of garbage recycled in Japan (Hershkowitz, 1987). Recyclables from Residential Waste generated in Chiang Mai Municipality, were priced from actual prices by category from a nearby local merchant in the recycling business, indicating an approximate annual value of 23,145,000 baht or $578,625.

Table 1: Garbage Composition of the City of Chiang Mai

1. Garbage density	0.22 ton/cu. m.
2. Garbage composition (% wet weight)	
2.1 Paper	6.8%
2.2 Cloth, leaf & wood	12.2%
2.3 Plant	66.1%
2.4 Bone & meat	3.4%
2.5 Ash & dust	3.2%
2.6 Metal	1.1%
2.7 Plastic & rubber	0.4%
2.8 Glass, tile, brick	0.2%
2.9 Particle less than 24 mm. in dia.	0.1%
2.10 Others	6.5%
3. Composition (% wet weight)	
3.1 Water Content	53.45%
3.2 Combustible material	30.92%
3.3 Ash	15.62%
4. pH value	7.72%

Source: Sanitation and Environment Division, Chiang Mai Municipality (Aug. 31, 1995).

Table 2: Value of Recyclables from Residential Waste in Municipality

Components	*50% of Weight Disposed Daily (Metrications)*	*baht/kg*	*Value/Day (in baht)*	*Value/Year (in baht)*
Plastic	12.325	3	36,975	13,495,875
Glass	7.83	1	7,838	2,861,016
Paper	8.98	1.5	13,470	4,916,550
Metals (using value for steel as it is the lowest)	1.71	3	5,130	1,872,450
Total			**63,413**	**23,145,891**

Source: Research, Chiang Mai, Thailand, July-August, 1998, J. Schultz and Z. Zivkovich. Disposal tonnage figures provided by Chiang Mai Municipality; prices provided by local Chiang Mai recycling businesses; also based upon assumption of 50% recyclable garbage as redeemable in Japan (Hershkowitz, 1987).

Table 3 calculations were made by using annual figures from all solid waste collection, including the surrounding suburbs. Calculations indicate an annual increase to 50,866,034 baht for recyclables that could be reclaimed by presortment from the landfill (approximately $1 million U.S.). Thus, the Municipality of Chiang Mai possesses a major potential to extract and to resell recyclables, particularly plastics as a material or fuel. Market potential for plastics is expected to increase by 10-12% over the next three years.

Table 3: Value of Recyclables from All Municipal Solid Waste Landfilled Daily

Components	*50% of Weight Disposed Daily (Metrications)*	*baht/kg*	*Value/Day (in baht)*	*Value/Year (in baht)*
Plastic	34.22	3	102,649	37,467,177
Glass	10.88	1	10,870	3,970,032
Paper	12.46	1.5	18,694	6,823,492
Metals (using value for steel as it is the lowest)	2.3793	3	7,138	2,605,333
Total			**139,351**	**50,866,034**

Source: Same as Table 2 above.

4. CHALLENGES FOR MUNICIPAL SUSTAINABLE SOLID WASTE

4.1. Composting Opportunities

The most realistic opportunity for shifting from traditional to sustainable solid waste management practices for Chiang Mai lies in moving away from land filling waste to composting and to pre-sortment with subsequent reuse of recyclables. A feasibility study on composting has recently been completed for the municipal government and could be acted upon. Approximately 55.6% of the municipal solid waste that is landfilled currently has the potential to be composted, including food waste that constitutes 54% of the total. The City of Chiang Mai currently landfills 90.7 metric tons of compostable wastes from within the city limits, and 125.9 tons from the surrounding suburbs. These wastes, on average, come very close to meeting the ideal 30/1 carbon/nitrogen ratios for compost, with an average of 34.4/0.79, with moisture content of 43.6%, an ideal range being 40-60% (Table of Solid Waste Statistics).

Chiang Mai is situated in an ideal setting for the location of a major composting facility, since it is in a subtropical latitude and is surrounded by a major agricultural region of Thailand in close proximity for use of degraded compost, and with low distance transportation from composted byproducts. "Agriculture was only surpassed four years ago by tourism as the number one source of income." (Interview #3). Neighborhood fresh markets are also a major source of vegetable matter that ends up in garbage, and these markets offer opportunity for major education on the function of composting.

4.2. Reduction, Reuse, Recycling

If Chiang Mai can concurrently develop a major formal plan for presortment removal of glass, plastic, metal, batteries, etc., the residual solid waste could be composted, greatly reducing its volume and shifting it into a resource category. An educational program could provide incentives for individual homeowners as well. However, recyclable materials collected must first find markets, and it would be useful if markets for these items were closer than Bangkok, which is 500 miles away. Transforming waste materials into useful products can also provide jobs within the local community, while reducing the need for shipping out of the community. A materials exchange newsletter and network would also be useful for reuse of industrial waste products. Time did not allow for a thorough review of the industrial waste treatment, but interviews suggested major improvement was needed. Recycling may also require transfer stations for materials gathered or material recovery facilities. Currently, several private businesses fulfill that role. Additional research is needed to assess whether private enterprise or municipal government can provide more efficient reuse and recycling, although it appears to be working well in the hands of entrepreneurs. Presortment must be the goal, and the best plan would be for the municipality to contract with existing businesses rather than in citing new facilities. Since packaging is a major component of solid waste, a persuasion campaign to encourage consumer reuse of store bags or bringing a mesh bag could be initiated. Finally, the existing landfill, which was slated for closing in late August of 1998, could eventually become a modest biogas facility. This could be leased to a biogas recovery firm, although the decay process requires a 20-30 year lead-time.

4.3. Economic Considerations

The primary challenge in the midst of the crisis facing Chiang Mai is coping with the current economic crisis in Asia, while obtaining the domestic and foreign capital to build composting and recycling facilities. However, an advantage of composting facilities is that they have a lower operating cost than landfills. Secondly, a formal program in presortment must be developed to remove glass, plastic, metal, batteries, etc., and to separate compostable waste. Third, incentives for local markets for recyclables need to be provided by the central or prominent government for the nearby industrial region, as well as developing a transfer facility for recyclable items. Fourth, a plan for transforming wastes into useful products could create local jobs, while reducing transportation and landfill costs. While the now closed landfill has the potential to become a biogas recovery plant, the process requires 20-30 years, and importantly, in the midst of the economic crisis, demands additional economic resources. Composting of waste and a comprehensive recycling-reuse plan would shift traditional solid waste garbage management to a resource recovery effort as part of a sustainable management strategy.

Clearly, Chiang Mai is a city facing major critical decisions and options. The city continues to expand in both permanent and transient population, and has been designated for additional economic and industrial development by the National Economic and Social Development Board of Thailand as the "educational and tourist center for Northern Thailand". However, a 1996 report on the urban environment, prepared jointly by the Office of the National Economic and Social Development Board and Danish Cooperative for Environment and Development found that Chiang Mai has the second most degraded environmental conditions in the country, emphasizing the need to make environmental improvements wherever possible (Aieka, 1997). The current economic crisis offers a brief, but advantageous and challenging, window to plan for new sustainable directions in solid waste management, at the historical intersect of population growth and increased solid waste generation.[1]

NOTES

[1] Appreciation is extended to Zachery Zivkovich, who provided excellent research support as an undergraduate research assistant.

REFERENCES

AIEKA Business Development Project. (1997). Analysis of Waste Management in Chiang Mai Municipality Thailand, June, 1997. Swedish International Development Cooperation Agency.

Berg, P. et al. (1990). *A Green City Program for San Francisco Bay Area and Beyond.* San Francisco: Planet Drum.

Blassingame, L. (1998). Sustainable Cities: Oxymoron, Utopia, or Inevitable? *The Social Science Journal, Jan. V35, n1 (13).*

Brooks, D. (1990). What Does Sustainable Development Really Mean? *IDRC Report 18, (4)*, 24-25.

Brown, L. and S. Postel, C. Flavin. (1992). From Growth to Sustainable Development. In R. Goodland, et.al., (Eds.), *Population, Technology and Lifestyle: The Transition to Sustainability* (pp.119-127). Washington, D.C.: Island Press.

Campbell, S. (1999). Green Cities, Growing Cities, Just Cities: Urban Planning and the Contradictions of Sustainable Development. In D. Banister, et.al., *Environment, Land Use, and Urban Policy* (pp.463-478). Northampton, MA: Edward Elgar Publishers.

Daly, H. (1993). Sustainable Growth: An Impossibility Theorem. In H. Daly and K. Townsend (Eds.), *Valuing the Earth: Economics, Ecology, Ethics* (pp. 267-274). Cambridge, MA: MIT Press.

Dolan, L.W. (1993). *Local Administration and the Provision of Environmental Infrastructure in Thailand.* Bangkok: U.S. Agency for International Development.

Dominski, T. (1992). The Three State Evolution of Eco-Cities-Reduce, Reuse, Recycle. In Walter, et. al. (Eds.), *Sustainable Cities* (pp. 16-18). Los Angeles, CA: EHM Eco-Home Media.

Drakakis-Smith, D. (1995). Third-World Cities, *Urban Studies, May, v32, h4-5,* 656.

Freeman, A. (1979). *The Benefits of Environmental Improvement: Resources for the Future.* Baltimore: John Hopkins University Press.

Haughton, G. and Hunter, C. (1994). *Sustainable Cities.* London: Regional Studies Association.

Hawken, P. (1993). *The Ecology of Commerce.* New York: Harper Collins.

Hershkowitz, A. and Salerni, E. (1987). *Garbage Management in Japan: Leading the Way.* New York: Inform, Inc.

Ikeguchi, Takashi. (1992). *Appropriate Land Disposal Technologies of Municipal Solid Wastes in Tropical Region: Application of Lysimeter Study Results to Landfill Operation.* Paper presented at 4th National Symposium of Environmental Technology and Management, June. Chiang Mai, Thailand.

Investment Plan for Chiang Mai Province. (1998). Executive Summary. Board of Investment, p.4.

Joint Report of the Office of the National Economic and Social Development Board of Thailand and the Social Development Board and Danish Cooperative for Environment and Development. (1996). Chiang Mai, Thailand.

Orr, D. (1992). *Ecological Literacy.* Albany, N.Y.: The State University of New York Press.

Rees, W. (1992). Ecological Footprints and Appropriate Carrying Capacity. *Environment and Urbanization 4 (2),* 121-130.

Saaowapon, C. (1996). *A Perspective on the Water Related Issues in the City of Chiang Mai.* Unpublished paper. Chiang Mai University.

Satterthwaite, D. (1997). Sustainable Cities or Cities that Contribute to Sustainable Development? *Urban Studies,* October.

Schultz, J.M. (1998). Survey of Municipal Government and Chiang Mai Residential and Business Sector for Value of Recycling and Solid Waste Management. Unpublished.

Table of Solid Waste Statistics, Chiang Mai Municipality.

Van Dieren, W. (1995). *Taking Nature into Account.* New York: Copernicus, Springer, Verlag.

Whitney, J. (1992). Asia. In R. Stren, et. al. (Eds.), *Sustainable Cities: Urbanization and the Environment in International Perspective.* (pp. 229-232). Boulder, CO: Westview Press.

Young, Y. and G. McGee. (1986). *Community Participation in Delivering Urban Services in Asia.* Ottawa: IDRC.

INTERVIEWS

1. Interview and Tour of Chiang Mai landfill: Kasana Harnchipibulgul, Engineer for Chiang Mai Municipality, July 13, 1998.
2. Interviews with Chiang Mai recycling businesses.
3. Interviews with officials from Sanitation and Environment Division and Public Works Division of Chiang Mai Municipality, Chiang Mai, Thailand, July 3, 1998.
4. Interview and Meeting with Chamber of Commerce President Danai Leosawathipung and Chairman Thauatcha Taeriyaphirom, July 3, 1998.

โทรศัพท์
0208
12

CHAPTER 11

LIN LIU
XINHAO WANG

BUILDING A GEOGRAPHIC INFORMATION SYSTEM

Abstract Sustainable development is often associated with a trade-off between social and economic development on the one hand, and environmental protection and enhancement on the other. Optimizing these conflicting goals requires a computing environment that integrates storage and analysis of spatial and statistical data related to both development and environment. Geographic information systems (GIS) provide such an integrated computing environment to facilitate the complex data analysis and decision-making for sustainable development. This article addresses issues surrounding a GIS for the sustainable development of Chiang Mai, the second largest city in Thailand. Based on the information collected from literature search, personal communication, and interviews with municipal and provincial officials, university professors, and local GIS professionals, the authors provide a systematic analysis, including the examination of the costs, benefits, and challenges for developing the system. The analysis suggests that most digital datasets needed by the GIS are available in various state government agencies. Tax collection, a pressing issue for the municipality, is used as a case study for illustration.

1. INTRODUCTION

Sustainable development deals with a host of societal sectors, including planning, economic development, natural resources, and environmental protection, just to name a few. It often involves compromises between conflicting goals of different sectors. Thus, integration and collaboration among multi-sectors is important in the decision making process of sustainable development. However, integration of multi-sectors is often challenging because most organizations are single sector-oriented. Geographic information systems (GIS) can be used as a computation and database environment to facilitate such integration. It provides powerful tools for integrating and analyzing spatial and statistical data from different sectors. This integrated environment can help improve communications among multiple sectors and generate quality information for better decision-making for sustainable development (FAO, 1998).

Murai (1995) pointed out several reasons why a GIS is needed for sustainable development in Thailand. First, data are inaccurate, inconsistent, poorly maintained and outdated. Second, there is inadequate data retrieval. Existing data are not sufficiently utilized. Third, there is no data standard, which causes difficulty for data exchange. Finally, there is no data sharing. He argued that GIS can help solve these problems.

Although Chiang Mai is often regarded as the second largest city in Thailand, it is much smaller than the largest city, Bangkok. The city, like other smaller cities, is

M. Romanos and C. Auffrey (eds.), Managing Intermediate Size Cities, 223—234.

confronted with a number of development issues (Tantayanusorn, 1998), some of which contradict each other. For example, further industrial development and expansion of the city are achieved at the expense of losing agricultural land; historical architecture and culture attract tourists, but development of tourism can play a negative role in the preservation of history and culture. Solving these conflicting goals is vital for the sustainable development of Chiang Mai. GIS is a powerful tool for sustainable development (FAO, 1998), but the City of Chiang Mai did not have a comprehensive GIS when the authors conducted fieldwork in the city. The city showed keen interest in GIS, but was not able to develop its own system due to various reasons. Hastings (1995) discussed obstacles to GIS development in Thailand, most of which are related to the organization of institutions. For example, should an institution have a separate GIS unit? What are the pros and cons of having a separate GIS unit? The following sections of this paper examine the important issues related to a GIS for the sustainable development of Chiang Mai.

2. THE ROLE OF GIS IN THE SUSTAINABLE DEVELOPMENT OF ASIA AND THAILAND

Asian countries realized the importance of GIS for sustainable development as early as 1995. During the GIS AM/FM Asia '95 conference in Bangkok, a special session was organized to discuss GIS applications for sustainable development (Shibasaki and Hashimoto, 1995). In the keynote speech of the session, Peter King from the Asian Development Bank outlined the role of the Bank in applying GIS and remote sensing to the sustainable development of Asia and the Pacific. The participants exchanged their experiences of GIS applications in the Asian region, in relating to the potentials of applying GIS and database technologies in development projects. They also discussed strategies to promote GIS technology in development projects.

There have been successful implementations of GIS in Asia. For example, Yeh and Li (1998) developed a GIS for sustainable land development for rapid growth areas in Southern China. In Thailand, GIS has been applied to watershed management (Omakupt, 1995), spatial decision support for land use planning within a watershed area (Mongkolsawat, 1995), and decision making for groundwater management in Bangkok (Gautam and Gupta, 1997). GIS has been applied in a number of research projects conducted in Chiang Mai. Examples of these projects include a GIS demonstration project for Mae Taeng watershed (Apavatjrut, 1993), the design of a spatial information system to facilitate agriculture land evaluation in Chiang Mai Province (Chupraphawan, 1993), the evaluation of land in a highland watershed using GIS, and a multicriteria decision making approach using GIS (Salakij, 1997). However, these GIS implementations are only for the data management of a single sector or a few sectors, and are not comprehensive enough for the sustainable development of Chiang Mai.

3. FEASIBILITY OF A GIS FOR CHIANG MAI

It is certainly desirable to have a GIS for the sustainable development of the City of Chiang Mai. But there are many questions yet to be answered: what are the costs and benefits of the GIS? What are the main challenges for developing the GIS? Will the GIS fit into the existing organization of the municipality?

3.1. Costs and Benefits of a GIS

Developing a GIS is not a trivial matter for many organizations (Obermeyer, 1999, p. 601). It requires substantial commitment in money, staff, and effort. Unless the expected payoff surpasses the cost, most organizations would not make such a commitment. Berhardsen (1999, 600) pointed out that GIS is rarely a once only, short-term investment. The computer equipment of a GIS can last two to five years, software can last three to eight years, and data can last ten to fifty years and more. Therefore, the costs and benefits must be applied over time in any cost-benefit analysis.

The costs and benefits of a GIS can be tangible or intangible. Tangible costs/benefits are those that are readily quantifiable, while intangible costs/benefits are those that cannot be easily quantified. Tangible costs typically include hardware, software, training, new staff, and additional space, as well as cost reduction, cost avoidance, increased revenues, and new market services (Obermeyer, 1999, 602). On the other hand, intangible costs may include internal personnel shifts and layoffs. Intangible benefits can be reflected on improved client relationships, better decisions, and improved morale (Obermeyer, 1999, 609). In general, evaluation of the intangible costs and benefits of implementing GIS is more difficult (Longley, et al, 1999, 587). Furthermore, the cost evaluation of a GIS must be assessed in comparison with alternate implementation scenarios (Wilcox, 1996). Tulloch, et. al. (1997) measured the benefits of using a land information system in efficiency (reduce/eliminate costs, avoid costs, faster, easier), effectiveness (better service, new service), and equity (equal/improved access, equal treatment, altered decision-making). They used a benefits matrix, in which rows represented access-oriented outcomes, cost-oriented outcomes, and application-oriented outcomes; and columns represented efficiency, effectiveness, and equity.

The costs of developing a comprehensive, multi-purpose GIS for the City of Chiang Mai consist of the following expenses:

Equipment purchasing: The city had only two PCs capable of running GIS software at the time the authors conducted the fieldwork. To build a comprehensive GIS, the city needs to purchase more computers and some peripheral devices such as digitizers and plotters.

GIS software purchasing: The city needs to purchase GIS software packages since it does not have a license for any GIS software.

Training: There are few people knowledgeable of GIS. City officials believe a GIS can be developed using technical people within the organization, but these people need to be trained first.

Database development: Although Chiang Mai has been digitized many times (Hastings, 1998), a comprehensive digital database does not exist for the city. Different data sets have been generated from various projects conducted in Chiang Mai. However, data sharing is not a common practice and there is no central depository for these data sets. Therefore, the city needs to secure existing digital data from other sources, and it may even need to input some analog data (buildings, for example) into digital format.
Organizational restructuring: This is an intangible but unavoidable cost. The current structure of the city is rigid and there appears to be little collaboration between different departments. Since cooperation is key for a successful GIS, re-organization is unavoidable.

The purchase of computer hardware and software accounts for only a small portion of the overall cost, while the cost of data acquisition and database construction can be significant. However, it is typically the intangible costs such as the restructuring of the municipal organization that determines the fate of a GIS. It must also be pointed out that a one-time investment is not enough for a successful GIS because the databases need to be continuously updated, computer software and hardware need to be upgraded, and the staff members need to be trained to keep up with the latest technology, all of which cost money on a continual basis. Therefore, the financial burden to an intermediate sized city like Chiang Mai can be significant if the benefits of a GIS are limited.

The potential benefits of the GIS to the city include:
Revenue increase: With a GIS, the city will be able to collect more tax (ADS, 1998).
Cost avoidance: With the current structure, departments have overlapping functions, which cause redundancy and inconsistency. A GIS can help avoid these costs.
New service: The GIS can provide many new services. For example, it can be used to generate updated maps for tourists.
Better decision-making: The GIS can help the city to make better decisions on various development issues. More importantly, the GIS will help solve the conflicting goals of various development issues to achieve sustainable development.
Better efficiency: The GIS can improve the efficiency of many city functions. To illustrate, the building permit application process can be made faster and build and parcels maps can be updated more efficiently.

It is difficult to draw a definitive conclusion on the exact costs and benefits of a GIS for Chiang Mai in the absence of a detailed analysis. However, if the claim that "65% of the city revenue are not collected due to lost tax" by ADS (1998) is credible, the benefit from recovering lost revenue alone may surpass all the costs.

3.2. Data

Data act as the 'fuel' for GIS. Thus, data availability and issues related to data are major factors in determining the cost and effectiveness of GIS (Rhind, 1999, 767). A typical GIS database consists of framework data and non-framework data. Framework data include those "which underpin all GIS operations, providing the geographical context and structure within which spatially manifested processes operate and on which other data are assembled" (Smith and Rhind, 1999, 655). For example, framework data in the United States consist of geodetic control, digital orthoimagery, elevation data, transportation, hydrography, the geography of government units, and public land cadastral information (FGDC, 1995). Non-framework data can be anything else. In Thailand, the main source of the framework data is the Royal Thai Survey Department. Access to this framework data is highly restricted.

There is not a single documentation that lists all available digital datasets for Chiang Mai. Some data sources are not even documented. Through personal interviews and literature search, the authors discovered that a number of projects had created several digital databases that cover the Chiang Mai area. The following paragraphs describe the databases developed by the Thailand Environmental Institute (TEI), the National Economic and Social Development Board (NESDB), the Department of Town and Country Planning (DTCP), and the Land Authority.

TEI created a database named *THAILAND on a DISK*. It covers the entire country. Framework data consist of contours, elevation, hydrological features, political boundaries, and transportation infrastructure. These layers are digitized from small-scale topographic maps (1:250,000), originally produced by the Royal Thai Survey Department. Non-framework data include land use for 1990, forest, soil, geology, rainfall, and socioeconomic data summarized at the tambon (sub-district) level. Geologic layers are created from 1:500,000 geological maps by the Department of Mineral Resources. Land use and soil layers are based on 1:500,000 maps by the Department of Land Development. These map layers cover the city of Chiang Mai, but are not useful because they do not provide enough details for the GIS for Chiang Mai.

A database was created by NESDB (1994) from larger scale maps for a Chiang Mai – Lamphun twin city development project. The twin city database includes the following layers:

- Base maps: administrative boundaries for districts and sub-districts, river, canal, contour (1:50,000 maps from Royal Thai Survey)
- Detailed land use in 1992 for Chiang Mai and Lamphun (1:12,500, 1:25,000 and 1:30,000 maps from Department of Town and Country Planning)
- Land use in 1995 for the study area (interpreted from airphotos at the scale of 1:100,000)
- Geology (1:100,000 maps from Department of Mineral Resources)
- Forest type and coverage
- Areas with potential for future development
- Infrastructure: road and planned road, railroads, river, reservoir, irrigation canal (1:50,000 maps from Royal Thai Survey), irrigation area, water supply service area, garbage collection service area
- Historical place
- Land suitability (1:100,000 maps from Department of Land Development).

Compared to *THAILAND on a DISK* by TEI, these layers are much more detailed. The NESDB database can be used in the comprehensive GIS for Chiang Mai.

Another detailed database was created by DTCP. DTCP uses GIS to increase efficiency of data analysis for developing and revising comprehensive plans, to update the data more efficiently and accurately, and to enhance data distribution and service to the public. It consists of four types of data: base maps, planning data, transportation data, and public facility/utility data (Table 1).

Table 1. Data layers included in the database created by Department of Town and Country Planning (DTCP)

Base Maps	*Planning Data*	*Transportation Data*	*Public Facilities & Utilities Data*
▪ Administrative boundaries ▪ Buildings ▪ Existing major land use ▪ Geodetic control ▪ Municipal and sanitary district boundaries ▪ 1:4,000 grid ▪ Hydrology ▪ Estimate land summary ▪ Aerial photo ▪ Railroads and rail edges ▪ Road centerlines and road edges ▪ Structures ▪ Topographic contours and spot elevations ▪ Wells	▪ Comprehensive plan boundary planning ▪ Flood zone ▪ Land subsidence area ▪ Industrial zone ▪ Slope ▪ Soil ▪ Population and employment data ▪ Direction of growth ▪ Development intensity ▪ Land suitability ▪ Development availability	▪ Traffic analysis zone ▪ Transportation information system – O-d survey – Road condition – Classified count – 24 hours count – Analyzed data – Trips interface ▪ Planned road	▪ Service area – School – Police station – Fire station – Recreation center – Public health center – Marketplace ▪ Water supply service area ▪ Electricity service area ▪ Dump site area ▪ Waste water area

Source: Summarized from a presentation by DTCP, 1998.

Because descriptions of DTCP's database are not made available to the authors, the source of the analog maps, from which these digital map layers were created, is not clear. But a presentation by DTCP's staff left the impression that the map scales are 1:30,000 or larger. It is also unclear whether all the listed layers are available. It is doubtful that a digital layer for all buildings in Chiang Mai is available. Nonetheless, DTCP's database is an important source of data to the GIS for Chiang Mai.

In addition to the above sources, the Land Authority is another important source for digital data sets for the GIS for Chiang Mai. The Land Authority aims to develop a database for land parcels, buildings, and streets which has not been completed.

Government agencies throughout the world have played important roles in collecting and generating digital datasets for GIS. The use of these datasets is primarily influenced by the policies of these governments (Rhind, 1999). Unlike the United States, where data collected by the government is freely available to the public, access to government collected data is rather restrictive in Thailand. Based on the observation of the authors, the Chiang Mai municipality does not seem to have access to the above databases for the Chiang Mai area.

3.3. Data Standard and Metadata

The GIS data standard and format in Thailand is apparently influenced by the Environmental Systems Research Institute (ESRI), as is suggested by the fact that the databases of TEI, NESDB and DTCP are saved in either ESRI's Arc/Info coverage format or ArcView shapefile format. Metadata, or data dictionary, is limited to the description of the coverages or shapefile themes.

In general, metadata provides descriptive information about the producer, content, quality, condition, and other characteristics of a given item (Guptill, 1999, p.677). There are many standards for spatial metadata (Salge, 1999). For example, the Federal Geographic Data Committee (FGDC) of the United States set the following standard categories of spatial metadata: identification information, spatial reference, status information, metadata reference information, source information, processing history information, distribution information, entity/attribute information, and contact information (FGDC, 1994). Technical Committee (TC) 211, Geographic Information/Geomatics, of the International Standards Organization (ISO) also set a standard for spatial metadata which includes mandatory and conditional categories: identification, spatial data representation information, and metadata reference are mandatory; data quality, spatial reference, feature and feature attribute, and distribution are conditional (ISO, 1997).

Among the above databases, TEI and NESDB have a data dictionary that provides basic information of the database. It contains a description of individual coverages, coverage names, feature classes and themes. DTCP may have a data dictionary, but it was not provided to the authors. The description is limited to Arc/Info coverages and shapefile themes. This may suggest that a standard for spatial metadata does not exist in Thailand.

3.4. Management and Organization

Organizational structure and management direction determine the type of GIS organization to be created (Sugarbaker, 1999, 611). In general, there are three types of GIS: enterprise, departmental, and project GIS. A comprehensive, multi-purpose GIS belongs to the category of enterprise GIS. Are the current organizational structure and management direction of the city compatible with an enterprise GIS for Chiang Mai?

The efficient utilization of a new technology in an organization often requires changes in work routines and chains of command, which often lead to the restructuring of the overall organization (Bernhardsen, 1999, 599). GIS is no different from any new technology. Implementation of a GIS in a municipal government like Chiang Mai will indeed change the structure and management of the city government.

4. STRUCTURED ANALYSIS FOR GIS

GIS is not only an assemblage of software packages and computer hardware, it is closely related to organizational structure, data, staff, and other factors. It is naive to think that "implementation of a GIS is something which can be achieved simply by following 'cookbook' instructions and which will then lead to inevitable success" (Campbell, 1999, p.621). Therefore, detailed planning and structured analysis (Marble, 1994) are important for any GIS implementation. A structured analysis starts with an initial design that examines the goals, the constraints, the costs, the benefits, and the feasibility of the system. The second step is a conceptual design for requirements determination and database design. Then, a physical (or detailed) design maps the conceptual design to a specific GIS. The final step is the implementation of the physical design.

According to Bernhardsen (1999, 589-590), most major GIS projects traditionally include the following phases:

1) Assessment and evaluation of the status quo situation; 2) Development of a business concept; 3) Identification and specification of user requirements; 4) Identification and acquisition of data; 5) Cost-benefit analysis; 6) Devising a strategic plan; 7) Choice of hardware and software; 8) Defining and obtaining the necessary expertise; 9) Choosing a GIS supplier; 10) System implementation; and, 11) Operation and maintenance of the system.

Although the above phases are summarized from western experience, they can also be applied to Southeast Asia. Data availability and access, organizational and institutional issues, staffing and training, and hardware and software selection and support are important anywhere.

For the city of Chiang Mai to develop a successful GIS, a structured analysis of the system by an experienced GIS analyst is crucial. Omitting any of the above steps/phases may lead to potential failure of the system.

5. A CASE STUDY ON TAX COLLECTION

Many municipal officials in Chiang Mai believe that the most pressing issue is tax collection. A large percentage of taxes is not collected due to incomplete information on signs, buildings, and land parcels. ADS (1998) reported that the total Chiang Mai city revenue was 101,928,451 Bhat in 1996, 92,450,000 Bhat in 1997, and 99,300,000 Bhat in 1998. If a GIS were used, the city would be able to collect the lost tax revenue and the total revenue would be 161,386,713 Bhat for 1996, 165,000,000 Bhat for 1997, and 165,000,000 Bhat for 1998.

Three types of taxes are collected in Chiang Mai: sign tax, building tax, and land tax. Sign tax is collected on all advertisement signs posted along streets. It is calculated by multiplying the tax rate and the size of a sign. Signs with only Thai characters have the lowest rate, while signs that include foreign languages have the highest rate. Collection of land tax is based on parcel maps. And, building tax is collected according to building maps.

If an integrated cadastral information system (a component of the comprehensive GIS) were available, new buildings and new transactions of land would be updated in real time. Thus, tax could be collected accordingly. But this is not the case in Chiang Mai. It is estimated that a large percentage of taxes are not collected due to the lack of complete information. Let us examine how land and building data are maintained in Chiang Mai.

The province maintains 1:1000 analog parcel maps. The city maintains 1:1000 analog tax maps that include buildings and parcel boundaries, which were originally copied from the provincial parcel maps. Parcels on the tax maps are coded differently than those on provincial parcel maps. It is difficult to link the changes on parcel maps to tax maps. As a result, changes made on the provincial parcel maps are rarely updated on the tax maps. Within the municipal government, the tax collection process involves three departments: the City Tax Mapping Department, the City Building Control Department, and the City Income Department. Building permits and boundaries are first approved by the City Building Control Department, then passed to the City Tax Mapping Department to update the tax maps, which are then used by the City Income Department to collect tax. Transferring data from one department to another is often delayed, due to the tedious updating on analog tables and maps. Figure 1 shows the relationship between the departments for tax collection.

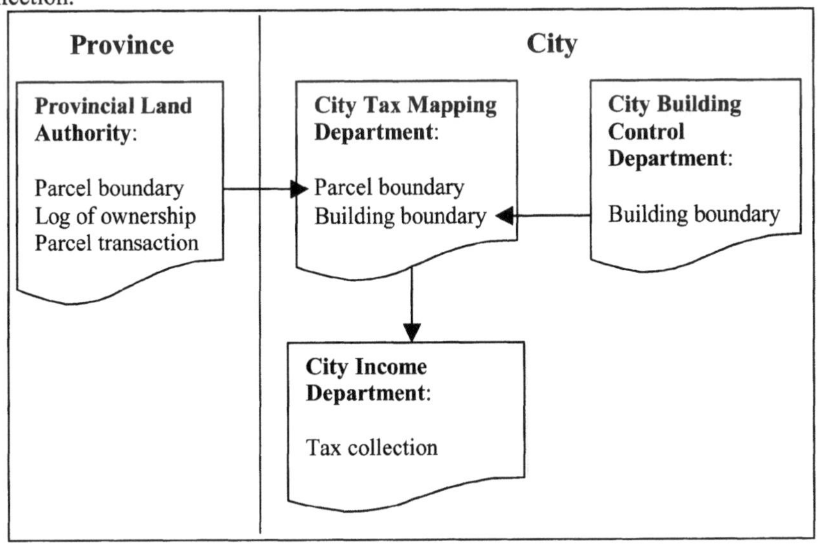

Figure 1: Tax Collection in Chiang Mai

GIS is a powerful tool that can be used to automate tax mapping and provide up-to-date information for tax collection. It requires all the related data be converted to digital format. All parcels and buildings must be uniquely identified in the database.

The parcel boundaries on the tax maps must be updated whenever changes are made on the provincial parcel maps.

There have been many successful implementations of GIS for this purpose. For example, York County in South Carolina uses GIS in departments of tax assessment, zoning, and voter registration (Jain, 1997); Dallas (Texas) Central Appraisal district implemented a cadastral GIS (Murchison, 1998); Chester County of Pennsylvania developed a computerized land records system which provides direct links among the Health, Planning, and Emergency Services, and Tax Assessment, Computing Information Services and Recorder of Deeds (Havan-Orumieh, et al, 1996); the City of Falls Church, Virginia, has a city-wide GIS for zoning, real estate assessment, and automated property owner notification (He, 1996); and land records modernization in Wisconsin (Tulloch, et. al., 1995, 1996). Obviously, it is technically feasible, so the remaining problem is whether the City of Chiang Mai can restructure the city's organizational structure to fit the new GIS technology.

6. SUMMARY, ASSESSMENT, AND POLICY RECOMMENDATIONS

Sustainable development calls for a comprehensive, multi-purpose GIS in Chiang Mai. A GIS can contribute greatly to solving some of the city's pressing problems, such as site selection for a new landfill, management of land records, and land use planning. The enthusiasm of the city's leaders towards GIS suggests that the city may implement a GIS in the near future. However, it must be noted that there exist significant challenges to implementing a GIS in Chiang Mai. Below the authors summarize major challenges and make relevant policy recommendations:

- A GIS brings significant changes to the organizational structure of the municipal government. The city's current departmental structure is rigid, and cooperation among departments seems to be rare. Successful implementation of a comprehensive GIS requires collaborative efforts from all departments. Thus, collaboration among the departments in the municipality needs to be improved, and a new department may be needed for coordinating the GIS development.
- A GIS is a complex information system. A structured analysis by an outside GIS analyst is key to the success of a GIS implementation. The view "we can do it by ourselves without outside support if we have computers" is naive and will inevitably lead to failure.
- Continued GIS education and training is vital for a successful GIS. This is an especially critical issue for Chiang Mai given the fact that few people in the municipal government are knowledgeable of GIS.
- It is important to realize that a GIS requires continued support and maintenance to achieve its full benefits. A long-term commitment from the municipality is critical for the success of the GIS.
- The city should obtain existing databases from the Department of Town and Country Planning (DTCP), the National Economic and Social Development Board (NESDB), and the Land Authority. These databases contain most of the datasets needed by the comprehensive GIS for Chiang Mai. Re-digitizing of

these datasets from original analog maps will be too costly, and thus should be avoided.

- The city should work together with local GIS experts. There are several professors in Chiang Mai University who have expertise in GIS. The ADS Survey & Computer Company in Chiang Mai has extensive experience in GIS.

If the city can deal with the above issues and challenges successfully, a comprehensive GIS for the sustainable development of Chiang Mai can become a reality in the near future.

REFERENCES

ADS Survey & Computer. (1998). GIS for planning, development, management and environment.

Apavatjrut, D. (1993). *Report on GIS demonstration project: Mae Taeng watershed.* Social Research Institute, Chiang Mai University.

Bernhardsen, T. (1999). Choosing a GIS. In P. A. Longley, M. F. Goodchild, D. J. Maguire, D. W. Rhind (Eds.), *Geographical Information Systems* (pp. 589-600). London: John Wiley and Sons, Inc.

Burke, L. (1995). Urban and municipal GIS applications in developing countries – the problems and the potential. *Proceedings of 1995 ESRI User Conference.* http://www.esri.com.

Campbell, H. J. (1999). Institutional consequences of the use of GIS. In P. A. Longley, M. F. Goodchild, D. J. Maguire, D. W. Rhind (Eds.), *Geographical Information Systems* (pp. 621-631).London: John Wiley and Sons, Inc.

Chiang Mai Provincial Statistical Office (CMPSO), National Statistical Office, Office of the Prime Minister, *Statistical Report of Chiangmai: 1996 Edition.*

Chupraphawan, S. (1993). *Designing spatial information system to facilitate agriculture land evaluation in Chiang Mai Province.* Masters thesis. Chiang Mai University.

FAO. (1998). Geographic information systems in sustainable development. http://www.fao.org/sd/Eidirect/GIS/intro.html.

FGDC. (1994). *Content standards for digital geospatial metadata.* Reston, Federal Geographic Data Committee Secretariat.

______. (1995). *Development of a national geospatial data framework.* Washington D.C.: Federal Geographic Data Committee.

Gautam, T.R., and A.D. Gupta. (1997). Integration of geographic information systems and multiple criteria decision making for groundwater management. *Proceedings of AM/FM ASIA '97 and GEOINFORMATICS '97*, Taipei.

Guptill, S.C. (1999). Metadata and data catalogues. In P. A. Longley, M. F. Goodchild, D. J. Maguire, D. W. Rhind (Eds.), *Geographical Information Systems* (pp. 677-692). London: John Wiley and Sons, Inc.

Hastings, P. (1995). GIS experiences in Thailand: institutional GIS issues. *Special session on GIS for sustainable development,* GIS AM/FM Asia '95, Bangkok, Thailand.

______. (1998). Personal Communication between P. Hastings and L. Liu.

Havan-Orumieh, M., M Massucci, M. Lavan. (1996). Reengineering the land records process in Chester County, PA. *Proceedings of 1996 ESRI User Conference.* http://www.esri.com.

He, P. (1996). GIS Implementation at the City of Falls Church, Virginia. *Proceedings of 1996 ESRI User Conference.* http://www.esri.com.

ISO. (1997). *ISO Standard 15046-15, Geographic Information – Metadata,* Version 2.0, 20 January.

Jain, C. (1997). Applying GIS to resolve issues in county government – York County, SC. *Proceedings of 1997 ESRI User Conference.* http://www.esri.com.

Longley, P. A., M. F. Goodchild, D. J. Maguire, D. W. Rhind (Eds.). (1999). *Geographical Information Systems.* London: John Wiley and Sons, Inc.

Marble, D. F. (1994). *An Introduction to the Structured Design of Geographic Information Systems.* (Source Book, Association for Geographic Information). London: John Wiley & Sons.

Mawby, D. (1997). City of Oshawa/City of Chiang Mai Technical Exchange. City of Oshawa.

Mongkolsawat, C., P. Thirangoon, and A. Sowana. (1995). *Spatial decision support system for landuse planning within a watershed area*, presented in GIS AM/FM Asia '95, Bangkok, Thailand.

Murai, S. (1995). *GIS for sustainable development*, presented in GIS AM/FM Asia '95, Bangkok, Thailand.

Murchison, S.B. (1998). Planning and implementing a cadastral GIS: Dallas central appraisal district. *Proceedings of 1998 ESRI User Conference.* http://www.esri.com.

National Statistical Office (NSO), Office of the Prime Minister, *Key Statistics of Thailand 1997.*

NESDB. (1994). *GIS Manual for the Chiang Mai-Lamphun Twin City Project.*

Obermeyer, N. J. (1999). Measuring the benefits and costs of GIS. In P. A. Longley, M. F. Goodchild, D. J. Maguire, D. W. Rhind (Eds.), *Geographical Information Systems* (pp. 601-610). London: John Wiley and Sons, Inc.

Omakupt, M. (1995). *Remote sensing and GIS application for watershed management*, presented in GIS AM/FM Asia '95, Bangkok, Thailand.

Rhind, D.W. (1999). National and international geospatial data policies. In P. A. Longley, M. F. Goodchild, D. J. Maguire, D. W. Rhind (Eds.), *Geographical Information Systems* (pp.767-787). London: John Wiley and Sons, Inc.

Roose, D. (1997). Customizing GIS for economic development. *Proceedings of 1997 ESRI User Conference.* http://www.esri.com.

Salakij, P. (1997). *Land evaluation in the highland watershed using GIS and multicriteria decision making approach.* Masters thesis. Chiang Mai University.

Salge, F. (1999). National and international data standards. In P. A. Longley, M. F. Goodchild, D. J. Maguire, D. W. Rhind (Eds.), *Geographical Information Systems* (pp.693-706). London: John Wiley and Sons, Inc.

Shibasaki, R. and T. Hashimoto. (1995). *Special session on GIS for sustainable development,* GIS AM/FM Asia '95, Bangkok, Thailand.

Smith, N.S., D. W. Rhind. (1999). Characteristics and sources of framework data. In P. A. Longley, M. F. Goodchild, D. J. Maguire, D. W. Rhind (Eds.), *Geographical Information Systems* (pp. 655-666). London: John Wiley and Sons, Inc.

Sugarbaker, L. J. (1999). Managing an operational GIS. In P. A. Longley, M. F. Goodchild, D. J. Maguire, D. W. Rhind (Eds.), *Geographical Information Systems* (pp.611-620). London: John Wiley and Sons, Inc.

Tantayanuson, N. (1998). *Development issues in Chiang Mai.* Chiang Mai Municipality.

Tulloch, D., B. Niemann, S. Ventura, and E. Epstein. (1995). Land records modernization in local governments and its outcomes. *Proceedings of 1995 ESRI User Conference.* http://www.esri.com.

______. (1996). Measuring GIS/LIS process in local governments: land records modernization and its outcomes. *Proceedings of 1996 ESRI User Conference.* http://www.esri.com.

______. (1997). Assessing the outcomes of MPLIS implementation: identifying and describing benefits. *Proceedings of 1997 ESRI User Conference.* http://www.esri.com.

Yeh, A.G., and X. Li. (1998). Sustainable land development model for rapid growth areas using GIS. *International Journal of Geographical Information Science.* London: Taylor & Francis Ltd.

PART IV

THE BUILT ENVIRONMENT'S ROLE IN SUSTAINABLE DEVELOPMENT

金
行

CHAPTER 12

BOBI B. SETIAWAN
DWITA RAHMI

LAND USE AND URBAN GROWTH IN CHIANG MAI: LESSONS FOR MORE EFFECTIVE URBAN DEVELOPMENT MANAGEMENT

Abstract Sustainable cities require appropriate management of the urban environment and an understanding of the dynamics of urban growth. Unfortunately, local urban problems are poorly understood in developing countries and require major research, development, and evaluation efforts to identify effective approaches to problem solving and remedial policies. This chapter identifies several issues of urban development in Chiang Mai and suggests policy interventions to address these. It is argued that to ensure the sustainability of Chiang Mai, it is of fundamental importance to properly manage and control the urban growth of the city. In order to be able to do this, it is necessary to better understand factors, patterns, and consequences of the growth itself. This chapter begins by describing the forces and patterns of growth in Chiang Mai. This is followed by consideration of the consequences of growth regarding both physical and socio-economic dimensions. By assessing the way in which growth and its related problems are managed, the need for increasing the capacity of the local government in managing urban development is highlighted. The chapter concludes by presenting a number of policy recommendations for improving Chiang Mai's capacity to manage its urban growth.

1. CHIANG MAI'S SIGNIFICANCE TO THAILAND

Chiang Mai, is the third largest city in Thailand, and plays an important role nationally as an intermediate size city and the designated service center for the highly productive northern agricultural region (Porananond, 1997). The city is one of the top tourism destinations in the north, with its famed walled ancient city surrounded by a moat in the center of its urban core. In addition, it has become the northern region's center of education, culture, business, finance, real estate, and industrial development. In 1994, there were 104 hotels, 14 hospitals, 79 academic institutions, 22 markets, 9 department stores, and 495 small industries within the city (Sanitation and Environment Division, Chiang Mai Municipality, 1994).

According to Rondinelli (1986), secondary cities in Asia, such as Chiang Mai, have at least four important roles. First, as a propulsive growth center in terms of economic agglomeration; second, as an absorptive growth center in terms of redirecting and diminishing migration streams to the major cities; third, as a service center for accelerated rural development; and fourth, as the service center for both the urban and rural development in terms of social and public facilities. These roles

M. Romanos and C. Auffrey (eds.), Managing Intermediate Size Cities, 237—250.

have been the factors that prompted the rapid development of Chiang Mai over the past two decades.

2. THE GROWTH OF CHIANG MAI

With a 1995 population of about 170,000 in an area of 40 sq. km., the population density within the city of Chiang Mai was 4,259 people per sq. km. (Saowapon, 1998). Table 1 shows the population and density for Chiang Mai in 1980, 1990, 1995, and projected population and density for 2005 and 2015. During the 1980-1995 period, the population increased by 70%, putting Chiang Mai among the fastest growing cities in Thailand.

Table 1. Population projection and population density

Year	*1980*	*1990*	*1995*	*2005**	*2015**
Population	100,146	160,143	170,348	197,481	229,184
Density (per sq. km.)	2,504	4,004	4,259	4,937	5,730

Source: Dept. of Local Affairs, 1990; Saowapon, 1998. * projected

The percentage of the Thai population living in urban areas (28.7% in 1988) is significantly below the average for developing countries (37%); yet, the rate of urbanization has increased greatly over the past two decades. In the Northern Region of Thailand, where the Province of Chiang Mai is located, the percentage of the population in urban areas (21%) is below the national rate, but the annual rate of urban population growth in the greater Chiang Mai area (2.16%) has been significantly above the national average (0.98%). In contrast, the population growth rate inside the city is only 0.64%. This slower growth in the city reflects the transformation of Chiang Mai's periphery to urban uses, as a result of both increasing population and the many emerging non-farm activities such as industry and tourism.

The urbanisation process in Chiang Mai is characterized by three important demographic factors. First, migration seems to contribute more than natural increase to urban population growth. Especially since the fertility rate in Thailand has fallen, migration has tended to become a more important component of urban growth. Second, seasonal migrants in the study area have increased although the absolute size of this group is unclear. This leads to the third demographic characteristic of Chiang Mai: more heterogeneous urban residents in terms of socio-cultural characteristics. If job opportunities and income levels in rural areas continue to be limited, while industrialization and tourism development continue in the urban areas, the disparity in wealth between these two areas will increase, and the three demographic factors of urbanisation will be major agents of change in Chiang Mai.

The urbanized portion of the city historically began within, and then slowly emanated from, the ancient walled city to the north and south. After a bridge was built across the Ping River in 1903, and Chiang Mai was connected to the national railroad in 1921, the areas of urbanization spread to the east and across the river

(Porananond, 1997). Between the early 1800s and 1935, the urbanized area grew dramatically, largely reflecting the eastward expansion. Between 1935 and 1981, the urbanized area expanded not only to the east, but to the north, south, and west, as

well. Commercial land use became more pervasive and became increasingly linear along major travel arteries.

By 1992, the commercial land use continues to expand, with residential development following a related sprawling pattern. Much of the land converted to urban use was highly productive rice land. Moreover, farmland has been replaced by development along major transport routes in a ribbon-like fashion, particularly in areas outside the city center. This sprawling development pattern can be expected to lead to higher levels of energy consumption and environmental resources in the future.

3. THE ECONOMIC CHANGE

Chiang Mai is typical of an intermediate size city in a developing country with a mixed economy. Western capitalism and semi-capitalism are manifested in various forms of formal trade, including manufacturing, banking, and tourism. At the same time, the informal economy provides a variety of jobs for a large portion of urban residents. However, several data sources also indicate a significant change in the economic structure in Chiang Mai. There has been an economic shift toward non-agricultural sectors, particularly trade and tourism, and it can be expected that these sectors will continue to increase their share of local economic activity in the future.

Despite the shift toward the non-agricultural sectors, the local labor force is still largely characterized by low levels of formal education and skills. While there is a small but growing middle-income population, most Chiang Mai residents have low incomes. The latter include: (1) about one-third of the total labor force dependent upon agriculture; and, (2) a growing number of informal sector workers. Data indicate that the number of farmers is in decline because of a number of factors, including the loss of agricultural land, decreasing quality of irrigation systems, and increasing costs of rice production. Since the growth of Chiang Mai appears inevitable, an important question is whether there will be appropriate alternative jobs for those who are forced to, or are willing to leave, the agricultural sector.

It is also important to note the increasing numbers of people working in the informal sector. Although there are no systematic data on the informal sector, evidence indicates that in the last decade the variety and number of informal sector activities have increased. This is significant because, as McGee (1982) argues, socio-historical backgrounds in Southeast Asia are likely to hinder the process of social mobilization from the poor to the middle class, as has occurred in Western societies.

4. LAND USE PROBLEMS AND CONFLICTS

4.1. Land Use and Spatial Structure

The spatial pattern of Chiang Mai's growth can be described by three characteristics. The first is urban sprawl, caused especially by the formal housing development mechanism to accommodate the middle- and upper-income households and by the infrastructure investment by the government. The second is transformation and densification of the existing villages both in the city center and periphery; these are caused especially by the informal housing development mechanisms to accommodate the middle- and low-income households. The third is the growing number of marginal settlements, in the form of squatters; these accommodate low-income households, particularly migrants in informal sector activities. This pattern has a number of implications for the city's natural, social, and economic environments, and therefore threatens the sustainability of the area.

As a consequence of the need for housing and other urban facilities, combined with the growing tourism industry and related services, developable land is becoming increasingly scarce in Chiang Mai. At least five types of land use conflict and stresses are related to these changes. These are: (1) significant loss of agricultural land; (2) uncertainty of the designated green zone; (3) increasing amounts of idle land; (4) increasing amounts of marginal land; and, (5) land fragmentation within existing settlements. These stresses are manifested in the haphazard and scattered pattern of urban growth on one side, and densification of existing urban settlement on the other side.

At present, it is clear that the pattern of recent urban growth in Chiang Mai has occurred along the transportation networks. Land along the major roads is developed to a depth of 10 to 200 meters from the road, and the land farther away from the road often remains unused, leaving pockets of empty land between the road networks. The land becomes developed only after roads are built through or access roads are built into these pockets of empty land. This type of ribbon development is common to urban areas of Thailand, and can be seen in other cities like Bangkok (Setchell, 1995).

4.2. Land Use Management and Land Values

Inadequate land use management has been linked to a number of negative socio-economic consequences. In Chiang Mai, six problems related to land reflect the distribution of wealth in the area, including: (1) the structural changing of land values and ownership; (2) large increases in land prices; (3) increasing land speculation; (4) pressure on traditional land; (5) uncertainty of the future for farmers; and, (6) more difficult access to land, particularly for the poor.

No detailed study has been found that explains the effects of urban development on the land ownership patterns in Chiang Mai. However, the large increases in the price of land have occurred not only in the tourist areas but also on the urban-rural fringe and the nearby rural areas. Table 2 shows the relationship between land use and land value in the city of Chiang Mai, and the change in land prices over 10 years, from 1979 to 1989.

Table 2. Changes in Chiang Mai land values by land use, 1979-1989

Type of land use	*Land value (Bath/4m^2)*		*Land price change*		*Rate of change per year*	
	1979	*1989*	*Bath/4m^2*	*%*	*Bath/4m^2*	*%*
Residential	458.93	11,312.42	+10,853.49	2,364.96	+1,085.35	236.50
High residential level	1,240.28	14,351.85	+13,111.57	1,057.15	+1,311.16	105.72
Medium residential level	450.44	11,287.28	+10,836.84	2,405.83	+1,083.68	240.58
Low residential level	298.73	9,182.93	+8,884.20	2,925.06	+1,338.74	292.51
Commercial	2,659.24	23,970.56	+21,311.32	801.41	+2,131.13	+80.14
Facility & utility	457.68	13,845.09	+13,387.41	2,925.06	+1,338.74	292.51
Public health	677.59	11,681.53	+11,003.94	1,623.98	+1,100.39	162.40
Industry	330.15	8,333.83	+8,003.68	2,424.26	+800.37	242.43
Government office	481.18	11,638.89	+11,157.71	2,318.82	+1,115.77	231.88
Education	204.46	8,662.97	+8,458.51	4,137.00	+845.85	413.70
Temple	692.39	19,426.43	+18,734.04	+2,705.71	+1,873.40	270.57
Public space	265.88	7,745.60	+7,479.72	2,813.19	+747.97	281.32
Agriculture	118.54	5,257.50	+5,138.96	4,335.21	+513.90	433.52

Source: Department of Urban and Regional Planning, Faculty of Architecture, Chulalongkorn University. Undated.

It appears that the lack of meaningful urban land policies have led to high levels of speculative investment that have burdened Chiang Mai's development. Large changes in land prices took place in Chiang Mai over the 10 year period. Land for commercial use increased by almost 800% or about 80% per year on average. Increases for other uses were even greater. This suggests that external capital has caused two unanticipated results: distortion of the land market and tremendous pressure to convert agricultural land to urban uses.

4.3. Housing: Second Homes and Informal Settlements

A number of housing projects for low, middle, and high-income persons have been built over the past two decades by the government and private sectors in the greater Chiang Mai area. The rising cost of land in the urban core has made most formal housing projects infeasible in that part of the city, and has led to the conversion of agricultural land, higher infrastructure costs, proliferation of squatter settlements, and the degradation of the natural environment through air and water pollution. The absence of adequate land use controls brings about a condition in which land development decisions are left entirely to market processes. Given that land is an essential factor in housing development, access to affordable land is an important determinant of the failures and the successes of housing development. In Chiang Mai, the failure of formal low and middle-income housing developments is the result of the uncertainty of land supply, while the strength of informal housing reflects flexible, informal access to land. However, Chiang Mai's current approach to informal housing is not sustainable, and includes a number of problems such as a growing number of marginal settlements, overcrowding, and inadequate urban services.

An assessment of housing development in Chiang Mai reveals how urban growth and housing development relate to and influence each other. Formal housing has followed a sprawl and scattered pattern of development, while the informal housing has led to intensification and fragmentation within the urban core.

Informal settlements are typically found in developed portions of cities. A study conducted by Community Architects for Shelter and Environment (CASE) (1998) counted at least 38 informal settlements in the various districts of Chiang Mai. These settlements housed some 7,500 people. About 30 of the settlements are relatively small, ranging in size from 10 to 70 households, while the largest settlement had about 200 households. Most of the residents are the urban poor. They earn less than the Thai minimum legal wage of 142 Baht per day, most working irregularly, on a daily-wage basis. They do not have the financial resources to construct basic infrastructure such as piped water connections, sanitary sewers, waste disposal, and drainage facilities. With little assistance forthcoming from outside the settlements, these facilities are often inadequate, resulting in unhealthy living environments. Continuing rapid urbanisation, coupled with growing poverty and income inequality, could give rise to higher rates of slum formation in the coming years, and thus expose even more Chiang Mai residents to unhealthy and degraded environmental conditions.

5. PUBLIC SECTOR MANAGEMENT

Urban land is a limited resource, and the use of urban land is shaped by many individual and group decisions. The problem is complex because the dynamics of urban growth exist as an overlapping of competing activities among diverse interests. Particularly for housing, conflicts among groups tend to exclude many people from affordable access to land. In this context, the key issue is whether land use should be managed directly or indirectly by the public sector planning and land use policy intended to ensure that land is used efficiently and equitably. Indeed, the challenge of urban management is to ensure a balance between efficiency and equity in the use of the urban environment (Bartone, 1991).

Since land is a basic resource for urban development, local government requires the ability to manage, to some degree, the ownership and use of land for the benefit of all parties. Unrealistic spatial planning, building regulations and standards, and inefficient public infrastructure provisions all indicate a lack of real understanding of local urban issues. The strategies and instruments for urban management that have been employed often seem to be formulated with little or no reference to the market. It is argued, therefore, as has also happened in other urban areas in Thailand, that the knowledge and capabilities necessary to intervene in urban development are inadequate at the local government level.

In Thailand, there is a serious problem with the lack of coordinated urban land management and the limited capacity of local governments to manage the complexity of urban development. In the case of the more complex urban problems, it also has been questioned whether local governments and planners understand clearly the dynamic nature of urban growth. There is a need for local governments to reappraise their roles, policies, programs, and instruments to assess whether they are able to meet the fundamental planning and management objectives for their community. As Devas (1993) argued, in theory, the government has a considerable range of tools to manage fast-growing cities. Public authorities or local governments must develop the will to evaluate the performance of the planning instruments they do adopt, while establishing new enabling mechanisms which will solve the problem of urban growth and its related consequences.

The practice of urban planning and management in Chiang Mai has been marked by confusion over the national policy and legal support for urban planning. This has caused uncertainties and overlap regarding the involvement of several government agencies in urban management. Many government agencies have difficulties sharing their mandates and obligations concerning urban development, and even seem to have a divergent vision and interpretation of their roles in urban development and management (Douglass 1989). In addition, the administrative status of Chiang Mai makes urban management practices much more complex since the provincial, district, and municipality levels have a shared role in urban management and development. (See Romanos, et.al., Chapter 2, Section 3.)

Under these circumstances, the local government has used several planning instruments such as zoning, location permit approval, building permit approval, land consolidation, and environmental impact assessment procedures. In practice, however, three obstacles have been encountered: (1) poor coordination and unclear mandates among local government agencies; (2) lack of human power; and, (3) lack of financial and technological support. There is the need for the local government to limit these obstacles, but perhaps of much greater significance is whether the local government could develop a better and clearer vision of the dynamics of urban development, or give more attention to the actual mechanism or urban growth and the land market. In other words, the lack of clear and appropriate objectives concerning government roles in urban development hinders efforts toward a better urban environment.

This has been especially true with respect to housing where most government interventions have not been directed to the issue of affordable access to land as a fundamental concern in housing development. Policies and programs to promote more efficient and effective use of land have been neglected. Rather, the existing policies tend to support, either intentionally or unintentionally, more urban sprawl and distortions in the price of land.

6. LESSONS FOR MORE EFFECTIVE URBAN DEVELOPMENT MANAGEMENT

In the context of sustainable development, managing urban growth is complex and there is no single means to solve the problems. As McGee (1982) suggested, the real challenge is how to convince the local governments of the need to adopt innovative policies that are connected to local problems and potentials. With this in mind, a number of policy initiatives have been organized into four areas related to urban management considered critical for the sustainable development of Chiang Mai.

6.1. Urban Development Management

Taking a broader view of urban issues within the National Development Plan
Considering the inevitable growth of the urban areas in Chiang Mai, with significant implications for the sustainability of Chiang Mai as a whole, a broader view of urban issues in future development planning is very important. The National Development Plan must give more attention to declining urban environmental quality while, at the

same time, encouraging the protection of the natural resource base from expanded urban impacts.

Balancing urban development in the regional context
In a regional context, more balanced and equitable development should be implemented with emphasis on: (1) promotion of more balanced development among sub-regions; (2) prioritization of urban development in Thailand based on the functional importance of the city in a regional framework; and, (3) maintenance of focus on small and intermediate cities. Thus, the trend toward developing "greater Chiang Mai" should be carefully re-examined.

Fostering and strengthening the capacity of local government
In this respect, strengthening existing local institutions is better than building new institutions. Education and training programs on urban management for local government staff should be emphasized.

Adopting more dynamic approaches to urban management
Urban management should not limit itself to the production of spatial plans. It should also consider a wide range of alternatives for urban management. For example, local governments could employ basic urban infrastructure investment or tax systems as effective tools for guiding urban growth. An incremental approach based upon degree of densification is needed for existing settlements. Here, household/private service can be replaced with government/public service in currently dense areas.

Involving people in the urban management and development processes
Urban policies and programs should enable the private sector and community organizations to contribute more to urban development and management. The role of local community groups could be incorporated, for example, to monitor and guide informal development mechanisms.

Developing innovative policies to increase local revenue
Decentralization of urban management could be fostered by maximizing possible sources of financing or mobilizing of local resources, such as betterment taxes, borrowing, local charges, and public service cost recovery.

6.2. Spatial Growth Management

Promoting more compact urban patterns
It is now accepted that compact development can make cities more liveable and environmentally sound. Urban development programs can reduce inefficient urban extension by simultaneously increasing the density of areas that are already planned and serviced. High density control, progressive taxation, and incentives toward using vacant land within the city could be implemented.

Fostering more integrated urban-rural linkages
Special attention should be given to the increasing numbers of temporary migrants and commuters. There should also be careful monitoring of their impacts on both urban and rural areas.

Acknowledging the possible negative environmental impacts
Urban development, as a means to increase economic development without paying attention to the possible negative impacts on the environment, should be avoided. The loss of productive agricultural land can be minimized by careful control of urban growth.

6.3. Urban Land Management

Providing for the more equitable distribution of urban land
Efforts should focus on providing more opportunities for the needy, while at the same time, protecting the rights of others. Programs of land consolidation, for example, should facilitate greater access to urban land by having on the market more small, serviced plots in locations suitable for middle- and low-income households.

Strengthening land administration while maintaining the flexibility of the land tenure system
Land administration needs to be improved. More efficient systems of land mapping and registration are crucial. This will require simplifying administrative, institutional, and legal procedures of land administration. Making affordable land available for middle- and low-income groups should be maintained.

Improving the land and building taxation systems
Since the present land tax system is not sufficient to implement such taxes in a rapidly expanding urban area such as Chiang Mai, it is suggested that the mechanism of land taxes should be improved (especially the price/value assessment). This will counter increasing land speculation and could be used to generate local revenue for financing services.

6.4. Housing Development

Developing housing that focuses not only on product, but also on process
In general, programs should aim to strengthen the essential needs of housing development, particularly access to land and finance. In particular, the incremental and self-help process of the informal housing development within the existing settlements should be encouraged.

Targeting specific groups with housing programs
The development mechanism of the formal housing for middle- and upper-income households should be re-examined. Limitation of the formal housing development is needed to avoid the urban sprawl that causes the loss of agricultural land and the degradation of the environment. Special attention should be given to the increased

housing supply for low-income households. Community credit systems for housing development that operate at a community level could be developed, and the process can be enforced by appropriate standards and regulations.

Increasing the flexibility and efficiency of formal housing development
With the growth of the middle-income group in Chiang Mai, the formal housing development sector should be supported by an environment that enables the public and private sector to be more flexible and to operate more efficiently. Programs that could be implemented include: modification of standards, simplification of administrative procedures, better coordination of urban infrastructure development, and the offering of incentives for in-fill development within the city.

Enhancing the rental housing system as a means of increasing housing supply
Programs such as credit systems, technical assistance, appropriate standards, and a simplified building permit procedure should be developed. At the same time, regulations to meet the minimum housing quality standard should be enforced. Builders should be encouraged to share the cost of a better urban infrastructure.

7. CONCLUSIONS

The integration of religion and culture in the daily life of the Thai people has drawn the attention of the world to Thailand. However, Thailand is now also undergoing rapid modernization and urbanization. This change presents Thailand with a conflict between a highly traditional culture based on a productive agricultural society and a modern culture based on urbanization and tourism. Clearly, Thailand faces some difficult choices – choices of development, choices of land use, choices of traditional vs. modern value systems and activities, and choices of public policy. From the point of view of sustainable development for the whole region, increasing urbanization, tourism, and modern economic development, an important question arises: in what ways does the rapid urban growth create negative impacts on both the physical/natural environment and the culture of the Thai people?

The evidence from Chiang Mai suggests that Thailand is being stressed as a result of urbanization. The combination of haphazard urban growth, land use conflicts, a deteriorating urban environment, and various socio-economic problems related to land threatens to disrupt the sustainability of Thailand. However, an assessment of the role of the local government in managing urban growth indicates that the problems are not just from rapid growth, but also from the way in which the growth is guided and managed. In addition to institutional and financial weaknesses, an important factor is the lack of a clear vision concerning the local government's role and involvement in urban management. The critical trends in growth patterns that endanger the whole region must be managed carefully. A sustainable region requires the proper management of the urban environment based upon a holistic understanding of the linkages between urban and rural environments. Clearly, there is a need for the central and local governments in Thailand to take a broader view of urban issues in the future.

REFERENCES

Bartone, C. (1991). Environmental Challenge in Third World Cities. *Journal of the American Planning Association, Autumn, Vol. 57(4),* 411-415.

Community Architects for Shelter and Environment (CASE). (1998). *The Invisible People: a Census of the Slum Dwellers of Chiang Mai.* Chiang Mai,Thailand: CASE.

Department of Local Affairs. (1990). *Some Information on Municipalities and Sanitary Districts* (in Thai), Bangkok.

Department of Urban and Regional Planning, Faculty of Architecture, Chulalongkorn University. Undated. *The Role and Impacts of Land and Housing Subdivision Development Projects on the Urban Change of Chiang Mai.*

Devas, N. and Rakodi, C. (1993). *Managing Fast Growing Cities: New Approaches to Urban Planning and Management in the Developing World.* New York: Longman.

Douglass, M. (1989). The Environmental Sustainability of Development: Coordination, Incentives and Political Will in Land Use Planning for the Jakarta Metropolis. *Third World Planning Review, Vol.11(2),* 211-238.

McGee, T. G. (1982). *Labour Markets, Urban Systems, and the Urbanization Process in Southeast Asian Countries.* Honolulu, Hawaii: East West Centre.

Porananond, U. (1997). Housing for the poor in Chiang Mai, Thailand. In Schneider, H., and Vorlaufer, K. *Employment and Housing: Central Aspects of Urbanization in Secondary Cities in Cross-Cultural Perspectives.* Brookfield, IL: Avebury.

Rondinelli, D. A. (1986). Metropolitan Growth and Secondary Cities Development Policy, *Habitat International, Vol.10(1/2)*, 263-272.

Saowapon, Chaithawat. (1998). *A Perspective on the Water Related Issues in the City of Chiang Mai,* unpublished paper.

Sanitation and Environmental Division, Chiang Mai Municipality. (1994). *A Survey for Garbage Collection Report,* November 4 (in Thai).

Setchell, C. (1995). The Growing Environmental Crisis in the World's Mega-Cities: The Case of Bangkok, *Third World Planning Review, 17(1).*

EXCHANGE

CHAPTER 13

BRENDA SCHEER
DAVID SCHEER

TOWARDS A SUSTAINABLE URBAN FORM IN CHIANG MAI

Abstract This chapter compares the morphology of the traditional central areas of Chiang Mai with the emerging morphology of its suburbs. In particular, the morphological analysis is used to compare elements that could affect long-term sustainability of this rapidly growing region. The thesis is that traditional settlement patterns, created when energy use was very low and when local production dominated, have much to offer as models of environmental and social sustainability. Traditional forms promoted social cohesion, walking, nearby agricultural work, local economic interactions, and an evolved response to climate. The paper compares the settlement patterns in qualitative and quantitative terms. It examines one particular central city neighborhood in particular, using principles of sustainability. At the same time, the analysis shows the patterns of change in the central city and suburbs that are a response to new development priorities. Historic preservation management is discussed as a tool of development, protection, and tourism.

1. A FRAMEWORK FOR SUSTAINABLE URBAN FORM

This chapter reports on the traditional and new urban forms of Chiang Mai, Thailand. The research was conducted over a six-week period in July and August of 1998, as part of the Joint Study. The purpose of our particular study was to compare the traditional urban settlement form of the city with the newer patterns that are now being built in response to its rapid growth. In particular, we were interested in their comparative sustainability.

Fast growing cities are particularly vulnerable to non-sustainable building practices: overtaxing of the infrastructure; construction of new buildings that require excessive resources to build and maintain; building in distant areas that are difficult to access and may destroy farmland or habitat; and destruction of traditional cultural settings (e.g. neighborhood coffee shops) that support a particular quality of life. Thus, a sustainable urban form is one that can adapt well to the requirements of growth and change without destroying natural resources and traditional culture in the process.

The idea that different urban forms can have very different degrees of sustainability is well-established (Frey 1999; Jenks et.al., 1996). It is also established that there is not a single model of sustainable urban form that is applicable in all situations (Guy and Marvin, 2000). In fact, there is some controversy regarding even the simplest assumptions about sustainable form; however, there are some widely

M. Romanos and C. Auffrey (eds.), Managing Intermediate Size Cities, 253—271.

accepted principles of sustainable urban form that might serve as criteria for evaluating particular urban form alternatives:

1) *Create and preserve higher density, compact forms.* Up to a point, greater density and compactness allows more efficient provision of urban services, increases the potential for walking and public transportation, and creates community cohesiveness and sense of place (Frey, 1999; Newton, 2000; Buxton, 2000).
2) *Preserve the urban region's agricultural land, water systems, recreational areas, and fragile ecologies.* This suggests an overall limit of urban development or clustered development that avoids sensitive land (Newton, 2000).
3) *Provide a mix of land uses.* Segregated land uses require more frequent and longer trips, isolate people without vehicles, create inefficient "day" and "night" districts, and inhibit the formation of complex neighborhoods that promote integration, social cohesiveness, and attachment to place (Masnavi, 2000).
4) *Preserve the existing built form.* The preservation of infrastructure and of buildings is more efficient than wholesale urban renewal and restructuring. It also preserves the sense of place and cultural continuity that is important for quality of life (Falk, 1993; Heath, 2000).
5) *Provide open spaces.* Parks and green spaces promote clean air, exercise and recreation, animal habitat, and urban cohesiveness. They help moderate local climate, and encourage the preservation of natural areas such as rivers and forests (Alberti, 2000).
6) *Encourage moderate parcel sizes.* Moderate parcel sizes impel small-scale changes that are less expensive, easier to implement, and require less disruption to the physical environment and social fabric than large scale change (Frey, 1999; Cuff, 2000). There is more inherent flexibility and adaptability of parcels that are a moderate scale (Scheer and Ferdelman, 2001).
7) *Limit buildings to a moderate size.* Although significant environmental technology (e.g. solar panels) and specialized design can mitigate these effects, relatively large buildings are usually less energy efficient, create urban temperature sinks, require more sophisticated technology to operate, and are built from materials that require high energy in manufacturing (McCarthy and Battle, 2001).
8) *Provide a mix of building types, sizes and ages.* A mix increases land-use adaptability, moderates social and economic stratification, and provides for economic flexibility (Jacobs, 1961).

These principles have generally been developed in the context of American and British cities, where the key issues of sustainable urban form often have to do with mitigating the effects of unsustainable, sprawling urban forms through redevelopment, densification, and creation of alternative centers. In Chiang Mai, sprawl is only one issue among many physical challenges arising from rapid urbanization. There is also an issue of rapid destruction and redevelopment of the historic core and the provision of adequate urban services such as water and transportation.

The major challenges of sustainable growth in intermediate-size Asian cities are reflected in the two trends of densification of form and dispersal over the countryside. Since growth is inevitable and likely to be very rapid, it is critical that a balance be achieved between these two poles. On the one hand, there is the danger of becoming severely over-developed within the city center, as seen in the model of the stages of development of Bangkok that Chiang Mai seems to mimic (Cohen, 1999). On the other hand, disbursed development in the western style is also unsustainable, requiring greater reliance on cars, encouraging single-use districts, and eating up the land in the countryside.

Most of the Thais we interviewed were horrified at the thought that Chiang Mai would become, like Bangkok, overbuilt, crowded, polluted, and unlivable. Concern for the dispersal of urban development (sprawl) was not much in evidence, although to our eyes it also was an issue, primarily because of its demands on transportation. Transportation options in Chiang Mai are limited. Few people have private vehicles and there is no mass transportation. Public jitney service (song tao) is widely available inside the compact city, but the most popular form of urban transport is the motor scooter.

In contrast to Bangkok, Chiang Mai is a much smaller urban agglomeration with much more tractable and predictable physical growth patterns. The provision of infrastructure and some limited measures of growth control would still be very effective sustainability tools for urban management.

2. TRADITIONAL FORM AND SUSTAINABILITY

The thesis of this study is that the traditional form of Chiang Mai has much to teach us about sustainability. Traditional settlements developed in eras where energy use and industrialization were significantly less than today and thus may be expected, on the whole, to support lower levels of energy consumption and lower levels of pollution (Dumreicher,et.al., 2000). Traditional forms promoted social cohesion, walking, nearby agricultural work, local economic interactions, and an evolved response to climate. On the other hand, the physical arrangements of the traditional environment are predicated on a smaller population living in more primitive conditions and in a more closed economic system than are possible in a modern city. Traditional forms did not need to support industrial or global economic activity. Traditional cities did not have electricity, telecommunications, or motorized vehicles, and they often did not have sewage treatment or even clean water piped to the house, all of which are necessary for a modern city. Therefore, the lessons of sustainability that traditional urban forms teach us must be understood in the context of their applicability to modern conditions, rather than as a return to old ways.

What is urban form? One of the basic theories of urban morphology (Conzen, 1968) is that the urban form is a product of the street formation, land use, buildings, and the land subdivision (lots). Others, most importantly Caniggia and Maffei (2001), hold that urban form derives primarily from building types, which in turn generate the dimensions of lots and streets. In either method, a key resource is the analysis of the urban plan, specifically the plan of lots, buildings, and streets. This kind of analysis involves using maps, plans, and field investigations as well as historical studies to establish current patterns of urban development as well as its evolution over time. Urban morphology provides a way to read a culture and its transformation through measurable, physical evidence. One intention of this work is to provide an analysis of the trends, so that appropriate new responses can be crafted.

Urban morphology takes into account a range of scales, from the overall urbanized area down to the individual building. It is particularly useful in identifying relationships between individual building types and larger-scale urban forms. Usually, a morphological study will look at a particular place in its evolution over time, using data from different time periods.

Like most Asian cultures, the Thais did not create accurate, scaled maps as the Europeans did, beginning as early as the Roman era (Winichakul, 1994). The first scaled map of Bangkok was completed by an Englishman in the late 19th century. Even today, accurate maps are difficult to come by and historical information on the physical form of Chiang Mai had to be gleaned from legends, twentieth century aerial photos, old photographs, pictorial representations, and interviews. Thus, our description of the "traditional" form is more a description of the parts of the city where the old plan still persists to a greater or lesser extent.

3. ELEMENTS OF URBAN FORM IN CHIANG MAI

We begin with the form of the city as a whole. Chiang Mai (which means "New City" in Thai) was established in the 13th century as the capital of the Lanna people, a culture that occupied an area of southern China and northern Thailand. The historic center of Chiang Mai was walled and nearly a perfect square, although the slight kilter of the square was apparently intentional (Witti, 1998). In the walls were five gates: a single northern, eastern and western gates, and two southern gates (Figure 1). It was, and still is, surrounded by a non-navigable moat, which was fed by the nearby river. Between the walled city and the river, it is likely that rice farming took place. The site of Chiang Mai, with a river to the east and rising hills to the west, was considered propitious and the story of its founding according to traditional principles of feng shui is recorded in *The Chiang Mai Chronicle* (Wyatt, 1999). The city suffered a cataclysm in the eighteenth century and was, for a time, abandoned. Re-populated in the 19th century, it has grown slowly until recent decades, when the growth has been explosive.

By 1954, the total urban conglomeration of Chiang Mai was between two and three times the land area of the walled city alone. Tremendous population growth and land consumption have prevailed since that time, and a freeway-like loop road was partially built by 1978. In the entire urban form today, the old city has been subsumed in an urban conglomerate, yet its walls and moat are still clear.

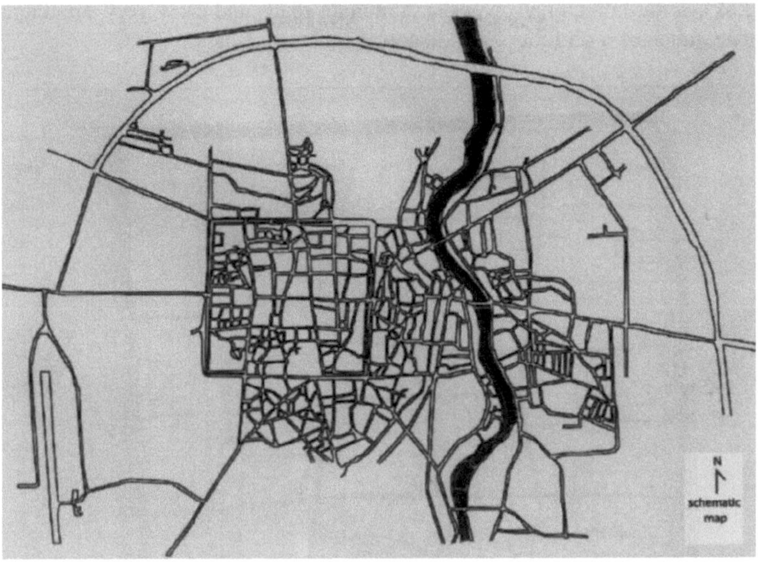

Figure 1: Chiang Mai Central City

3.1. Traditional Street and Block Form

The city shows definite signs of being influenced by Chinese capital city planning standards, which are described by Steinhardt (1999). Following the Chinese tradition, the palace and important religious center were nestled against the northern wall of the city, along the central north-south axis (Figure 2). In this location in Chiang Mai, there was a palace that was destroyed and a temple (or Wat) that partly remains. The grounds have been replaced by a school and a prison. Many of these changes were imposed by the Thai regime centered in Bangkok after it assumed dominance over the Lanna kingdom in the 19th century. On the west side of Chiang Mai, the most important temple of the city, Wat Phra Sing, was built where the ideal Chinese capital city model would call for a large market. As the grounds of a Wat can often function as a community center, gathering space and market, the original role of this space in the city has not been entirely lost.

Except for a few main streets connecting the gates, the Chinese capital city is not usually a gridded city. The streets (soi) of the city branch off the few main streets, usually connecting to another main road, but somewhat erratically. Branching off these minor streets are even smaller dead-end streets providing access to the interior of the block. The pattern of streets inside the walled city and directly outside the walls follows this tradition. Cohen (1985), in his study of Bangkok, argues that the development of multiple, branching soi off of linear main arteries and highways is the traditional method of urban expansion in that Thai city. In Chiang Mai (and

Bangkok), soi are much narrower than the main roads and carry very small amounts of traffic, much of it on foot or motorscooter rather than cars.

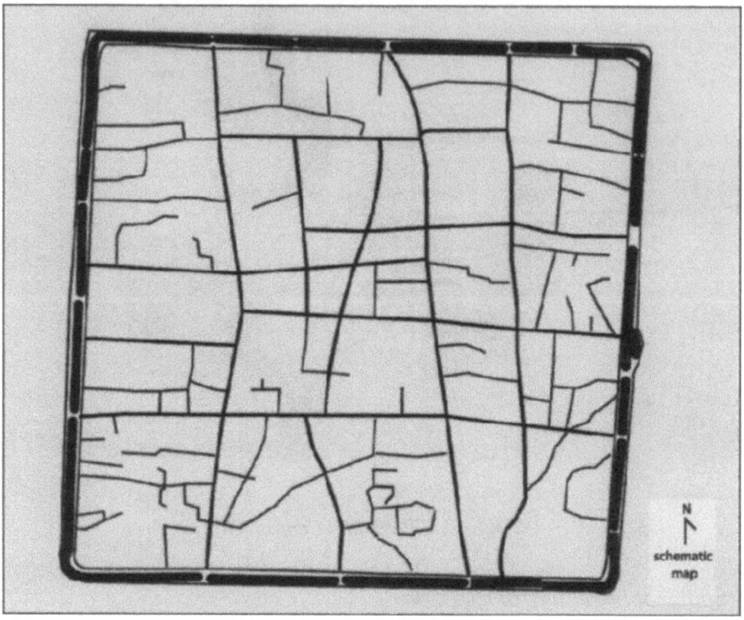

Figure 2: Walled City of Chiang Mai

Because streets are very persistent urban formations, we can assume that many of the current streets are almost as ancient as the city itself. Chiang Mai's present mix of institutional, commercial, and residential uses evolved after the 14th or 15th century. Before that time, only religious and governmental buildings and housing for the aristocracy stood within the city walls (Vitti, 1998). In comparing aerial photographs from 1956 to 1989[1], the most striking change is the replacement of small footprint buildings with larger buildings and paving and the concomitant loss of vegetation. In 1982, perhaps in recognition of the damage to the traditional fabric from these changes, a Cultural Conservation Zone was established within the walls (Narumit-rekagarn and Tansukanan, 1988). This and subsequent regulation determines a maximum height (four stories) and land use allowed and requires buildings to maintain local architectural identity, which has come to mean the superfluous addition of a galare (an ornament derived from crossed roof rafters) on the roof of otherwise modern buildings.

For a closer look at traditional form of the building types, streets, lots and block, we carefully examined a small neighborhood in the Chiang Mai Night Market District directly outside the walls of the city, between the city and the river. This neighborhood exhibits a pattern of street and lot development that is clearly similar to all the traditional lot and street patterns within the walls of Chiang Mai. Fortunately, we were able to obtain an accurate and complete map of this neighborhood, which we used for subsequent analytical work (Chiang Mai Municipality, 1998). The neighborhood has undergone change in the last twenty or thirty years, yet most of the newer structures fall into categories that might be termed "traditional types".

Figure 3 locates this neighborhood in the context of its immediate surroundings, the Night Market District. We see the old wall of the city and the now familiar pattern of major streets and soi. A clear division in urban pattern occurs near the river, where larger blocks are apparent. This is the newer pattern of development near the old city. It contrasts strongly with the traditional, organic form, with its tiny lots, small footprint buildings, and wiggly soi. In this area are large market halls and hotels, structures that are not possible within the tightly woven fabric of the old city. These patterns suggest that the small creek visible today may once have been the true boundary of the urbanized area and that the tight, traditional development pattern never existed between it and the river, probably because of the flooding that occurs regularly.

The map series shown in Figure 4 shows the layout of the neighborhood itself. Along the main road (Thae Pae Road), the shop houses are concentrated, with only a few breaks in the continuous street front wall. Shop houses have been in this location for one hundred years or more, but these were built as replacements for wooden structures of the same type, which are visible in photographs of the street from the 19th century. This street is the major east-west connection through Chiang Mai and connects to one of the few bridges, so it is always full of traffic. The shop houses create a solid urbanized wall, concealing the true nature of the neighborhood behind it from the tourists and taxi drivers.

The smaller streets, soi, are known by a number and the name of the major street from which they run, an indication of the development of the street pattern, i.e., growing out from a few planned streets. Turn the corner from the traffic, noise, and pollution of the main road and you have reached a quiet eddy in the apparent bustle. Only steps from the shop houses and hotels, life goes quietly on in the neighborhood behind. The soi are narrow streets with dead end sub-soi branching off them. On the inside of the block, undeveloped land still exists in the heart of the city. The soi and sub-soi are not arranged in an orderly system because they were never planned systematically. Further development of the interior of the block, for example, awaits the development of a new sub-soi to serve this land, which is subdivided but has no access.

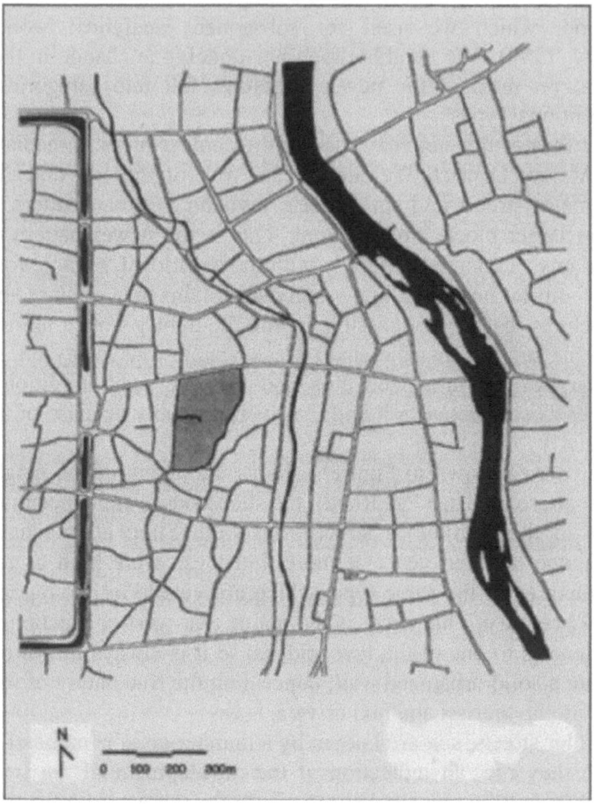

Figure 3: Neighborhood Location near Night Market District, Chiang Mai

As the maps in Figure 4 indicate, there is a great variety in the houses and land uses of the study area. None of the lots is a regular rectangle, and all are rather small compared to the footprint of the buildings that sit on them. In many places, the lots contain more than one building, in the fashion of a rural compound. In newer construction, the buildings are larger, but they tend to honor the close fit of the lot and the building.

3.2. Traditional Building Types

Traditional Lanna Thai house types provide the basis for most private residences in our study area. These types are well documented in Thai architectural literature (Vadhanasindhu, 1999; Charernsupkul, 1978). The most common type of traditional urban house is built of wood and is raised off the ground six to eight feet (Figure 5). The house has a simple peaked roof. The most beautiful examples are built of teak post and beams, with infill of teak panels. Window frames and shutters are often beautifully handcrafted. Houses of the less well-to-do are similar in layout but are constructed of bamboo. In the countryside, the ground level of the house is often completely open to the weather and provides a shaded working space for the family. In the city, it is more common to find that the ground level has been walled in with concrete block or other permanent material. Sometimes this area under the house is left open. It provides a handy garage for the ever-present motor scooter or small car.

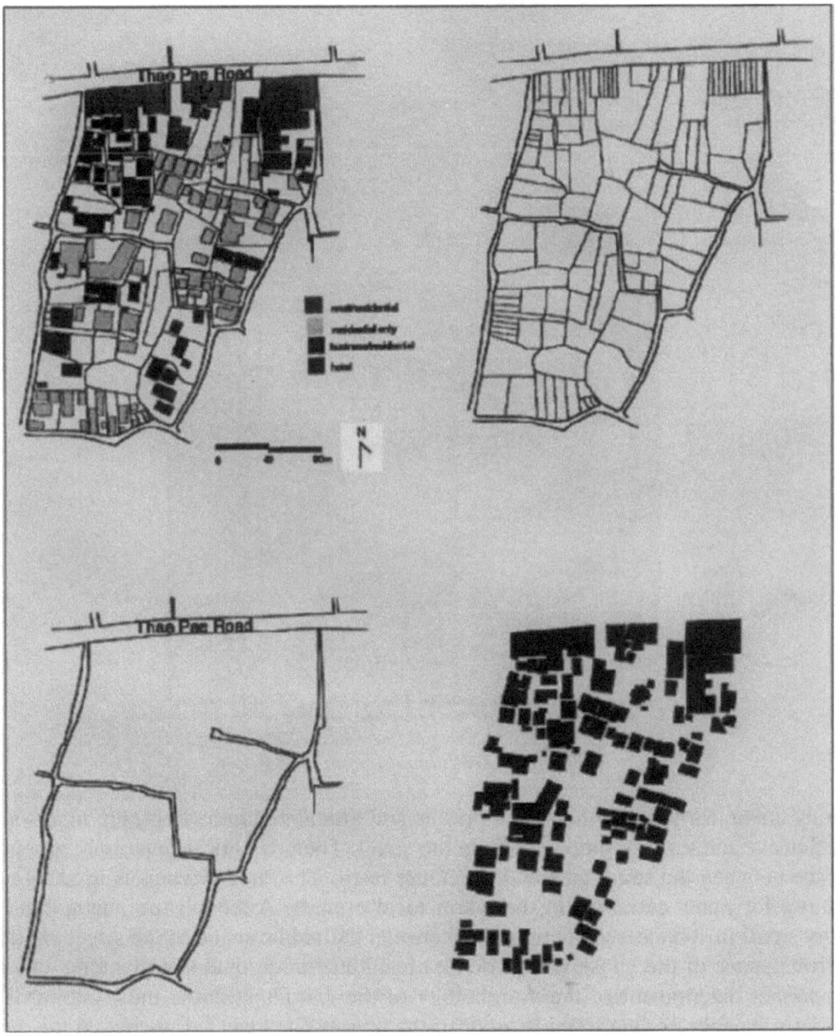

Figure 4: Study Area Neighborhood
Showing: Land Use (upper left); Lots/land subdivision (upper right);
Street (Thae Pae Road) and soi (bottom left);
Figure Ground –building footprints (bottom right)

In the country, the Thai house often comprises two buildings: a regular home and a "rice house", which has both ceremonial and storage functions. Compounds of several houses occupied by a large extended family are also quite common.

Of particular interest for the morphology of the study area is the planning of the typical house compound. The Lanna Thai types that are the basis of the houses in the

Figure 5: Traditional Lanna Thai House

study area clearly recall their rural origins and articulate Lanna domestic traditions. A fence separates the compound from the street. There is almost invariably an open space between the street gate and the house itself. This area (*kuang*) is used by the family for many activities in the warm local climate. Although the kuang can be very small in some cases, its presence between the public space of the street and the private space of the house is an essential feature of traditional site planning and an important determinant of the morphology of the district. It is the most public of a series of spaces leading to the interior of the house. The kuang may contain the well and will usually feature one or more shade-giving trees. The house itself is positioned to form an edge of the kuang and usually presents an oblique view to the visitor arriving at the gate.

The planning of the house itself extends the sequence established by the kuang. After traversing the kuang, one ascends a stair leading to a terrace (*chaan*) that is several inches below the main floor level of the house. This partly-covered terrace is where the family entertains guests. Adjacent to the chaan is another covered terrace at the main floor level (*tern*). The tern serves as the common space of the house, where the family eats and socializes and where young male members may sleep. Visitors are rarely invited onto the tern. Only the bedrooms are enclosed. The kitchen is usually housed in a separate part of the house.

Another common type, found only in larger settlements, is the "shop house" (Figure 6). This type is not of Lanna origin, but is associated with immigrant Chinese and has a century-old history in Chiang Mai as well as in most Asian cities (Cohen, 1985, 12). Early photos show the Chinese shop house type well established along Thae Phae Road. Today's shop house is a three or four story, narrow building with a broad opening on the ground floor (the shop). Passage to the upper floor is usually available only within the shop. The upper floors are usually exclusively residential. The shop is entirely flexible in use – within our study area it was used for active retail, office, parking, and as a kind of extended living room. These shop houses, which are built side-by-side with party walls, represent a more densely built up fabric than the small houses.

Figure 6: Two Different Sets of Shop Houses

These two types and their variants comprise the major vernacular, traditional types in the city. Another important urban type is the *wat*. These are not single buildings, but walled compounds comprising many distinct building types including the *wihan* or worship hall, library, dormitories and conical or pyramidal burial monuments (*chedi*). Wats vary greatly in size and usually contain a great deal of open space. As urban elements they act as public spaces, serving not only religious purposes but also as community centers and primary schools. Since they are never traversed by streets, they block the street grid.

An emerging building type is the small Thai office building, characterized by its slight setback off the street, modern construction in steel, glass and concrete, height of three to five stories, and extremely small footprint (Figure 7).

In our study area, there were also several modern style small hotels and apartment buildings. One of them was the tallest building in the block, about eight stories tall. These modern types were wedged into the traditional neighborhood, some more comfortably than others.

Figure 7: New Office Buildings

3.3. New Urban Forms in Chiang Mai

Rather than acknowledge the traditional form, Chiang Mai's rapidly developing suburbs seem to be modeled on western suburban forms. For example, Figure 8 is a photo of new housing that is being developed north of the city. These houses are set in a subdivision not unlike that of an American city – even the scale is similar (although the lots tend to be smaller). Figure 9 is a same-scale comparison of the layout of this subdivision with the traditional plan of Chiang Mai and an American suburb. The new subdivision exhibits many of the characteristics of the American suburban type, including single family lots and an area next to the highway reserved for commercial use. It is located off of the main highway north of the circumferential highway and is intended for well-to-do families with cars. Even farther north is a golf-course community.

Figure 8: New Housing on the Outskirts of Chiang Mai

New office developments in the periphery are composed of the small-footprint office buildings, set in a single-use office park landscape that recalls an American single-family housing subdivision. Immediately outside the walls of old Chiang Mai, a large retail shopping mall attracts young Thais for entertainment and shopping, just as in the western world.

Within the already urbanized area, urban change also takes new forms. As in Cohen's study of Bangkok, we find low-scale residential buildings replaced by larger apartments or hotels. Comfortably-scaled neighborhoods like the study area already have larger scale buildings intruding on them. Within the night market district, for example, a large western hotel developer has apparently leveled several blocks of tightly-scaled traditional urban fabric in order to create a lawn and garden for the high-rise hotel. Ironically, a remnant, traditional house is preserved on the site as an icon.

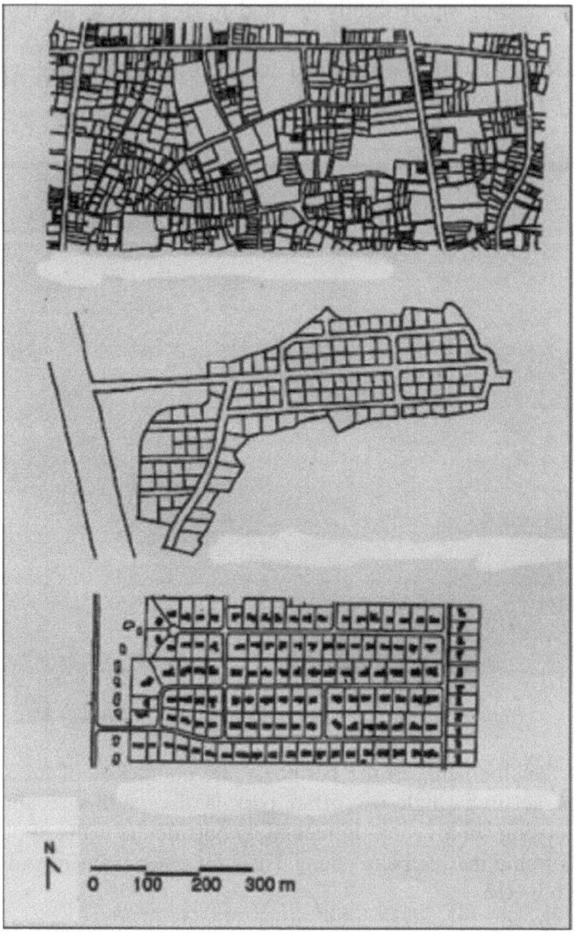

Figure 9: Comparison of Urban Form

4. THE SUSTAINABILITY OF TRADITIONAL FORMS

The traditional forms of Chiang Mai seem to provide a possible model for balance between the unsustainable physical extremes of Bangkok-like over-development and western-style suburbs. Characteristics of sustainability in the traditional form that might serve as a guide for new development include the following.

4.1. Dense, But Low Scale Development Patterns

Except for the shop houses, all the structures in the neighborhood were single buildings separated by small yards from their neighbors. Only a few were taller than two stories. Lush vegetation was evident everywhere that a building was not actually standing, giving the neighborhood the feeling of a small village. Even the shop houses and the new office buildings were low-scale compared to western "urban" development. With a height limit of about four stories, this neighborhood was far denser than the new Chiang Mai residential subdivision (35 dwelling units per hectare compared to 9 units per hectare), and it easily meets the recommended minimum density figures for efficient provision of mass transportation. In compact settings, urban services can be provided less expensively, especially utilities. Likewise, a dense but low-scale environment responds to the traditional village character of old Chiang Mai.

4.2. A Mix of Land Uses

This small neighborhood contains retail shopping, offices, services such as a bank, laundry and massage, hotels, small snack shops, and restaurants. Nearby schools and temples complete the urban ensemble. Even so, the primary land use is residential. Nearly every building that contains another use also contains a residential use and many houses also contain workshops or other work places. Residents may work, shop, and recreate within walking distance, which avoids most of the pollution and waste of resources necessitated by personal autos or other motorized transport. Neighborhood facilities and services draw residents closer together and create personal connections, which is another sustainable quality. Children can be tended to close to home, or even in the work place, which we observed as a common custom in Chiang Mai.

4.3. A Wide Variety of Incomes and House/Lot Sizes

The older neighborhood contains both very wealthy people, living in large, secluded, gated houses, and very poor people, living in an intriguing settlement only a few steps away. In between are prosperous shop owners in their shop houses, apartment dwellers, and ordinary working people. Unlike segregated American communities, people of different incomes living in proximity insure that the vast, pathological urban slums common in America cannot develop. Areas where poverty is the only model have proven to be breeding grounds for crime and urban decay, the very definition of unsustainability.

4.4. Adaptable Forms and Patterns

Like many older environments, the traditional neighborhood of Chiang Mai is built to accommodate change easily. In effect, the small highly varied lots are constantly undergoing change without significantly damaging the overall sense of place in the neighborhood. For example, one house was abandoned and rapidly deteriorating. It was easy to imagine another large house or two small houses replacing it, or even a small hotel. The scale of the lot restricts the size of the eventual building, as long as two or more lots are not joined together. Such adaptability is commonplace in traditional fabrics around the world, where lots act as a brake and a guide for development, while allowing tremendous variety. The traditional Thai building types are themselves highly adaptable to changes in use or family type.

4.5. Traditional Building Types

The traditional Lanna Thai house responds very well to its locale (Figure 10). Extensive use is made of outdoor spaces for domestic functions, allowing houses to be relatively small and keeping impermeable surface area to a minimum. This is particularly important in this region where rainfall is heavy at certain times of the year. The steeply sloped roofs shed the heavy rains as well as creating storage space between the supporting trusses. The open areas surrounding the houses also provide space for plants which create a cooler microclimate and give privacy despite the high density of development. The houses' open ground floor and woven or latticed wall panels of the upper floor provide natural ventilation.

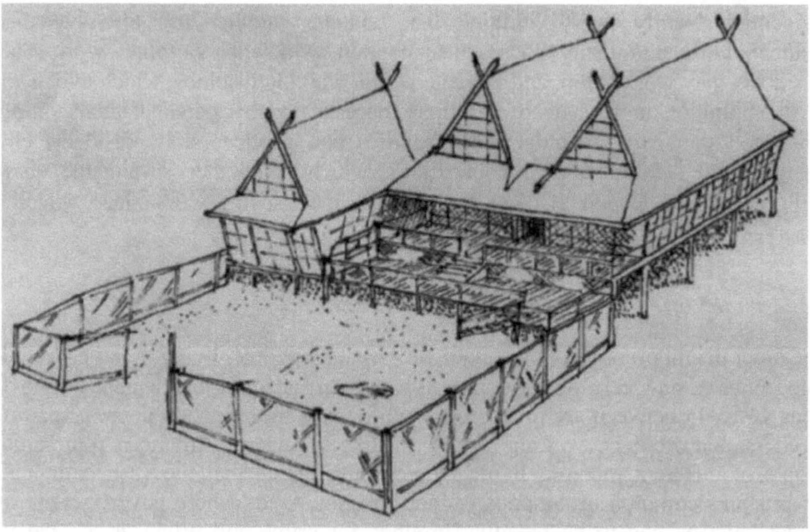

Figure 10: The Traditional Lanna Thai House

The houses in our study area were built of local materials, but this may have to change in new construction. The older houses were built of local woods, primarily teak. Teak naturally resists moisture and rot, and is thus well-suited to building in the local climate. Unfortunately, due to overcutting, the Thai government has severely restricted the harvesting of domestic teak and most teak now is imported from neighboring Myanmar, making it expensive. We observed that most modern houses are constructed of concrete block and stucco. This change in materials need not impose any change in site planning or interior spatial arrangements. Most of the new houses we saw did not adopt these traditional forms; however, nor were they as well adapted to the climate. The new form responds to a desire of young Thais to reject the traditional form as a way of declaring their modernity (Vitti, 1998). Many climatically appropriate aspects of the traditional house, such as allowing natural ventilation and incorporating outdoor space, can easily be adapted to modern construction techniques.

5. ASSESSMENT

The new construction in Chiang Mai is happening in two important mechanisms. In the central city, even in the protected area within the walls, low-scale and fine-grain development characterized by traditional small lots, small buildings, and multiple streets is being slowly supplanted with larger buildings and a greater amount of paved area. Immediately outside the protected area, this is accelerated with high-rise buildings and larger developments like the shopping mall supplanting older patterns. In the suburbs of the city, entirely new patterns of development are growing, often in master-planned projects. These offer modified western-style low-density form, with single-use office parks and single-family subdivisions instead of the traditional mix of uses. Neither of these development patterns has the sustainable characteristics enjoyed by the traditional form, and neither offers the unique traditions of the Lanna-Thai.

This suggests two important points: First, where traditional urban patterns exist, they should be zealously guarded. Second, these patterns may be used as a model for new development, rather than saddling Chiang Mai with the worst kind of western development.

It is important to distinguish between preserving an urban pattern of lots and blocks and streets versus preserving individual buildings or houses. In the western tradition, conservation or preservation is usually directed at particular buildings as unique forms. The traditional Thai house, while very beautiful, is a relatively fragile construction built to last only a generation or two, at which point it is demolished and rebuilt on the same site. With a few exceptions born of the rapid disappearance of traditional houses, Thais do not seem to value the individual house as an historic artefact (unlike temples, which are venerated); but they do value the traditional way of building. In fact, the original type was designed to be dismantled, moved, and re-erected. Preserving the urban form would allow the traditional types to remain in an appropriate and supportive setting, but not fix individual artifacts in place. This provides a flexible and living city, which is ultimately more sustainable.

It also makes a city that is more traditional, more unique, more "Thai" and thus more appealing to tourism. We believe it is a grave error to destroy the traditional physical fabric of this town to build hotels, shopping centers, office buildings, subdivisions, and condo towers that would look at home in Los Angeles. Not only are these forms inherently more energy intensive than traditional types, but the urban pattern that supports them calls for relatively low density single use districts, segregated incomes, and vast transportation networks.

The current preservation policies in the core of Chiang Mai have been concentrated inside the walls and have been limited to a few architectural guidelines coupled with a height limit. Although it has been important in directing high-rise growth away from the center, it has not been entirely successful in maintaining the historic fabric. Instead, this policy has so far led to four story buildings with much larger buildings that can be overwhelming in scale and impact to the delicate fabric of a traditional block and soi network.

In the aggregate, controlling the specific design and layout of master planned, infill and redevelopment projects has important effects on the region's sustainability. Yet, it is equally important for policymakers to control the overall physical distribution of growth in the region. If past patterns are an indication, the total population growth within the region of Chiang Mai will be rapid and extensive over the next ten years. On the scale of the region, it is well worth evaluating the tendency to sprawl that is apparent in the current development patterns. Western development teaches us that the time to plan more sustainable urban form is prior to expansive growth – not after unsustainable patterns have already been promoted. The planners of Chiang Mai, both at the national level and locally, have the means to direct growth into reasonable locations that can preserve valuable agricultural land, environmentally sensitive areas, and areas of cultural or aesthetic value. The future location of mass transit must also be determined, so that outlying town centers can be established. Such locations should not be built at the ultra-low densities that resist transit options and are more expensive to serve with infrastructure, rather, the moderately high density of the traditional urban form should be used as a guide.

By the same token, planning for the areas already urbanized should recognize the tendency of redevelopment to destroy the existing urban fabric and build much more intensely, in a manner that eventually leads to the kind of overcrowding that is the scourge of Bangkok. Whereas rapid change is necessary to the growth and development of the city, those changes must be placed in a sustainable framework that preserves the scale and livability, if not the actual old buildings, of Chiang Mai and Lanna Thai traditions.

NOTES

[1] These aerial photographs are part of the collection of the University of Chiang Mai library and are not reproducible because of their poor quality.

REFERENCES

Alberti, Marina.(2000). Urban Form and Ecosystem Dynamics. In Williams, et.al (Eds.), Achieving Sustainable Urban Form (pp. 84-96). London: E&F Spon.

Battle, Guy and Christopher McCarthy. (2001). Sustainable Ecosystems and the Built Environment. London: Wiley-Academic.

Buxton, Michael. (2000). Energy, Transport and Urban Form in Australia. In Williams, et.al (Eds.), Achieving Sustainable Urban Form, (pp. 54-63). London: E&F Spon.

Caniggia, G. and G. L. Maffei. (2001). Architectural Composition and Building Typology: Interpreting Basic Building. Translation from Italian by Susan Fraser, with revisions by Karl Kropf and Brenda Scheer. Florence, Italy: Alinea.

Charernsupkul, Anuvit. (1978). Northern Thai Domestic Architecture. Unpublished paper. Chiang Mai University.

Chiang Mai Municipal Government Planning Office. (Undated.) Maps and aerial photographs.
Cohen, Erik. (1985). A soi in Bangkok – the dynamics of lateral urban expansion. Journal of the Siam Society, Vol. 73, no 1, 1-35.
Conzen, M.R.G. (1968). The Use of Town Plans in the Study of Urban History. In Dyos, H.D. (Ed.), The Study of Urban History (pp.113-130). New York: St. Martins Press.
Cuff, Dana. (2001). The Provisional City. Cambridge, MA: MIT Press.
Drumreicher, H., Levine, R., Yanarella, E., and Radmard, T. (2000). Generating Models of Urban Sustainability: Vienna's Westbahnhof Sustainable Hill Town. In Williams, et.al (Eds,), Achieving Sustainable Urban Form (pp.288-298). London: E&F Spon.
Frey, Hildebrand. (1999). Designing the City: Towards a More Sustainable Urban Form. London: E&F Spon.
Guy, S. and Marvin, S. (2000). Models and Pathways: The Diversity of Sustainable Urban Futures. In Williams, et.al (Eds.), Achieving Sustainable Urban Form (pp. 9-18). London: E&F Spon.
Jacobs, Jane. (1961). The Death and Life of Great American Cities. New York: Random House.
Jenks, M., Burton, E., and Williams, K (Eds.). (1996). The Compact City: A Sustainable Urban Form? London: E&F Spon.
Masnavi, Mohammad-Reza. (2000). The New Millenium and the New Urban Paradigm: The Compact City in Practice. In Williams, et.al (Eds.), Achieving Sustainable Urban Form, (pp. 64-73). London: E&F Spon.
Narumit-rekagarn, S. and Tansukanan, P. (1988). Problems and Proposals on Conservation: the Case of Chiang Mai. Paper presented at the UNESCO/SPAFA Workshop: Principles of Preservation Applicable to the Ancient Cities of Asia.
Newton, Peter. (2000). Urban Form and Environmental Performance. In Williams, et.al (Eds.), Achieving Sustainable Urban Form, (pp. 46-53). London: E&F Spon.
Phanichphant, Vithi. (1998). Professor of Art, Chiang Mai University, 1998. Interview: August 5, 1998.
Scheer, Brenda and Daniel Ferdelman, (2001). Destruction and Survival: the Story of Over-the-Rhine. Urban Morphology, Vol. 5(2).
Steinhardt, Nancy. (1999). Chinese Imperial City Planning, (pp. 1-28). Honolulu, Hawaii: University of Hawaii Press.
University of Chiang Mai Library, Aerial Photographs from 1954, 1955, 1956, 1978, and 1989.
Vadhanasindhu, Pongsak. (1995). Local Knowledge in Physical Design and Planning: A Case Study of Chiang Mai. Ph.D. Thesis. School of Community and Regional Planning, University of British Columbia.
Williams, K., E.Burton, and M. Jenks. (2000). Achieving Sustainable Urban Form. London and New York: E & F Spon.
Winichakul, Thongchai. (1992). Siam Mapped: A History of the Geo-body of a Nation. Honolulu: University of Hawaii Press.
Wyatt, David and Aroonrut Wichienkeeo. (1999). The Chiang Mai Chronicle. Bangkok, Thailand: Silkworm Books.

CHAPTER 14

CHRISTOPHER AUFFREY

THE INFORMAL SETTLEMENTS OF CHIANG MAI: LESSONS OF SUSTAINABILITY FOR OTHER INTERMEDIATE SIZE CITIES

Abstract This chapter uses a sustainable development framework to assess the status of the informal settlements in Chiang Mai, an intermediate size city in the north of Thailand, and considers the lessons from these settlements that may be useful for the sustainable development of intermediate size cities in Asia. Informal settlements serve a critical role in providing affordable shelter close to jobs for poor urban residents, yet impose substantial costs on those living in and near the settlements as a result of significant externalities. A sustainable development framework is well suited for the issues of informal settlements that encompass social, cultural, economic, and environmental impacts. This is especially true in intermediate size cities, where most Southeast Asian urbanization is currently focused. Such cities are more readily able to adopt the decentralized public management approaches needed by their informal settlements than their mega-city counterparts. The lessons for sustainable development include: recognition that residents of informal settlements are a community asset; eviction and relocation programs should be approached with caution; community networking and cooperation should be nurtured; social service provision must be made a priority; and, environmental infrastructure has tremendous potential for return on investment and need not impose a financial burden.

1. INTRODUCTION

In developing countries, where most of the world's population growth is occurring, the number of people living in urbanized areas is rapidly increasing (UNCHS, 1997). This escalating urbanization has been accompanied by a proliferation of informal settlements without access to basic infrastructure, safe water, and sanitation, with associated health and property risks (UNEP, 1997). It has been estimated that informal settlements comprise 30 to 70 percent of the urban housing stock of developing countries, reflecting the pressures of population growth, the relatively low wages paid unskilled workers, and the ineffective systems of housing finance and land use (Pugh, 2000; Choguill & Choguill, 1996; Choguill, 1995; Yap, 1995; Peattie, 1987; Habitat, 1982). The informal settlements serve a desperate need for affordable shelter among the poorest urban residents yet impose substantial social and economic costs on both the poor and others as a result of significant environmental and social externalities (Kirby, 2001; Pugh, 2000; Habitat, 1996; Main, 1990).

M. Romanos and C. Auffrey (eds.), Managing Intermediate Size Cities, 273—292.

The concept of sustainable development has emerged in the development literature over the past two decades as a means for balancing issues of economic and population growth with social, cultural and environmental concerns, and as such, would seem especially appropriate for addressing the multidimensional and complex issues of informal settlements (Coomer, 1979; WCED, 1987; Hossain, 1995). If cities throughout the developing world are to effectively utilize sustainable development frameworks for improving the quality of life for all their residents over the longer term, the issues of informal settlements must be made a fundamental part of the overall development process.

This is especially true in intermediate size cities where the environmental, social, and economic impacts of the informal settlements' externalities may be accentuated by the cities' relatively small area, population and access to resources. Such cities have become the major recipients of urban population growth in Southeast Asia over the past decade, and the issues of their development are receiving increasing attention (Rondinelli, 1991; also see Chapter 2, Romanos & Auffrey, this volume). It has been argued that they may offer a greater likelihood for the timely and successful adoption of the decentralized public management approaches useful for responding to the needs of their informal settlements than their mega-city counterparts (see Chapter 8, Romanos, Wagner & Thadaniti, this volume), and therefore may have better prospects for successfully implementing sustainable development approaches. This chapter uses site visits, structured interviews, and reports from public and private organizations to assess the informal settlements in Chiang Mai, an intermediate size city in the north of Thailand, and considers the lessons to be learned from these settlements that may prove useful for the sustainable development of other intermediate size cities in Asia.

2. CHIANG MAI'S EXPERIENCE WITH INFORMAL SETTLEMENTS

The rapid economic expansion from the mid-1980s until the Asian economic crisis beginning in 1996 led to uneven economic and social development, and a number of serious environmental and social problems in Chiang Mai (Phongpaichit & Baker, 1998, 2000; Dixon, 1999; Parnwell & Arghiros, 1996). The growth of informal settlements of low-income residents along the municipality's canals and other vacant public and private spaces was among the most visible result of this "maldevelopment" (Bell, 1996, 49). It is argued here that efforts to promote the sustainable development of Chiang Mai must explicitly address the problems in the city's informal settlements. As an initial step in this process, the historical context of these settlements and the prior efforts to document their existence must be understood.

2.1. Chiang Mai's History of Housing Low-Wage Rural Migrants

The presence of poor residents living on land they do not own is not a new phenomenon in Chiang Mai, though the character of the early settlements was fundamentally different than those currently found. During the period of the Chiang

Mai kingdom (1774-1874) laborers were allowed to build small houses on the land of the ruling class so they might be able to provide their services. Later, as the economy of the city developed and the need for labor increased, it was not uncommon for new workers from rural areas who had come to the city "to seek their fortune" to live rent-free near factories or mills, and for wealthier Chiang Mai residents to offer temporary or even permanent housing to workers migrating from the rural village of their ancestors (Tan-Kim-Yong, 1979).

In more recent times, the need to provide low-cost housing for new workers from rural areas continued, albeit in a somewhat different form. Rural workers continued to move to Chiang Mai during the decades following World War II as the city grew at a moderate pace (Goldstein, 1972). The increased business activity attracted a new wave of rural migrants who, unlike the earlier groups, were unable to find adequate housing with the established families from the same home villages, or in employer-provided units near the factories and mills. Given the shortage of affordable housing, some of the workers new to Chiang Mai found housing in an informal settlement to be one of their few options for establishing low-cost housing close to job opportunities (Tan-Kim-Yong, 1979). Provided vacant property could be found, a shelter could be simply built without permission of the landowner. If the piece of property was large enough, it could be informally subdivided and pieces of land rented to subsequent settlers, though without the agreement of the property owner. In some cases, the settlers might directly contact the property owner, who having accepted that the land was occupied and other more lucrative uses were not immediately available, would allow others to live there as renters (Yap, 1995).

In the 1960s and 70s, Chiang Mai continued to grow and emerged as the commerce and administrative center of the northern region with expanding trade and factory production facilitated by improvements in communications, transportation, and finance. It was during this time that Chiang Mai University was established,[1] and increased resources were provided for Chiang Mai's technical and teacher training colleges, and the Institute for Agricultural Technology. Many rural workers seeking higher wages and a better life responded by migrating to Chiang Mai, but soon found affordable shelter near jobs to be in short supply. As a result, informal settlements sprouted near the city's outer wall,[2] along vacant private land, near the markets, and on the municipal cemetery. This was especially a problem near major construction projects, where many of the workers and their families built residences on nearby public land. The city's official policy with respect to these settlements was no policy at all, provided that others in Chiang Mai did not demand that the city take action against what were ostensibly illegal acts[3] (Tan-Kim-Yong, 1979).

Beginning in the 1980s, the pace of development in Chiang Mai quickened, especially for the tourism sector, and related construction and services, creating opportunities for many unskilled and semi-skilled workers. Room cleaners, laundry workers, kitchen staff, and gardeners were needed in hotels and restaurants. Numerous employment opportunities in Chiang Mai were created for those who could drive a small transit pickup truck (song-tao), tourist van, or "tuk-tuk" taxi. Sales clerks were needed in stores and the Night Market, and laborers were needed at the many construction projects. Similarly, workers were needed in the production of the handicrafts, jewelry, and textiles for which the city and region are famous (CASE, 1998).

2.2. Documenting the Informal Settlements of Chiang Mai

It was not until relatively recently that the informal settlements of Chiang Mai were formally documented. This is significant because without credible documentation, authorities would likely continue to ignore the settlements. Such a strategy may avoid short-term conflict, but is clearly not sustainable given the increasing magnitude of the settlements' potential externalities. Thus, it is not surprising that with the rising numbers of informal settlements and the potentially heightened sensitivity to their visibility that accompanied the city's emergence as an international tourist destination, a number of academics, non-governmental organizations, and community groups undertook efforts to better understand the nature of these communities. These efforts intended to make informed decisions in regards to controlling the proliferation of the settlements, improving the quality of life of the residents, and mitigating the impacts on other Chiang Mai residents.

One of the first studies was conducted in 1979 by Chiang Mai University to assist planning by the National Economic and Social Development Board (NESDB). This effort found nine separate informal settlements within the municipality on both public and private property, mostly along the city's canals. These communities were found to include 909 households with 4,090 persons. Among the nine settlements, none had adequate water supply, sewerage or drainage, though all communities had electricity (though only available to 40 percent of households). Wells and canals were used for all manner of waste disposal, including cases where toilets were built above the same waterway used by others for bathing and cleaning. Municipal solid waste disposal was available to only some of the residents (Tan-Kim-Yong, 1979).

The study concluded that development pressures would increase the likelihood that many of the informal settlements on private land eventually would be cleared and profit-generating developments built. It recommended that the informal settlements should become a higher priority for those concerned about the city's growth and development. Most importantly, there was concern that important historic sites (the ancient outer wall) were being destroyed, and that immediate action was needed to protect these. Further, it was strongly recommended that policies must be clearly established to more effectively provide for low-income housing, and that difficult choices must be made from among the alternatives of public housing, community development, site and services provision, and relocation (Tan-Kim-Yong, 1979). It is unclear what response, if any, resulted from the report. If the NESDB did use the information in their plans, it apparently did not result in substantive action. Subsequent studies only refer to the municipality's inability or unwillingness to develop a policy for low-income shelter during the period following the 1979 report.

Seventeen years after that first study, another project was undertaken, as the issues of informal settlements in Chiang Mai had not gone away. Community leaders, municipal officials, academics, and NGOs joined together in 1996 to conduct a survey of 58 communities in order to collect information about the problems of shelter for low-income residents. Unfortunately, the resulting Thai-language "Report of the Urban Communities in Chiangmai" (reported in CASE, 1998) failed to clearly define the physical, tenure, services, and income issues of the communities. As a result, municipal officials were unable to determine which of the communities were in most need, and there was no change to the official list of 21 communities eligible for annual financial assistance (CASE, 1998).[4]

Nonetheless, this second study was considered to have been successful in mobilizing the residents of several of the communities to take action that would elicit the attention of city officials. While the city did not officially accept the study's results, for the first time there was an informal acknowledgement of the existence of a considerable number of informal settlements, and the need for the municipality to address their major problems such as the lack of access to basic services.[5] Also, at this time the central government instructed municipalities that all citizens[6] are entitled to registering their residence in the official housing register.[7]

Largely in response to what was perceived to be the weak official response of the municipality and other levels of government to the 1996 survey and follow-up mapping of the previously identified communities, another municipality-wide survey of poor communities in Chiang Mai was conducted in 1997-98 by the NGO Community Architects for Shelter and Environment (CASE) to assist efforts for legal recognition by the residents of informal settlements. This time, special attention was given to those communities with housing, land tenure or shelter-related problems. This new study was designed to clearly define the physical characteristics of Chiang Mai's poorest communities using an accepted methodology that would provide valid and verifiable results. Of the 58 communities included in the 1996 study, 38 were identified as "slum communities" populated by low-income residents with significant and widespread problems of tenure, housing, and basic services (CASE 1998, 13). Map 1 shows the distribution of these settlements surrounding the ancient walled-city in the center of the municipality.

3. CHARACTERISTICS OF INFORMAL SETTLEMENTS IN CHIANG MAI

As part of the Joint Study[8] field project involving Chulalongkorn University, Gadjah Mada University and the University of Cincinnati, and with collaboration with Chiang Mai University and the Municipality of Chiang Mai, a series of intensive site visits and interviews were conducted with city officials, community residents, and NGO representatives to assess the current conditions and issues of Chiang Mai's informal settlements. In addition, a number of government and private sector reports on the settlements and related topics were collected.[9] Results from the 1997-98 census provide a demographic overview of the communities (see Table 1).

Chiang Mai's informal settlements face many of the same problems found in urban areas throughout the developing world and, as such, the city faces challenges for confronting these problems in their pursuit of sustainable development that are common to many other intermediate size cities in Asia. The issues of these informal settlements are manifold and complicated, though in the context of developing countries generally, and Thailand more specifically, they fall into one of five areas: location and access to employment, environmental infrastructure and utilities, poverty, social services and community organizations, and land tenure and threats of eviction. The following sections report on the characteristics of the informal settlements in Chiang Mai with respect to these issues, and relate these to the core issues of sustainable development.

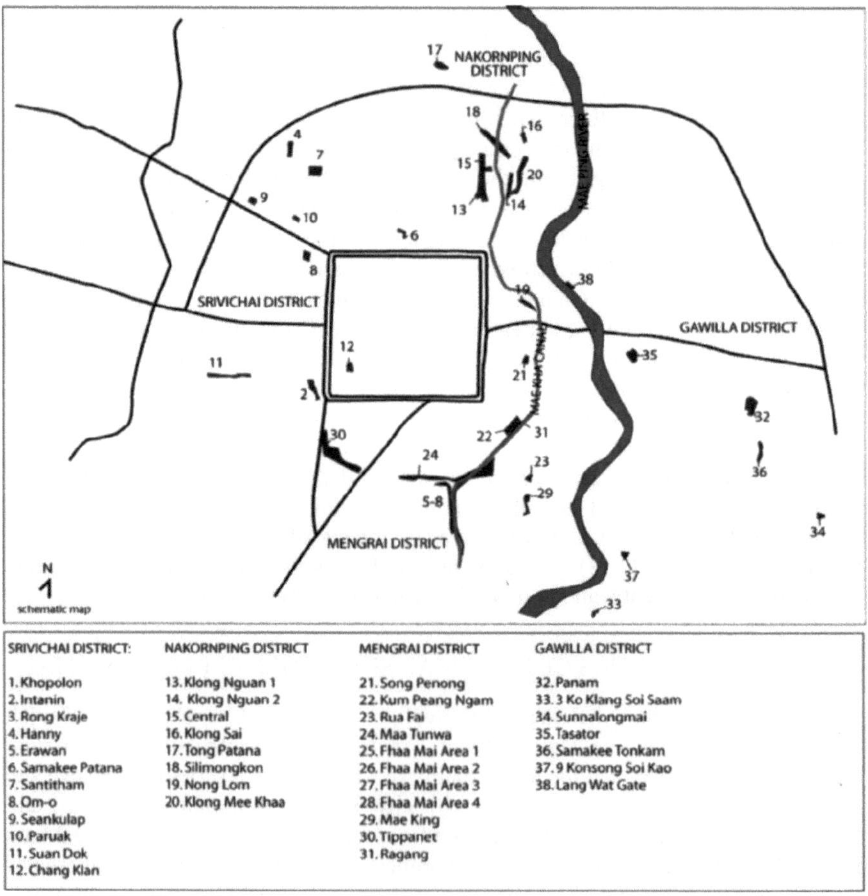

Map 1: Informal Settlements in Chiang Mai
Source: CASE, 1998.

Table 1: Demographic Overview of Chiang Mai's Informal Settlements, 1997-98

Settlements by District	No.	Reasons in Settlement (%)	
Srivivchai	12	Near workplace	40
Nakornping	8	Near relatives	18
Mengrai	11	There for long time	24
Gawilla	7	Low cost/no other place	41
Total	38	Eviction	3
Population		Years at Present Site (%)	
Total	6,346	Less than 1	3
Male	3,108	1 to 5	14
Female	3,238	5 to 10	14
		10 to 20	25
Age Distribution (years)(%)		20 to 40+	44
Less than 5	9		
5 to 16	24	Land Status (%)	
16 to 60	67	Own	4
Greater than 60	8	Squatter	62
		Rent	34
Education (%)			
Illiterate	13	House Condition (%)	
Less than grade 5	50	Satisfactory	48
Less than secondary	63	Needs repair/replacement	52
Occupation (14+ yrs) (%)		Water Supply (%)	
Unemployed	15	Legal meter	39
Unskilled	33	Well	29
Housekeeper	9	Hand pump	20
Self-employed	23	Illegal connection	12
Student	10		
Other	7	Electric Power (%)	
		Legal meter	62
Distance from job (%)		Illegal connection	31
No commute	72	No electricity	7
1 to 10 minutes	14		
More than10 minutes	14	Solid Waste Disposal (%)	
		Municipal service	70
Monthly Income (%)		In community	5
Less than 4,000 baht	41	Burn	25
4,000 to 5,000 baht	10		
More than 5,000 baht	49	Sewage Facilities (%)	
		Own toile/latrine	94
		Neighbor's toilet/latrine	3
		No toilet/latrine	3

Source: CASE, 1998.

3.1. Issues of Location and Access

The issues of location and access are concerned with access to jobs, exposure to hazards, and pedestrian and vehicle circulation networks. Access to jobs is for most residents the single most important factor for taking up residence in an informal settlement. Residents may have better quality housing available at locations significantly more distant from major employment areas, yet choose to remain in lower quality housing in the better employment location.

In Chiang Mai, both the collected reports and our interviews confirm the high priority that informal settlement residents place on location. Most informal settlement residents in Chiang Mai have sought affordable housing close to the areas where they earn their living (see Table 1, Reason for Staying in Settlement). Most cannot afford the time and expense of traveling by vehicle from more distant locations where property and housing is more affordable. Street vendors with heavy carts that must be pushed by hand to vending locations each day are at a competitive disadvantage if they live any distance from the center of the city. Thus, vacant properties along the canals near the center of the city have become attractive locations for informal settlements for low-wage workers.[10] The center-city locations minimize transportation time and costs, and the informal nature of the settlements allows the provision of very basic shelter at a lower cost than would be possible in a legal housing development where building codes, infrastructure charges, permits and taxes would add expenses. Efforts to direct these workers to public housing projects away from the selling areas may present an additional problem since they fail to provide a place to store the carts at night.[11]

3.2. Issues of Environmental Infrastructure and Utilities

The availability of adequate environmental infrastructure to all segments of a community is an essential condition for its sustainable development. Issues of economic growth, environmental preservation, and social cohesion have links to the need for clean water, safe disposal of sewage and solid waste, and stormwater drainage.

The issues of environmental infrastructure and utilities are primarily concerned with access to clean drinking water and electricity, sanitary sewage disposal, removal of solid waste, and drainage of stormwater. Most of the residents of the informal settlements in Chiang Mai have self-built their homes (78%) without government authorization and have no official access to municipal water (61%). Providing safe drinking water to a crowded urban settlement is typically most effective through piped connections to treated municipal supplies. Twelve percent of households reported an illegal piped water connection to a neighbor. Wells may offer a lower-cost alternative to piped water systems, yet these are vulnerable to contamination, especially in urban areas where sewage and solid waste disposal, or stormwater drainage is an ongoing problem. Nearly one-third (29%) of residents reported a well as their source of water. The remaining 20 percent use a hand pump, though the source is unclear.

Houses built over Chiang Mai's Mae Kha Canal

Most of the residents use cesspool or septic tank sanitation systems, and 94 percent report having their own toilet or latrine. Such private systems can be effective if properly designed and maintained, though the evidence from site visits showed that many were ill maintained and a major source of environmental contamination.

Municipal solid waste collection was reported by 70 percent of residents. This is accomplished by taking the household waste to centralized collection sites distributed throughout the city. The collection sites were observed to be ill kept in many cases with garbage strewn around the receptacles. One-fourth (25%) of the residents reported burning their solid waste. This is particularly troublesome given the density of the settlements and the potential for ash and residue to be washed into the nearby canals.

Household wastewater drains into an open canal

Flooding and drainage were identified as the two highest priority issues among all the informal settlements in Chiang Mai. The lack of effective drainage systems for stormwater leads to frequent flooding, often with property damage, land erosion, and increased transmission of diseases, spreading of waste, and proliferation of odors. Flooding was identified as a priority for 43 percent of residents, while 36 percent cited drainage as a priority (see Table 2). Given that many of the settlements in Chiang Mai are located along the Mae Kha canal on land that was vacant, at least in part, because of its susceptibility to seasonal flooding, it is not surprising that these issues rank so high.

The lack of officially authorized electrical connection has lead to makeshift connections that can overload circuits and cause potential fire hazards. Nearly one-third (31%) of the residents reported having unauthorized electrical connections from a neighbor. This presents a fire risk in an area where houses are made of inexpensive flammable materials and access for fire fighting equipment is extremely poor. In 1997, a fire in the Nong Lom settlement destroyed nearly all of its 32 houses (CASE, 1998, 9).

Table 2: Community Priorities of the Informal Settlements in Chiang Mai (percent of residents selecting, with multiple choices)

Flooding	43%
Drainage	36%
Access/walkways	35%
Poverty/unemployment	33%
Water provision	31%
Eviction	24%
Housing conditions	23%
Fire risk	23%
Electricity	21%
Drug use	18%

Source: CASE, 1998.

3.3. Issues of Poverty

The issues of poverty are most concerned with the ability to earn an income sufficient to support oneself, lack of financial resources for improving one's quality of life, and the lack of savings and the need to borrow to take advantage of business opportunities. Many of the residents of informal settlements are initially brought to urban areas by the push of rural poverty and the pull of increased opportunities for jobs and income in the cities. Yet, despite the higher wages, many of these workers are unable to find affordable and accessible housing in the formal housing market. In Chiang Mai's informal settlements, 41 percent of households reported monthly incomes of less than 4,000 baht (approximately US$150 at the pre-Thai financial crisis exchange rate), and one-third (33%) reported poverty and unemployment a priority issue for their settlement (see Table 2).

Chiang Mai's newest informal settlements provide shelter for its poorest residents

Residents of the informal settlements are generally unskilled. In Chiang Mai, 33 percent of workers classified their occupation as unskilled, and another nine percent were housekeepers. Nearly one-fourth (23%) were self-employed as vendors or making handicrafts. The remainder were unemployed (15%) or students (10%). Low levels of education also were the norm. Thirteen percent reported to be illiterate, with 50 percent having less than a Grade 5 education, and 63 percent having no secondary education.

Further, because informal settlement residents have little or no savings, when investment opportunities arise that could lead to a higher standard of living, one can only turn to high interest rate moneylenders or forego the opportunity. Nearly three-fourths of the residents (72%) reported no savings, and more than one-half (54%) of them were in debt (see Table 1).

3.4. Issues of Access to Social Services

The issues of access to social services and community organizations concerns the ability to utilize basic public services such as free education, medical care or childcare, and the development of strong organizations that can effectively advocate for the community. Low levels of health care and education and lack of knowledge about personal hygiene and nutrition compound the other problems of informal settlements. With one-eighth of the residents illiterate and nearly two-thirds having only a primary school education (see Table 2), access to education might be considered to be a problem, yet residents did not list education among their priorities. This may reflect the perception that "education," as defined for the CASE study, is strictly for children, and adequate levels of primary education are currently available. Likewise, the perceived educational needs of the adults (workforce training) may be imbedded in the high priority that residents placed on poverty and unemployment issues, and that workforce training is not considered to be a social service, per se. Health care also ranked relatively low among resident concerns with only nine percent identifying this as a priority. Again, it may be perceived that the Thai national health care system is meeting the basic needs of the residents. Visits to community health centers indicated that basic health care was widely available to those with citizen status and able to pay the minimal annual premium for enrollment in the national system. Unfortunately, Chiang Mai's poorest residents are from among the hill tribes people whose citizenship is often questioned and may have insufficient income for any enrollment premium.

Often social services are provided to those communities that are able to effectively lobby public and private agencies through strong community organizations. The informal and often chaotic nature of settlements may hinder the formation and nurturing of organizations that could represent the communities with a stronger voice. Community participation was identified by 15 percent of the residents as a priority, ranking behind the problem of drug use at 18 percent, and ahead of gambling at 13 percent.

3.5. Issues of Land Tenure and Eviction

The issues of land tenure and threats of eviction concern the insecurity resulting from the lack of title to the land where one lives, and the potential that one will be forced from the property and one's residence demolished. Since informal settlement residents generally have no title to the land they occupy, they are put in the position of having to invest in buildings from which they can be summarily evicted. In Chiang Mai, 62 percent of the settlement residents reported they were "squatters" on the land their house occupied, with this number varying substantially from settlement to settlement. In Nakornping District, for example, 97 percent of the residents were squatters, while in Mengrai District, 52 percent either owned or rented the land.

Even when residents are financially able, site visits and interviews indicate they are justifiably reluctant to make the long-term social and financial investments necessary to improve the living conditions in their communities, given the ongoing threat of eviction. Over one-half of residents (52%) reported that their house needed repairs or to be replaced. About one-quarter of residents (24%) cited concern of eviction as a priority. The investments by residents could potentially have wide-ranging effects such as decreasing fire risk, and reducing crime and drug abuse (see Table 1).

4. LESSONS FOR OTHER INTERMEDIATE SIZE CITIES

Based on the results of this exploratory research, it is clear that there are serious threats to Chiang Mai's sustainability as a result of the unresolved issues in the community's informal settlements. Some steps taken by central and local government agencies have succeeded, yet it is unlikely that the sustainable resolution of the issues identified here can be accomplished without new and innovative approaches for securing the additional policy and financial resources necessary to effectively improve conditions. Nonetheless, several lessons emerge from this research that may prove useful for other intermediate size cities in Asia that are faced with the issues common to informal settlements.

Lesson 1: Recognize Residents of Informal Settlements as Assets

Too often local officials simply view the residents of informal settlements as parasites on the system, only taking and contributing very little. The lesson from Chiang Mai is the overwhelming evidence that these residents serve vital economic, social and cultural resources that directly and indirectly benefit the entire city. These residents are among the lowest paid workers that provide goods and services to all Chiang Mai residents on a daily basis. Further, they provide a portion of the unskilled and semi-skilled labor for the crucial tourism, handicrafts, and manufacturing sectors of the Chiang Mai economy. Their low wages contribute to lower production costs and make Chiang Mai's products less expensive for local consumption and more competitive when exported outside the region. The workers are able to accept such low wages, in part, because their housing costs are low. Without these workers, Chiang Mai may lose a significant portion of its economic base.

In addition to their economic contributions, the more recent migrants from rural areas may serve to maintain better linkages with the community's social and cultural heritage. Chiang Mai, like many of the intermediate size cities of Asia, especially those with significant international tourism, is experiencing pervasive influence from western consumer products and lifestyles. The youth are widely exposed to western music, television and movies, and there is concern that the community will lose a sense of its unique Lanna heritage. The more recent migrants from rural areas come to the city with a stronger sense of that heritage and could be better utilized as a resource to cultivate and maintain it throughout the community.

Lesson 2: Proceed Cautiously with Eviction and Relocation Programs

Too often the official response to informal settlements is no response until a combination of forces (usually economic or social) applies sufficient pressure to force government action. Then bulldozers are ordered to the site, an announcement is made that residents have so many minutes to leave, and all the structures are cleared away. In other cases, more notice may be given, but with the same results. In still other cases, negotiations may take place, and residents may be offered a portion of the land, payments for resettlement, and/or alternative locations where they can either build new housing or move into existing public or private units. The lesson from Chiang Mai suggests that eviction and relocation programs will not achieve their objectives if they fail to consider the reasons for informal settlements in the first place. Issues of location and access to employment are just as critical as the cost of housing. Those evicted without formal alternatives that accommodate their location and access needs are likely to resettle into new informal settlements nearby. The experience in Chiang Mai has shown that many of the residents who had been evicted from one informal settlement may have simply added to the overcrowding and associated problems at another. Good public or semi-public transportation can compensate for some of the access problems, though a street vendor may have no place to store a cart in a public housing complex and is unable to put it on a bus.

Further, the threats of eviction serve as a strong disincentive for investment in one's home and community. The housing conditions were clearly worse in those Chiang Mai settlements where issues of tenure and eviction were most threatening. This suggests the value of policies that reduce uncertainty and unequivocally communicate to residents the future status of their settlements.

Lesson 3: Strengthen Efforts to Build Community Networking and Cooperation

Resolving the issues of informal settlements will require assistance from central, provincial, and local government agencies, as well as national and international NGOs. The lesson from Chiang Mai shows that significant resources exist within the communities that can and should be utilized. Tapping these resources may require efforts to build community networks and promote cooperation among residents. The city's Community Development Department had assisted efforts in a number of informal settlements to organize resident associations. The purpose of these groups is to identify local priorities and work with city and NGO staff to design appropriate solutions. In several cases, the residents contributed both time and modest sums of money for infrastructure improvement. In others, where the residents were unable to contribute financially, residents contributed labor in self-help projects. While the improvements in infrastructure were significant, important other benefits to the community included the establishment of leadership, and a sense of pride and capability to take part in shaping their own destiny.

Lesson 4: Extend Social Services

Social services address real needs that, if ignored, will generate significant future problems. The lesson from Chiang Mai is that the investment of these services can provide large returns in terms of a healthier, better-educated citizenry essential for a sustainable society. In the more established of Chiang Mai's informal settlements, especially where residents have become organized and have opened formal lines of communication with city agencies, the issues of education for children, health care, crime, drug use and gambling are less pressing and a lower priority than in the newer, less organized settlements. An educated citizenry and workforce is essential to a sustainable society, and it is especially important that those coming from the poorest communities be given the opportunity to reach their full human potential through education and opportunities to participate in self-determination. Workforce training[12] also is critical if a local economy is to effectively compete in a global economy. Of course, among the essential prerequisites for taking advantage of educational opportunities are good health and freedom from threats to one's personal security. Preventive health education is far more important than curative medical care in terms of the sheer numbers of lives improved and saved, but access to health care provides a safety net, especially against the communicable diseases endemic to densely populated urban areas with poor environmental infrastructure.

Lesson 5: Extend Environmental Infrastructure

The lack of adequate environmental infrastructure is the source of most of the acute problems of the informal settlements. The lesson from Chiang Mai is that, without infrastructure to divert human, household and commercial waste products, the waterways, ground water, and soils will become toxic to residents of the settlements as well as surrounding areas. In most cases, the contamination is not quickly remediated through natural processes, and contaminated areas may be unsafe for long periods even after the sources of contamination have been removed. Communicable diseases resulting from such contamination kill millions of urban residents worldwide, especially the very young and very old uneducated poor. While specific mortality and morbidity data were not available for the informal settlements in Chiang Mai, it was clear that health issues were especially prominent in those areas with the worst problems of contamination. Perpetual drainage problems during the monsoon rains exacerbate the already bad situation.

While the environmental and health problems are critical, there also are important economic reasons for improving environmental infrastructure. Those living in or near highly contaminated environments make unproductive workers. This is especially true in the tourism sector where hygiene is especially important. An outbreak of cholera could cost an international tourist destination such as Chiang Mai untold millions in lost revenues. Low-cost technology is readily available that can prevent most of these problems. In this light, any investment in improved environmental infrastructure has the potential to be highly cost effective.

Indeed, it can be argued that the lack of adequate urban environmental infrastructure is not primarily a budgetary problem, but rather, a political and management problem. Political leaders, concerned that by providing infrastructure they will be perceived to be endorsing the formation of informal settlements, would rather pretend the settlements did not exist, or at least by allowing the residents to suffer they are providing disincentives for the formation of new settlements. The management problem concerns using already-available institutional and financial resources to devise appropriate self-help schemes that use resident labor to construct and maintain functionally simple but effective systems. Usually this involves mobilizing residents so they are able to direct savings from fewer purchases of health care, medicine and potable water from vendors, less sick time away from their jobs, and more productive time on the job to the needed investments in their homes and communities. Adopting a sustainability framework for an entire city may help resolve both sets of problems, because it makes an explicit linkage between the welfare of the larger community with that of its component parts.

5. CONCLUSIONS

In the course of the research upon which this chapter is based, it became evident that the informal settlements in Chiang Mai exist because they meet a real and ongoing need for low-income housing in close proximity to employment opportunities. When each of the settlements was originally established, some over 40 years ago, the residents sought to balance their need for shelter with their need to earn sufficient income in order to acquire the other basic necessities of urban life. They did not have explicit preferences for locations with drinking water, sewerage, drainage and solid waste disposal problems, where access to education, health care and other social services would be difficult, or where they would be under a continual threat of eviction. Yet, they accepted these conditions as best they could, and in most cases, contributed to the development of thriving settlements that have made significant contributions to the larger Chiang Mai community. In some cases, particularly where the settlements are more recent or are in especially vulnerable locations, the settlements continue to have problems that present significant dangers to the health and welfare of the residents of the settlements as well as nearby areas. Their economic, environmental, and social problems transcend the boundaries of their settlement and detract from the sustainable development of greater Chiang Mai. In other cases, the municipality has stepped forward and provided much needed financial, technical, and organizational assistance to effectively address many of the issues of location, environmental infrastructure, poverty, and social services.

It is clear that the de facto policies of the past, whereby government agencies pretend these settlements do not exist, or downplay their significance and attempt to hide them behind walls from the view of other residents and visitors is not a sustainable option. Nor are piecemeal approaches that simply react when political and economic pressures reach a critical level as a result of social, environmental, or health crises. Rather, more proactive, integrated and comprehensive measures are needed. Thailand's Sixth (1987-91), Seventh (1991-1996) and Eighth (1997-2001) National Economic and Social Development Plans have contained numerous items designed to fight poverty, provide low-income housing and upgrade slums, though with limited success, as the informal settlements in Chiang Mai demonstrate. The lack of financial resources is not entirely responsible for the limited impact. Innovative approaches to self-help, and community organization have been shown to particularly effective. Even more important is the strong leadership needed to make the issues of informal settlements a community priority. The use of a truly comprehensive and innovative sustainable development framework that explicitly acknowledges the existence of the city's informal settlements and the reasons why they exist will go a long way toward improving the quality of life for all Chiang Mai residents.

NOTES

[1] In 1964, Chiang Mai University was established as Thailand's first provincial university, bringing both prestige and resources to the community.
[2] Originally two rings of walls surrounded Chiang Mai. A wall made of brick with an exterior moat surrounds the ancient inner city in an almost perfect square. A mostly earthen wall (with brick in some portions) was built as an outer protection.

[3] At least four times between 1968 and 1996, the residents of informal settlements in Chiang Mai were forcibly evicted from informal settlements. In one case, the settlement had existed for 10 years (CASE 1998).

[4] As of June 1998, the municipality of Chiang Mai officially recognized only 21 communities as eligible for receiving the special financial support available for poor communities.

[5] Information from an interview at the Chiang Mai Municipality, Department of Social Welfare, Office for Community Development, July 27, 1998, with Mr. Chalor Bunthong, Director, Mr. Somsak Upapan, and Mr. Jedsada Phromkhan.

[6] There are nine major groups of hill-tribes or upland ethnic minorities that live in upper Northern Thailand, comprised of approximately 400,000 persons. While these people were born in Thailand, they are ethnically and culturally different than the majority Thais, and many have been denied citizen rights, including the rights to education, health care, and social services (Ritchie, 1996).

[7] Such registration is necessary in order to legally obtain water, sewer and electric connections. While it has been reported that a number of the residents of informal settlements in Chiang Mai have attempted to register their houses, at least one official had refused, indicating that the registration of "squatter families" was inappropriate (CASE 1998, 10).

[8] See the Preface for a more complete explanation of the Joint Study and its participants.

[9] The combination of the information from the Joint Study site visits and interviews with the details of the CASE census is especially significant because it provides current information to link the plight of the informal settlement residents to issues of sustainable development, both within their communities and to the city as a whole. By viewing the characteristics of these communities within the context of integrated economic growth, social/cultural development, environmental preservation and natural resource conservation, clear patterns and lessons emerge.

[10] The canals surround and radiate out from the old walled city, with a number of the major streets crossing over them. In many cases, small settlements are located along the canals near theses major thoroughfares.

[11] Information from an interview at the National Housing Authority, Chiang Mai Office, August 10,1998.

[12] Other chapters in this book emphasize the workforce training needs in two of Chiang Mai's most important economic sectors. Barry emphasizes the need for better workforce training for the manufacturing sector in Chapter 5, and Romanos similarly calls for better training of workers in the tourism sector in Chapter 7.

REFERENCES

Bartelmus, P. (1994). *Environment, Growth and Development: The Concepts and Strategies of Sustainability*. New York: Routledge.

Boonyabancha, S. (2001). Citizen Networks to Address Urban Poverty: Experiences of Urban Community Development Office, Thailand. Asian Coalition for Housing Right. Accessed at http://www.achr.net/network2.htm.

CASE. (1998). *Report on Chiang Mai Slums*. Chiang Mai: Community Architects for Shelter and Environment.

Choguill, C. (1995). The Future of Planned Urban Development in the Third World: New Directions. In B. Aldrich and S. Ranvinder (Eds.), *Housing the Urban Poor: Policies and Practices in Developing Countries*. London: Zed Books.

________ and Choguill, M. (1996). Toward sustainable infrastructure for low-income communities. In C. Pugh (Ed.), *Sustainability, the Environment and Urbanization*. London: Earthscan.

Coomer, J. (1979). The Nature of the Quest for a Sustainable Society. In J. Coomer (Ed), *Quest for a Sustainable Society*. Oxford: Pergamon Press.

Goldstein, S. (1972). *Urbanization in Thailand, 1947-1967*. Institute for Population Studies, Bangkok: Chulalongkorn University.

Greenberg, M., and Schneider, D. (1996). *Environmentally Devastated Neighborhoods: Perceptions, Policies, and Realities*. New Brunswick, NJ: Rutgers University Press.

Hossain, K. (1995). Evolving Principles of Sustainable Development and Good Governance. In K. Ginther, E. Denters, and Paul J.I.M. de Waart (Eds.), *Sustainable Development and Good Governance*. Norwell, MA: Kluwer Academic Publishers.

International Council for Local Environmental Initiatives. (1997). *Local Agenda 21 Survey: A Study of Responses by Local Authorities and Their National and International Associations to Agenda*. Toronto: ICLEI. Accessed at http://www.iclei.org/la21/la21rep.htm.

IRC International Water and Sanitation Centre. (1997). *Operation and Maintenance of Sanitation Systems in Urban Low-Income Areas in India and Thailand: Report on a joint research program 1989-1993*. Delft, The Netherlands: IRC International Water and Sanitation Centre.

Kirby, A. (2001). Editorial: Housing in the Developing Economies. *Cities. 18:(4)*, 211-212.

Main, H.A.C. (1990). Housing Problems and Squatting Solutions in Metropolitan Kano. In R. Potter and A. Salau, (Eds.), *Cities and Development in the Third World*. London: Mansell Publishing Limited.

Milton, D. and Satterthwaite, D. (1994). *Cities and Sustainable Development*. Manchester: Manchester City Council.

National Housing Authority. (1991). *Slum Development*. Bangkok: National Housing Authority.

Peattie, L. (1987). Shelter development and the poor. In L. Rodwin, (Ed.), *Shelter, Settlement and Development*. Boston: Allen & Unwin.

Phantumvanit, D. (1988).Thailand: Degradation and Development in a Resource-Rich Land. *Environment, 30(1)*.

Pugh C. (2000). Squatter settlements: Their sustainability, architectural contributions, and socio-economic roles. *Cities, 17(5)*, 325-337.

Ritchie, M. (1996). *A Model for Successful HIV/AIDS Education in Upland Minority Communities of Northern Thailand*. Project Evaluation Report. Chiang Mai: The Health Project for Tribal People.

Rondinelli, D. (1991). Asian Urban Development Policies in the 1990s: From Growth Control to Urban Diffusion. *World Development. 19(7)*,791-803.

Suphapodok, A. Tridech, S., and Fongsatitkul, P. (1992). Sustainable development initiatives in Thailand, *Canberra Bulletin of Public Administration, 69*.

Thailand Development Research Institute. (Undated). *National Urban Development Policy Framework, Final Report Area 2: Urban Population, Employment Distribution and Settlement Patterns*. Bangkok: Thailand Development Research Institute.

United Nations Center for Human Settlements (Habitat). (1982). *Survey of Slums and Squatter Settlements*. Dublin: Tycooly International Publishing Limited.

United Nations Center for Human Settlements (Habitat). (1996). *An Urbanizing World: Global Report on Human Settlements 1996*. Oxford: Oxford University Press.

United Nations Center for Human Settlements (Habitat). (2002). Area and population of formal and informal settlements. Accessed on the Internet at esl.jrc.it/envind/un_meths/UN_ME048.htm.

Urban Environmental Guidelines Project. (1994). *Slum Improvement.* Bangkok: Local Government Development Division, Department of Local Administration.

Van den Berg, J. (1993). A framework for modeling economy-environment-development relationships based on a dynamic carrying capacity and sustainable development feedback. *Environmental and Resource Economics, 3 / 4.*

Wang, X., and Auffrey, C. (1998). Analyzing Spatial Correlations Between Hazardous Waste Sites and Mortality Rates, *Environments, 26 (2),* 39-57.

World Bank. (1996). Changing the City for Good: Developing a Focused Urban Agenda. *Environment Matters, Summer.*

World Bank. (1997). World Development Indicators 1997. Washington, D.C.: World Bank.

World Commission on Environment and Development. (1987). *Our Common Future.* Oxford: Oxford University Press.

World Health Organization. (1998). *The World Health Report 1998: Life in the Twenty-First Century: A Vision for All.* Geneva: World Health Organization.

Yap, K. (1995). Slums and Squatter Settlements in Thailand. In *Housing the Urban Poor: Policy and Practices in Developing Countries* (pp. 261-276). London: Zed Books Ltd.

CHAPTER 15

HUGO RINCON
DWITA RAHMI

PRESERVATION OF CULTURAL AND HISTORIC HERITAGE AS A TOOL FOR SUSTAINABLE DEVELOPMENT

Abstract Cultural preservation is becoming an urgent concern of urban residents worried about the loss of their cities' cultural identity, history, and heritage. The conservation of this urban heritage and the preservation of the social and economic activities that make use of it, are now seen as part of the community's sustainable development effort. This chapter addresses this approach to development while exploring the conditions for the preservation of historic buildings and urban spaces in the city of Chiang Mai. The methodology used included reviews of national and local regulations, assessment of the findings of previous studies, structured interviews with residents, businesses, community leaders and local government officials, field work, and a ground survey for the Tha Phae Road case study. The study's findings are: (a) that Chiang Mai cannot avoid the inevitable influence of cultural and economic modernization; (b) that it is unclear at this point if the efforts of citizens and government can contribute to the long-term sustainability of the city, despite their enhanced awareness of the need for preservation of the local cultural heritage; and, (c) that to become a more important component of the development process, preservation in Chiang Mai should be institutionalized as part of a strategic urban development plan, including documentation, legal and financial support, and community participation.

1. CONSERVATION AND DEVELOPMENT

Some cities are static, but most will change physically as their economies undergo transformations (Loo, 1996). In this process of physical change, cities and their communities are in danger of losing their sense of place, their social connectivity and fabric, their environmental integrity, and their cultural foundations. Physical change means the alteration of buildings, of the space encompassing them, and of the interactions they determine. By destroying this physical complex of structures and interactions, change affects the essence of the community, and the way the people in that community live, interact, and react to external stimuli. These changes raise issues of protection of the architectural heritage. In recent years particularly, the loss of cultural heritage as a result of globalization, massive reconstruction, and rapid redevelopment of the urban fabric is becoming recognized as a pressing concern in developing world cities (Yeah and Huang, 1996), especially in the secondary cities of Asia. Thus, cultural conservation, which just a few years ago was not even considered part of the development priorities of developing nations, is becoming an urgent concern of urban residents worried about the loss of cultural

M. Romanos and C. Auffrey (eds.), Managing Intermediate Size Cities, 295—318.

identity and severance of the cultural linkages between the new generations and their history and heritage.

The concept of urban conservation includes the preservation of the historic urban character and form of a community, the architectural quality of its structures, its street patterns, its outdoor furniture and objects, and its other physical features and characteristics. It also covers the traditional fabric of the community and its successful adaptation to changes inherent in the urban development process. This preservation in the past was primarily based on public or private action relating to the protection and maintenance of ancient buildings and monuments of national or regional historic and architectural significance or interest. But today, it covers much more, because communities are beginning to pay major attention to many aspects of their past, their ethnic heritage, and the events that shaped them, regardless of the archaeological significance of the artifacts and their national as opposed to local role.

Conservation is concerned with the wise and beneficial use of natural, built, human, and financial resources. As such, for historic and cultural preservation to make social and economic sense, it must be considered as a part of a community's development program and plans, that is, integrated into a single and unified process following agreed planning policies. For example, the conservation of buildings and monuments is always facilitated by making use of them for some socially useful purpose (what is called adaptive reuse). If the desire is to preserve the character of a structure or an urban complex, attention has to be focused on the urban ensemble rather than just on the architectural merit of individual buildings. However, in order to remain living structures, historic buildings must continue to fulfill their original or adapted reuse functions (Gonzalez, 1977).

Conservation has a protective character as well, that includes, besides the preservation of artifacts, the control of conflicting uses and architectural forms in a given area. The historic, artistic, archeological, and aesthetic interest of historic buildings not only call for the maintenance of the character of a given structure but also that of the architectural and aesthetic quality of the ensemble to which it belongs. Further, safeguarding historic areas does not only mean the preservation of a valuable heritage, but also its enhancement and the harmonious integration of the old and the new, as part of the effort to enhance the overall character of the community.

Recognizing, therefore, the need to preserve a community's cultural and historic heritage, but also the need for scarce resources to be allocated efficiently, the need for old structures and complexes to play a role in today's community life, and the positive or negative impacts that such area conservation can have on the entire community, makes it obvious why, in the urban areas of developing countries, the balance between development and conservation is the subject of major and intense policy and design debate.

In this context, conservation of historic areas as part of a community's sustainable development program refers not to the static conservation of a significant site, but to a dynamic process in which the preservation of the urban heritage becomes a means and an integral component of a community's efforts at social and economic development. The public sector has an indispensable role in developing the institutional mechanisms needed to balance these interests and in promoting collaboration among the many and diverse community and national actors (Rojas, 1999). But at the same time, the search for more effective management of the urban heritage may contribute to an increased consensus about the need to incorporate the private sector and civil society into public sector initiatives. The preservation of the historic built environment may aid in the search for a diverse, pluralistic, and sustainable model of urban and environmental development that would be economically feasible, socially accessible, politically participatory, and ecologically sound (Coyula and Lapidus, 1994).

2. CULTURAL AND HISTORICAL HERITAGE OF THE CITY OF CHIANG MAI

2.1. A Brief History

The city of Chiang Mai, also known as Chiengmai or Kiangmai, is located in northern Thailand and is the capital of Chiang Mai Province. Situated on the banks of the Ping River (Mae Nam Ping), it is about 130 km (80 miles) from the border with Myanmar (Burma). One of Thailand's largest cities, Chiang Mai is the chief economic center for the northern part of the country and the northern railroad terminus. The many 13th and 14th century temples and monasteries on the western bank of the river, in the old section of the city, and the nearby Wat Phra Dhat Doi Suthep hilltop temple complex, believed to contain relics of Buddha, are major visitor attractions to the area. Besides tourism, the economy of the area is based on agricultural production and on handicrafts, mainly traditional silverware, lacquer ware, pottery, and woodcarving. (See Chifos and Looye, Chapter 6, in this book).

Chiang Mai today is the cultural center of Northern Thailand. From its foundation in A.D. 1296 by Phya Mengrai, ruler of the Lanna Kingdom, the city has experienced both the apex of achievement as capital of the Kingdom of Lanna and virtual annihilation from the ravages of wars. Though modern Chiang Mai no longer is the seat of a kingdom, the city preserves its ancient royal identity with great and natural dignity. Traces of its heritage are present in the numerous ancient chedi[1], the walls and fortifications, and the other brick structures scattered around the old city and dotting the landscape around it.

In its early years as the capital of the Lanna kingdom, Chiang Mai took advantage of its strategic location between the Kok and the Ping river basins to become the economic and defense center of a broad region containing several smaller outlying towns. Its dominance of the north-south trading route on the Ping River helped it become a major commercial hub. In addition, the vast and fertile valley along the Ping River established the region as a significant center of agricultural production (Promsao, 1996, 14). This absolute dominance ended in 1327, when the center of the Lanna kingdom administration was moved to the new city of Chiang Saen, and Chiang Mai became the second city of the realm. The Lanna kingdom reached its peak between 1442 and 1487 and, by the time of the reign of Phaya Ket Chesatharaj (1525-1538), began to decline. In 1579, Burma conquered the city and took control over the region for the following two hundred years. Then in 1763, Burma lost the city to the Siamese Thai. Under Siam, Chiang Mai maintained a certain degree of independence, but had lost most of its prominence and entered a period of decline and abandonment which lasted until 1782, when the deserted city was repopulated and rebuilt. Modernization of the city began with the introduction of rail service between Bangkok and the province in 1902. Then, in 1933, Chiang Mai was officially absorbed by Thailand and became one of the country's northern provinces. But even with the construction of roads connecting it with Bangkok and other major cities of the country, and the establishment of an airport in the following years, Chiang Mai remained a small and remote town until the 1970s. Improved transportation, however, helped the city begin to move out of its isolation, and its many Buddhist temples and its renown handicrafts -- gold and silver jewelry, lacquer ware, baskets, woodcarving, pottery, and textiles – gradually began to attract visitors in great numbers. Since 1982, the city has been designated as a center for commerce and services as well as for tourism and handicrafts.

2.2. The City's Heritage

The historic section of Chiang Mai has a square plan of approximately 2.5 square kilometers. The urban plan consists of a walled compound with walls facing the cardinal points, and with one gate in each of the four walls, except for the southern wall, which has two gates. Outside the walled area, the old city spreads eastward to the river and northward along the old highway for about one kilometer.

Chiang Mai is renowned for its numerous monastic temples. A survey conducted in 1897 found 100 temples within the walls of the old city and another 51 outside the walls (Promsao, 1996, 30). Temple architecture varies significantly from one area of Thailand to another. In northwestern Thailand, many temples have a clear Burmese architectural influence, with Buddha statues painted in white and construction of the temple building -- the "wat" -- often using teak wood. In addition to the wat, temple compounds often contain several other buildings: the "bot", a main building which contains the Buddha statue, the "kut", which contain the rooms of the monks, the crematorium, easily recognizable because of its tall chimney, and the library. Other structures include the "sala", a big pavilion for ceremonies, and a building where the monks eat their meals. The "chedi" is the highest building in the wat and is used to store relics of Buddha -- in the most famous temples -- or relics of important monks -- in more ordinary temples. Roofs are often made of red tiles, and the number of roof layers -- "stacks" -- is always odd. Woodcarving is typically used for decoration and for temple offerings.

Most of the existing traditional architecture in the city are Buddhist religious structures; a good many of which are located inside the walled city The rest of them can be found scattered throughout the urban area. Local culture and climatic conditions have determined the living conditions of the Lanna people. Dwellings are made of natural materials, and are simple in style. The typical Lanna house is strongly built. It is usually a small building on high stilts covered with a gabled roof.

The wealthier people live in larger houses built with more attention to details. At the peak of the gabled roof, which may be tiled with slate or wood, an intricately carved wooden Ka Lae is placed for decorative purposes. Another traditional feature of the Lanna house is the religious altar attached to the eastern wall of the terrace, and an altar for the ancestral spirits found on the eastern wall of the bedroom above the bed.

Figure 1: Traditional Chinese Shop House

The Buddhist temples and the Lanna houses are not the only historically significant structures in the city. Traditional Chinese shop houses are important features in the urban landscape, especially in the commercial areas of the city (Figure 1). Shop houses represent the traditional architecture of Chinese migrants to Thailand. These rectangular two- and three-story buildings have a store in the lower level (with corridor and colonnade in the front) and a house above (with a balcony). In the early years, these shop houses were built using local teak timber. Later, at the beginning of the 20th century, these were usually replaced with brick and concrete structures, although they usually preserved the use of wood for the upper walls, railings, and the side-gabled or hipped roofs. Several clusters of older wooden shop houses and complete districts where such structures have been preserved can still be seen in Chiang Mai, and in other places in the region, such as in Lampung (Figure 2).

Figure 2: Older Wooden Shop Houses

3. CHIANG MAI'S PRESERVATION FRAMEWORK

3.1. National Level Regulations

Several legal instruments are in place to guide historic preservation in Thailand at the national level. Most notable among them are:

- The City Planning Acts 2518 (1975) and 2535 (1992). These two acts stipulate that comprehensive and land use plans should be developed to maintain and preserve places and structures of historic value or significant natural beauty. The acts also direct municipalities to define such areas, and determine the types of development that would be allowed in them.
- The Building Code Act 2522 (1979). This act provides for building controls to eliminate negative impacts on the community and empowers municipalities to issue regulations to that effect.
- The Ancient Sites, Antiques, Artistic Materials Act 2504 (1961) and the National Museum Act 2535 (1992). These two acts protect structures in historic sites that have been declared national monuments by the Department of Fine Arts of the central Ministry of Education[2].
- The National Tidiness and Orderliness Act 2535 (1992). This act protects buildings, trees, and natural environments from vandalism and abuse, and directs owners to maintain such resources in good condition.

However, these laws do not appear to be effective in controlling the uses of land in the proximity of historic structures or sites, in setting rules and guiding modifications to historic buildings, and in general, in protecting the cultural heritage. Furthermore, these laws have been initiated and are being administered by a host of different central government agencies. These agencies are often in conflict with each other, and such conflicts cannot be solved easily under the complex system of the Thai bureaucracy (Pimonsathean, 1999). And yet, municipalities have made considerable efforts to protect and preserve their heritage by implementing a combination of rules and regulations, and by taking advantage of the powers vested in them by the national laws.

3.2. Preservation Efforts in Chiang Mai

According to a 1987 survey conducted by the Department of Fine Arts of the Ministry of Education, the central government agency responsible for Thailand's cultural heritage, there were 109 old buildings and sites in the Chiang Mai district, 59 of which were considered significant archeological sites[3]. But, according to Narumit-Rekagarn (1988), only about half the historic buildings and sites of the city were included in this inventory. The buildings in the inventory were ancient and medieval structures, complexes, and temples. No newer significant buildings such as the former city hall, the courthouse or any traditional residences and shop houses were included. The value of this inventory is that structures included in it receive some protection from the city. A boundary is established around them, prohibiting construction and unauthorized activities within it. This boundary is very limited -- usually a 3-meter fence around a building -- and not enough to protect a site from construction infringement or from visual intrusion (Narumit-Rekagarn, 1988, 3). Also, some funds are available for basic maintenance of the buildings, but it is irregular and inadequate to cover all the necessary repairs and restorations needed. On top of all this, there is a serious shortage of trained craftsmen to undertake the restoration work, and of specialized architects and engineers to provide the necessary studies.

The Chiang Mai municipal administration has been relatively successful in planning for cultural preservation, mainly by establishing clear land use rules and regulations as part of its comprehensive plan. Unfortunately, it has not been able to enforce the provisions of these rules effectively, and has been especially weak in encouraging and promoting the preservation of significant buildings and important complexes and neighborhoods. For example, the land area outside and surrounding the old city moat is zoned as a historical and archeological reserve, and parts of it are characterized as an environmental conservation reserve. But the city has not been able to enforce these designations effectively, so construction is taking place in land parcels where supposedly it is prohibited. Elsewhere, developers find ways to build structures regardless of location or proximity to culturally sensitive buildings or complexes. Multi-story buildings are being constructed in various parts of the city using modern, non-local materials such as glass and steel. And, old structures are being demolished and replaced with modern buildings which radically alter the character of the community.

The Municipality has been more effective in the protection of areas within the enclosure of the city walls and moats. There, regulations are much more explicit, and are being enforced via the issuance of building permits requiring that "Buildings shall not be higher than three stories or 12.5 meters. The roofs shall be in Lanna style, and include the "Ka Lae" symbol, for conservation of Chiang Mai's arts as the unique characteristic of Chiang Mai..."(Local Government, Chiang Mai Municipality).

As mentioned above, one of the first attempts to carry out a conservation plan in Chiang Mai was introduced by the Comprehensive City Plan of 1982. The plan made the area within the city wall a cultural conservation zone, regulating the types of buildings, their heights, and the land uses. Following this designation, a special building code and construction regulations for the area were issued in 1983, requiring all new buildings to follow local architectural characteristics. These regulations also prohibit new construction or alterations to existing buildings within a distance of 20 meters from the ancient wall of the city (Figure 3).

Figure 3: New Construction

Despite these efforts, preservation in Chiang Mai has not been particularly successful to date. With the exception of the temples, old buildings have been deteriorating, collapsing, and continue to disappear. In the absence of government regulations or controls, many shop houses are being altered and modernized. The Chiang Mai landscape has changed rapidly from that of a traditional Lanna city of wooden residences and Chinese shop houses, to an unplanned and poorly designed collection of high-rise residential and commercial buildings and modernistic suburban communities. The architectural styles of these new buildings do not observe any of the elements of traditional local architecture, neither in terms of form nor in terms of materials[4].

One of the trends that has helped save some old buildings in the central city is the preference of tourist establishments for older structures. That is true for shops and galleries, as well as restaurants, bed-and-breakfasts, and small hotels. This adaptive reuse has at least provided an opportunity to preserve some traditional houses. As a result, guesthouses, restaurants, tourist agencies, tourist and antique stores using restored old buildings are scattered around the city. In most of them, restoration of the old structures seems to have been successful in preserving traditional architectural features. But construction techniques and materials used have in the past been a problem. For example, the timber used in the original buildings tends to deteriorate due to climatic conditions, making imperative, in most cases, its replacement. In some of the preserved buildings, new wood is used to replace the damaged parts. But more often, alternative materials are used, greatly affecting the building's integrity (such as tiles of different styles, prefabricated elements, metal and glass, etc). Wood actually is at the center of the materials integrity fight between preservationists and developers. Its cost is increasing continuously as more and more restrictions are imposed on tree harvesting. But most of the structural and other construction elements of the old residential and shop house buildings were from wood. So the tendency is to replace wooden components, such as beams, roof gables, and shingles with artificial materials which are cheaper and require less maintenance. But these materials change the appearance of the buildings as well as compromise their building integrity.

4. FACTORS AFFECTING THE PRESERVATION OF BUILDINGS, MONUMENTS, AND SITES

Climate, in all its aspects, is one of the fundamental causes of building decay through failure of materials, which in turn affects the support structure. In Thailand, radiation from the sun, seasonal temperature changes, occasional flooding, wind and moisture, as well as pests and insects may affect buildings, which thus require continuous maintenance. Wood has already been discussed in the previous section of this chapter. Other materials, such as bricks, when they deteriorate, are replaced by new materials that affect the integrity of the building and its mechanical reaction to elements.

The use of cement can be the enemy of historic buildings, especially when used in the form of concrete. Cement is not designed for use in mortar or plaster on historic buildings. Its use is not reversible without damaging the original structure. But concrete has been extensively used in restoration work in Thailand. In one case in Wat Phra Chedi Luang, figures representing elephants originally built of bricks as well as the ramps, were recently "restored" to the original condition using cement mortar which certainly is not the material that was used hundreds of years ago.

However, it is not in all cases that the use of cement has been a mistake. An example is the restoration of the ancient ruins of Sukhothai, an eight hundred-year-old city located in central Thailand, and the capital of the first Kingdom of Siam in the 13th and 14th centuries. The site of a number of notable monuments, this city illustrates the beginning of Thai architecture. In the restoration of Sukhothai, concrete has been used to stabilize old structures in a way that the visitor could perceive the difference between the old and new materials. This archaeological site is included in UNESCO's World Heritage List, and that allows the Thai government to receive funding for preservation and maintenance.

Figure 4: Influence of Different Cultural Forms

4.1. Impacts of Tourism and Economic Prosperity

Cultural transformation. The cultural, architectural, and natural attractiveness of Chiang Mai and its surrounding areas draw visitors throughout the year and entices residents of other Thai cities to resettle in this city. A major impact of this attractiveness is the gradually changing social and cultural composition of the population. As the old residents of the city move out of town, and new residents come in, many of them of non-Lanna ethnic origin, the cultural mix of the city changes, and with it the nature and value of its festivals, ceremonies, theater and other cultural performances, and the entire ambience of the city's cultural life and expressions. Tourism visitors exacerbate these trends. While visitors come to Chiang Mai to observe and experience these cultural patterns and stimuli, their demands transform the local expressions of culture into commodities with a monetary value. This commercialization of culture, in turn, weakens its social context and alienates the people whom it represents. At the same time, such transformations take place among the less traditionalist local residents. The locals are more and more exposed to different cultural forms from the media, from visitors or from other cities, like Bangkok where they are educated or with which they interact through their work. All these influences are bringing about significant, if gradual changes in attitudes and values (Figure 4).

Built environment transformation. The demand for tourist services and accommodations has resulted in a growing number of tourist establishments built, such as hotels, resort communities, golf courses, and related real estate developments. This construction is altering important characteristics of the local cultural landscape, adding taller buildings and new land uses to the older parts of the city. More and more, traditional architecture is giving its place to structures designed in international styles, emulating European or American ways of life. Local scholars have been critical of the increasing "disorder" of the city's skyline, the shapes, sizes and scale of new buildings, the perceived discord of colors and materials, and the lack of effort to incorporate the distinctive architectural features of the indigenous architecture and building methods (Figure 5). The increased number of high-rise structures, which in many cases overshadow historic buildings, has especially affected the cultural ambience of the older neighborhoods. Furthermore, the planning and the design for significant new hotels, office and institutional buildings, department stores, and housing projects is done by Bangkok design and engineering consultants, who often lack the knowledge, interest, and even sympathy for indigenous architecture and the local cultural environment.

Figure 5: "Disorder" of the Cultural Landscape

Natural environment transformation. Increasing prosperity has also brought about the expansion of the Chiang Mai metropolitan area, reducing the amount of farmland devoted to wet rice cultivation, and further altering the visual and cultural character of the urban region. The hillsides surrounding the city are being denuded of their forest, and the tribal groups living in the hill and mountain regions around Chiang Mai are being resented by the locals because of their living conditions, their control of the forests, the perceived damage to the forest caused by slash and burn agriculture, and the differences in cultural characteristics between the Lanna and the rest of the tribal groups.

4.2. Cultural Loss

"Preservation is not in our culture. They (the Thai people) have never known about it" (Vithi Panitchaphan, Thai Art Department. Chiang Mai University).

Buddhism teaches that everything on earth is temporary. In this sense, following their religious beliefs, Thai people should not be attached to belongings or property, and should look forward rather than backwards in their lives. This way of thinking may explain, at least partially, the Thai attitude towards the physical and cultural heritage and the relative lack of interest in what western preservationists would consider significant historic entities worthy of preservation for the continuity of culture and the benefit of future generations. In short, many Thai people do not place too much value on the importance of their heritage. Some feel that they need to fix up old temples and houses just to make them look nice for visitors. This is combined with a tendency towards the modern. Many residents who have moved to Chiang Mai in the last twenty years would like to see the city become a newer and cleaner version of Bangkok. There is also a lot of interest in western life styles, cultural norms, and architectural forms. More and more, the suburban homes of the wealthy resemble western villas, some of them complete with statues of Greek or Roman gods[5].

Preservation and sustainability. The combination of radical change in the old city brought about by rampant construction activity, and the transformation of the rural landscape surrounding the city, is already producing an urban environment very different from the traditional setting that made Chiang Mai such an attractive traditional cultural center. The question then is, how significant is the effort to restore and preserve some isolated structures and even some small urban complexes, if the rest of the urban and urbanized area is changing in appearance and character so dramatically (Figure 6). It is unclear at this point if the preservation efforts in Chiang Mai are contributing to a long-term sustainability of the urban fabric itself, or are simply another component supporting the city's tourist sector. While in either case preservation can be considered as contributing to the city's sustainable development, the preservation outcomes are diametrically different. A coordinated effort at comprehensive preservation of the urban fabric will carry with it the cultural and historic continuity that would make for a meaningful sustainable development path for the city; while the isolated and fragmented restoration efforts would simply produce a stage-set of architectural samples for tourist consumption. At this point, it is an open question how much of the character of an old, significant cultural center is critical to preserve in order to consider that a community is achieving long-term sustainability.

Figure 6: Preservation versus Urbanization

4.3. Public Management of Historic Preservation

In the highly centralized administration of the state of Thailand, most decisions regarding what to preserve and restore are made in Bangkok. Especially in the area of cultural and historic preservation, there is no efficient management of human and material resources, nor coordination among agencies and between central and local governments. The fragmentation of responsibilities among several central agencies and the limited training and expertise of local authorities make the understanding of the process of preservation extremely confusing. In addition, financial resources allocated to the preservation efforts of the country are extremely limited. For example, in Phuket and Chiang Rai we observed magnificent complexes of Chinese shophouses abandoned and in an extreme state of deterioration, because the funds approved for their structural supports several years earlier, never materialized. City officials in both cities confided in us that they had repeatedly requested permission from the Department of Fine Arts to spend their own funds for the partial restoration of these structures, but were denied permission because these buildings had been designated as historic, and therefore, fell under the jurisdiction of the Department. But, according to Narumit-Rekagarn (1988), the Fine Arts Department is weak in authority and administrative support, and has the lowest budget among the

Departments of the central government. In addition, the agency itself is not an advocate for broader preservation efforts. It has traditionally focused its attention only on ancient buildings, including temples and old city fortifications and gates. Very little funding is available for the restoration or maintenance of houses and other common buildings[6].

At the danger of generalizing, our investigation under the Joint Study has found evidence that preservation in Thailand lacks government concern. Thai government officials, both at the central and the local levels, associate the concept of preservation mostly with the restoration and maintenance of religious buildings - temples and monasteries - major archaeological sites and official buildings of historic importance for the nation, such as palaces and other government structures. The rest of the old architectural heritage, both residential and commercial, is not considered of major importance, and not worth investing funds in. Even at the local level, where this heritage is encountered daily, the local residents are often concerned with it, and tourism is dependent on its existence, preservation is not one of the priorities of local agencies. When, for example, we asked the city architect's office in the municipality of Chiang Mai to explain to us how preservation is done in Chiang Mai, we were directed to solicit this information from the local office of the Department of Fine Arts, the agency of the Ministry of Education responsible for the designation of historic structures and complexes and the funding of restoration efforts. The ignorance of the status and the procedures for historic preservation in the city indicated to us that the municipal agencies had little concern for, information about, or interest in the conservation efforts taking place – or rather being neglected – in their city.

5. LESSONS FROM THA PHAE ROAD

In order to understand how the development of the city affects historic preservation efforts, and how business and preservation interests can be connected, our team embarked on a study of Tha Phae Road, one of the major thoroughfares of the old city of Chiang Mai (Map 1). Tha Phae Road is a one-way street that connects the Old City of Chiang Mai with the Ping River and the eastern section of the city through the Tha Phae gate[7]. It is a major commercial district containing a variety of activities – hotels, restaurants, shops, banks, temples and travel agencies. The area is densely built and heavily used by pedestrians. Traffic is constant and congestion occurs frequently. Air pollution from the passing cars is evident on the façades of the surrounding buildings. Sidewalks are irregular, and there is no significant street furniture besides street lighting.

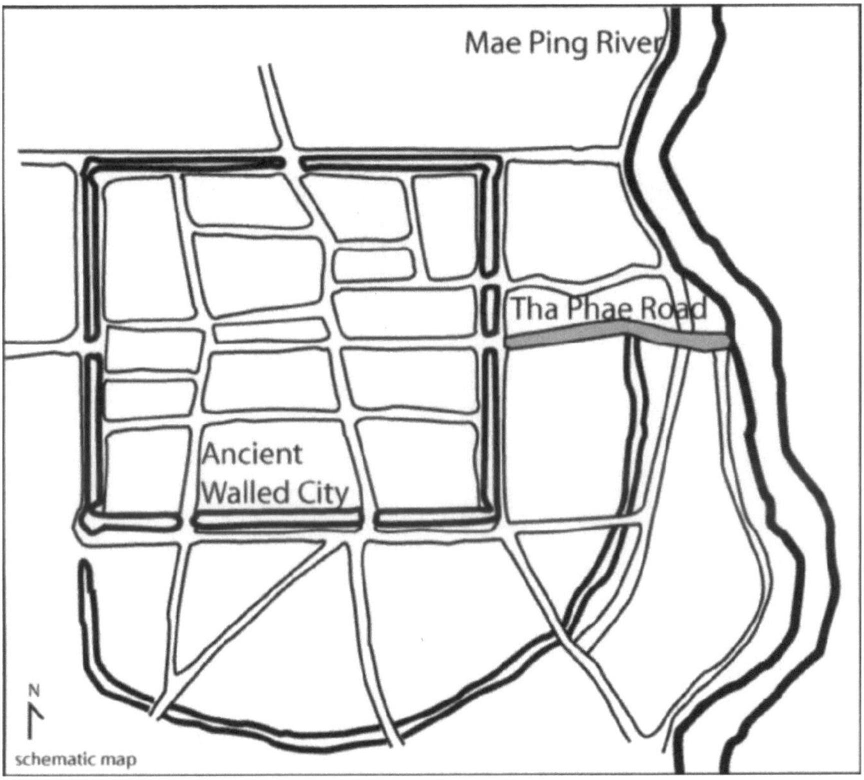

Map 1: Tha Phae Road
Source: Promsao 1996: 17

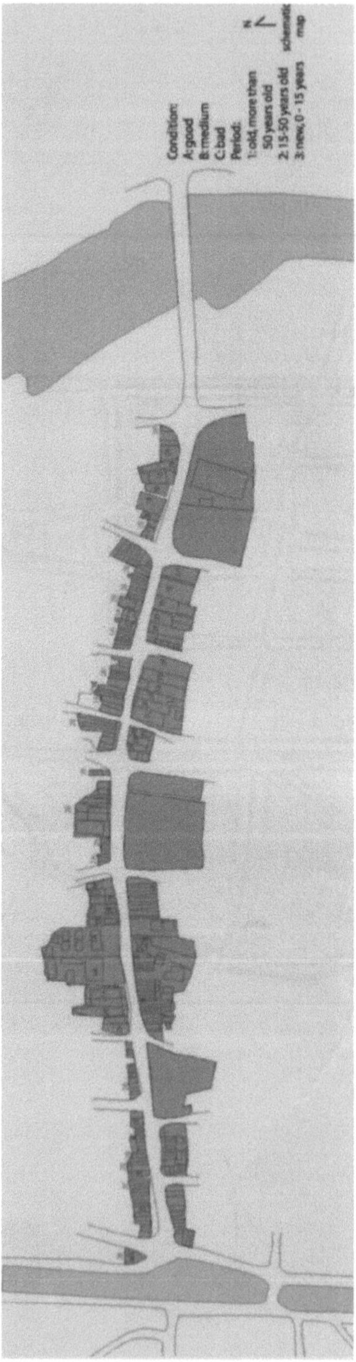

Map 2: Tha Phae Road Structures and Use

The first task of the study was to identify and count the types of buildings along the road, determine their uses/functions, describe their architectural characteristics (number of floors, height), and estimate their age. The findings of the survey are shown in Table 1 and in Map 2.

Table 1: Summary Description of Tha Phae Road Structures and Use

NUMBER OF INDIVIDUAL STRUCTURES	76
BUILDING FUNCTIONS	
banks	8
offices	9
single family houses	3
temples	4
hotels	2
travel agencies	20
restaurants and pubs	17
clinics and drug stores	7
clothes shops	16
craft/antique shops	33
other stores	36
Total Functions	**155**
BUILDING ELEVATIONS (Excluding temples)	
1-story height	4
2-story height	37
3-story height	19
4-story height	8
5-story height	2
6-story height	2
Sub-Total	**72**
Temples	4
Total Number of Structures	**76**
BUILDING AGE	
More than 50 years old	13
15 to 50 years old	50
0 to 15 years old	13
All Buildings	**76**

Source: Rincon and Rahmi, 1998

The next task was to survey the physical conditions of these buildings, in order to establish levels of maintenance. From the total of 72 buildings, again excluding the 4 temples, 22 were found to be in good condition, 45 were in mediocre condition, and 5 were in deteriorated condition. Once that was established, the team attempted to identify structures that had already been preserved or were worthy of preservation. In addition to the 4 temples, 13 of the 72 buildings on the Road were identified as historic structures that have maintained their architectural integrity over the years and have significant cultural or historic value. The remaining 59 buildings are either older structures that have been drastically modified over the years, or are relatively new additions to the streetscape. Some of the older structures are traditional Chinese shop houses, and are in relatively good condition, so that they would only require minimum maintenance to be preserved as part of the city's cultural heritage. The new buildings, on the other hand, invariably were designed with no regard to the traditional scale of the older buildings, paid no attention to their surroundings, and used materials alien to the traditional materials prevalent on the Road.

The team then set out to identify interventions to the historic buildings along the road and find out how changes in the appearance, the condition, and the role of the buildings had affected the urban landscape in the neighborhood. We found out that some older houses and commercial buildings on the Road, built in the traditional Thai designs of the late 19th century and the beginning of the 20th century, have been restored and transformed into tourist and other shops, tourist agencies, restaurants and bars. Private owners have invested in their restoration and have in several cases tried to use original materials and preserve the architectural features and construction details of the structures. In some other cases, restoration of the buildings took place simply to strengthen and protect the structures so that they could be used for commercial purposes. In these cases, the façades of the buildings had usually suffered drastic changes and had lost much of their traditional character.

The four temple compounds have preserved their architectural style intact, in spite of the numerous modifications and add-ons to them, most of which were built using reinforced concrete. The new additions, built to accommodate new needs and expanded numbers of monks, follow the Thai tradition of leaving the older structures in place and intact.

Some of the older structures remain empty or are used as storage of merchandise, while the sidewalks in front of them are used to display goods for sale. In one such case, we identified a house of Thai-Chinese architecture, with impressive woodcarving, wood tile roofs, and ka lae decorations on the roof tops. The building was empty but not in very poor condition. It would not require a lot of effort and expense to restore it and make it economically profitable.

The Road is dominated by advertisements and billboards, behind which great buildings are often hidden. Sometimes complete metal or wood structures have been erected in front of the old buildings to modernize the facades, changing drastically the appearance of the street at that location. In general, the use of non-traditional materials in the construction of the new buildings creates very strong visual intrusions. New buildings are having an even more powerful effect, because they use setbacks to accommodate parking on the front of the lots (in the case of office and bank buildings, for example). Finally, some of the new buildings are much wider than the regular lots because they are built on two or more lots connected after the demolition of the old structures. The break in the continuity of the street façade, combined with the modern design of the new buildings and with the wider lots, produces impressions of disintegration of the streetscape.

Finally, we found a perfect example of what can go wrong with restoration in the Tha Phae Gate which frames the intersection of Tha Phae Road and the moat surrounding the old city wall. The gate was restored by the city in the mid-1980s, but the supervising official had his own ideas about how the gate should look. So the wall and the gate were rebuilt using new materials and without regard to the original designs. Today, aggressive landscaping by the city manages to minimize the effect of concrete and other new materials on the medieval structure. In fact, the creation of a small park there has been one of the favorite spots for local residents, who assemble in the evenings to enjoy the open space and be inspired by the ancient walls and their cultural heritage.

This small case study, through the interviews we conducted with residents and shop owners and users along Tha Phae Road, revealed a number of things about several aspects of preservation in Chiang Mai. First and foremost, it showed that government initiative in the area is all but non-existent. Neither the Fine Arts Department of the central government, nor the archaeological service, nor the municipality of Chiang Mai has done much to preserve the character of the Road and conserve at least some of its key buildings. The temples are being maintained by the monks with funds from public alms, but that is a religious, not necessarily a conservation effort. Furthermore, restoration and protection of the old buildings on the Road from collapse is done by their owners or tenants without special preservation permits or supervision and enforcement of regulations by any government agency. Third, substantial old buildings have been demolished and have been replaced with new structures without any objections being raised by responsible local agencies.

On the other hand, many of the old structures on the Road have been preserved through private initiative. Owners of these buildings, and often the tenants themselves, attempted to preserve the traditional architecture and even replace traditional materials with similar elements, in an effort to preserve the character and ambience of the place. Businessmen have found that restoring an old building and then putting it to some tourist use is good business, because visitors appreciate the effort and enjoy shopping, eating, or even staying in a traditional setting. Also, the small but active groups and NGOs that have emerged in recent years in Chiang Mai for the purpose of protecting the city's environment and support the preservation of its monuments and buildings, have been actively pursuing and helping the efforts of these businesses, and have in the process persuaded the municipality to approve building permits and even extend some funding to assist building owners. Thus, the picture that emerges from this investigation is that while the central government and the municipality have not done much to enhance the traditional treasures of the city, the private initiative, both individual and group, has been at least partially successful at both raising awareness and helping save a number of buildings from destruction.

6. A PRESERVATION STRATEGY FOR CHIANG MAI

It is clear that the citizens and the government of Chiang Mai are becoming more and more aware of their cultural heritage and the need for its preservation. As modernization and the inevitability of global economic forces become more obvious, so do the people and organizations of the city recognize the urgency of protecting their historic buildings and sites before they disappear under the surge of new construction and of contemporary mores. As such, strategies for the preservation of these buildings and sites will be needed, if the city is to incorporate cultural conservation and preservation into its sustainable development program. Below we outline the basic elements of such a strategic plan.

1. Identification and documentation. Information and documentation should be collected and organized as part of the city's Geographic Information System, for each type of cultural and historic heritage resource, especially historic buildings, complexes, and sites. For buildings, this inventory would include building type by a predetermined list of categories, function, location, age, architectural style, history, ownership, market value, and structural conditions.
2. Preservation priority classification. Preservation priority ranking would be assigned to the buildings in the city inventory according to their relative value and conditions. Top priority structures and sites would have to be preserved at all cost. Medium priority buildings could be adapted for new uses, where that option would be possible. Buildings classified as low priority would be protected to the extent possible, but could be altered or even removed/demolished if they represent a public hazard or if they intefere with significant government projects.
3. Institutionalization of preservation effort. Preservation of historic buildings and sites should be done through collaborations of all segments of the urban society, including central and local government agencies, private institutions, non-government organizations and citizen groups, as well as individuals. Mechanisms should be established to ensure this collaboration, based on enabling legislation and capitalizing on the academic and cultural institutions of the city.
4. Legal, administrative, and financial support. Laws, rules and regulations aimed at the protection and preservation of the city's cultural, historic, and architectural heritage should be established. Where existing laws and regulations are insufficient for the purpose of protecting and enhancing the city's heritage, they should be supplemented by appropriate legal instruments at the national, provincial, and local levels. In order to carry out policies and programs of conservation, properly staffed administrative services should be established. Preservation of the architectural heritage should be encouraged by suitable forms of financial aid and incentives, including local tax measures. Property tax exemptions or reductions should be granted to private owners and developers to create incentives for private participation in the conservation effort of the city.
5. Participation of the community. Conservation planning and the management of the city's cultural resources, should involve and engage all components of the community, public and private. The public should be properly informed, because citizens are entitled to participate in decisions affecting their environment. Interest in heritage buildings and sites should be stimulated by every kind of activity capable of improving public knowledge and appreciation of preservation and its benefits. Activities could take the form of lectures, seminars, exhibitions, special educational training, architectural preservation campaigns, traditional architecture

awards, publications including works designed for the general public, use of the media, etc.

6. Involvement of local heritage conservation groups. Non-Government Organizations (NGOs) or Non-profit Organizations (NPOs) could promote the idea of preservation in their communities and could mobilize the community in support of specific conservation efforts or projects, as well as could lobby the local and central agencies for the development of appropriate preservation policies.
7. Technical support. In Chiang Mai there are too few architects, engineers, technicians, specialized firms and skilled craftsmen who can respond to the restoration and preservation needs of the community. It is necessary to develop training facilities and increase prospects of employment for the relevant managerial, technical, and manual skills. The local building industry should be urged to adapt itself to these needs.

7. CONCLUSION

Chiang Mai cannot avoid the inevitable influence of cultural and economic modernization and globalization. But such trends bring with them the danger of deterioration of old elements of the community that do not meet contemporary standards of efficiency, design, or comfort. Despite the continued improvement of the quality of life for the city's residents, long-term sustainable development would be incomplete if it materialized at the loss of the region's cultural heritage and historic continuity. That, in turn, makes preservation and conservation of the remnants of the past, be they buildings, archaeological ruins, or other forms of cultural expression, a significant element of a sustainable development plan for Chiang Mai.

The historical and cultural heritage of the city, in the form of traditional buildings and sites, both religious and secular, is an important asset for Chiang Mai. That heritage can be identified as comprising:

1) Significant symbols of northern Thai culture and identity;
2) Representative items of the northern region architectural /cultural style;
3) Symbols of the northern Thai history; and,
4) Items with outstanding historic, artistic, religious, or scientific value.

To date, preservation efforts in Chiang Mai have been weak and low key, while demolition of historic buildings, alteration of the city's skyline, and intrusion of new structures in the Old City fabric, have been prevalent. This study found that some of the reasons for the weak preservation efforts in Chiang Mai are related to two factors: inadequate management of the preservation effort, and lack of interest in preservation among the public and government agencies.

Inadequate management:

- Absence of information, maps, plans, and documents describing buildings and sites.

- Lack of expertise among architects, engineers, and technicians regarding methods, techniques, procedures, and materials for building and site conservation.
- Unavailability of traditional building materials for the replacement of decaying building components, especially timber, other wood components like shingles, and hand-made hardware.
- Building codes and regulations to protect old buildings and sites are inadequate, leave too many holes for developers and other violators, and are weak on penalties; enforcement is minimal, if it exists at all.
- In restoration projects, fragmentation of responsibilities exists because of laws being enforced by several different agencies of the central government. Lack of coordination and communication among agencies creates inefficiencies in the execution of restoration projects.
- Conservation consultants and contractors are chosen at the national level, resulting in lack of communication among the contractor, who is paid by Bangkok, the central government agencies, which make decisions about Chiang Mai without consulting with the city administration, and the local municipal agencies.

Lack of interest in preservation among the public and government agencies:

- Buddhism encourages people to look to the future, not be concerned with the past; and to not focus on the preservation and expansion of material wealth and property.
- The Thai people are not interested in preservation and do not like to live in old structures; they like modern buildings and homes.
- The Thai people do not recognize the importance of their cultural heritage and the need for its preservation; they consider the preservation of buildings of interest mostly to the tourists.
- There are few heritage associations and organized groups , and the ones in existence lack power and resources.

Both government and the public sees historic preservation mostly as needed for the protection and preservation of religious monuments, temples, and significant government buildings; common structures like homes and stores, including Chinese shophouses, are not considered worthy of conservation at the expense of government funds.

But there are some positive things happening in Chiang Mai in the area of historic preservation and conservation. First, citizens have began to organize into cultural and environmental preservation support groups, and have already mobilized in support of worthy causes, such as the protection of threatened temples from new construction in their proximity, and the prevention of construction by developers infringing on the river banks, both times successfully. Second, the city administration has shown strong interest in taking responsibility for the city's preservation efforts in its hands and away from the agencies of the central government. While the city is empowered to take such initiatives, the resources for the implementation of its preservation programs have in the past been controlled from Bangkok, and that has been an impediment that needs to be overcome. Third, Chiang Mai organizations as well as local government agencies have already advocated for better and more effective government programs at both the local and the national level. These would include appropriate building regulations, incentives for preservation and disincentives for the demolition of old buildings, as well as decentralization of preservation related decisions, and allocation of resources for preservation programs to the municipalities. Fourth, emerging interest in architectural heritage preservation has begun to create demand for Thai style homes and stores among educated Thai households and businesses. As a result, developers are beginning to pay more attention to the possibility of preserving older buildings. Last but not least, education and information programs are sprouting in the city, as in the rest of the country, promoting the protection of the national heritage. Thai people are proud of their national heritage and of their history, but to date did not relate these two components of their social continuity to preservation of buildings and other archaeological sites. As they begin to make this connection, Thai scholars at the University of Chiang Mai and Chulalongkorn University believe that there will be a major movement in support of preservation efforts, similar to the movement, about thirty years ago, that alerted citizens to the importance of the country's artistic heritage, and helped stop the massive exportation of art treasures from Thailand.

NOTES

[1] Chedi is a Thai or Ceylonese-style pagoda, the most important and imposing structure in a temple or monastery compound. The Chedi has a very powerful symbolic organization, in which the base of the structure represents Hell, the middle represents Humanity, and the top is a representation of the state of Nirvana.

[2] The Ancient Sites Act establishes the conditions for houses to be included in preservation programs. Houses must be more than 50 years old. Newer houses with character and local architectural significance or important historical background can also be included in this category.

[3] That same year, another inventory of historic buildings was conducted inside the Chiang Mai city walls. The study classified buildings into five categories according to their "merits", and an order of priorities for conservation was set up (Narumit-Rekagarn, 1988, 3).

[4] Chiang Mai's exposure to outside influences has ushered it into the era of modern development. There have been many cultural changes in the city in recent years, and they have had major impacts on the life styles of the old residents. Many of these residents have been abandoning the city as a result, preferring to live in relative seclusion in the countryside, away from the crowds, the noise and the clatter. Respondents to our survey commented that the construction that is constantly taking place in the city, even around temples and in proximity to archaeological and religious sites, greatly upset the "natives" who see in all this activity the destruction of the old city and its living conditions. Combined with the tendency of the Thai middle class to prefer living in modern homes, this reaction helps explain the large scale suburbanization of Chiang Mai and the gradual transformation of its central city into a mostly commercial and tourist district.

[5] A positive sign for the future of preservation in this city is the growth of non-government organizations and citizens groups concerned with preservation. As an example, a campaign to save the old post-office from demolition was recently waged successfully by a group of university faculty and students, architects, and Chiang Mai citizens.

[6] The Fine Arts Department has developed procedures describing how to rebuild, restore, and preserve older structures. These procedures, however, are only implemented in cases where the Department or its local offices are directly involved in a restoration project, and then only if funding is made available for it. The Department has no power or authority to prevent construction changes or to instigate improvements to privately owned buildings.

[7] This gate originally was called Tha Phae Nok (Tha meaning landing, Phae meaning raft, and Nok meaning outer), probably because there was a boat landing for traders on that location of the river bank (Promsao 1996 15).

REFERENCES

Bangkok.com. (1999). Thai Temple or Wat. (www.bangkok.com/misc/thaiworld/wat/wat.htm). Nov. 24.

Chiang Mai Newsletter. Vol 7. No. 7, July. (1998). Trisila Company Limited. Chiang Mai, Thailand.

Coyula, Mario; and Lis Lapidus. (1994). Historic Preservation in Cuba and Latin America. *Other Americas. Spring/Summer. Desigh Book Review. 32/33*, 43-45.

Dix, Gerald. (1990). Conservation and Change in the City. *Third World Review, Vol. 12. No. 4, Nov.*, 385-406.

Gonzalez, Jose M. (1977). Architectural conservation and enhancement of historic towns in America. Barcelona, Spain: Blume.

Jumsai, Sumet. Naga. (1997). *Cultural Origins in Siam and the West Pacific.* Bangkok, Thailand: Chalermnit Press & DD Books.

Local Government Chiang Mai Municipality. Public Relations Section. (Undated). Chiang Mai Municipality: Chiang Mai Dararat Ltd. Thailand.

Loo, Sim. (1996). Urban conservation policy and the preservation of historical and cultural heritage. *Cities, Vol. 13. No. 6*, 399-409.

Narumit-Rekagarn, Sirichai and Pranom Tansukanan. (1988). *Special report of Thailand. Problems and proposals on conservation. The case of Chiang Mai, Thailand.* UNESCO/SPAFA Workshop. Chiang Mai, Nov.

Saowapon, Chaithawat. (1998). *A Perspective on the Water Related Issues in the City of Chiang Mai.* Department of Civil Engineering. Faculty of Engineering. Chiang Mai University, Thailand.

Pimonsathean, Yongtanit. (1999). *Zoning and Land Use Codes for Historic Preservation in Thailand : The Case of Phuket.* Document presented at The Economics of Heritage: UNESCO Conference on Adaptive Re-use of Historic Properties in Asia and the Pacific. Penang and Melaka, Malaysia. http://www.unescobkk.org/culture/pdf-files/T6Pimonsathean.pdf.

Promsao, Kannika. (1996). *Chiang Mai: Seven Hundred Years*. Chiang Mai, Thailand: Within Design Co.

Rojas, Eduardo. (1999). *Old Cities, New Assets: Preserving Latin America's Urban Heritage.* New York: Inter-American Development Bank.

Yeah, Brenda; and Shirlena Huang. (1996). The conservation-redevelopment dilemma in Singapore. *Cities, Vol. 13. No. 6*, 411-422.

INTERVIEWS

Doyasa, Tasana. Fine Arts Department (Government Agency). July 28, 1998.

Narumit-Rekegarn, Sirichai Arch. Chairman, The Siamsun Co. Lecturer, Faculty of Fine Arts. Chiang Mai University. July 8, 1998.

Panitchaphan, Vithi. Professor. Fine Arts Department. Faculty of Fine Arts. Chiang Mai University. June 16, 1998.

JON HUGHES

A SENSE OF PLACE

Visual Documentation of a City and Its Prospects for Sustainability

1. THE CHALLENGE

The challenge of visually documenting specific issues in a very different milieu is the ideal assignment for the photojournalist interested in capturing a sense of place. It was the interdisciplinary nature of the project, the objective, and the location that motivated me to respond to a university-wide call for participants. Working with a team doing fieldwork in Chiang Mai was an extraordinary opportunity to document an existing situation, the city, being quantitatively and qualitatively studied for its sustainable prospects. This was also a unique opportunity for me to work day-to-day with individuals who moved from concrete particulars to the conceptual and who are qualified to make specific recommendations based upon their fieldwork findings. The images compliment the research as a document, a kind of qualitative evidence, assisting the researcher in making conclusions that may lead to recommendations which directly affect a community. This is unlike conventional deadline driven photojournalism, that at best impartially records a situation with images that are rarely considered as evidence for change by policy makers.

2. PHOTOJOURNALISM VS. DOCUMENTARY PHOTOGRAPHY

While the documentary photographer is a photojournalist, the photojournalist is not necessarily a documentary photographer. Both are photographers and the difference between the two is related to the assignment. In this context then, the rather minor differences between the photojournalist and documentary photographer are in time spent on a shoot (breaking news vs. long-term deadline), intended use of the images (news vs. feature), objective (report vs. inform), and, possibly, for the documentary photographer, the opportunity of art giving shape to the presentation of reality. While I consider myself a photojournalist, my objective is *documentation, not* covering news events.

Objectivity is a goal for the photojournalist, who, working with a subject for an extended period of time, must especially be resistant to the temptation of validating dogma in the guise of information. The photojournalist should not be a policy maker nor a philosopher determining what is best for a people. Rather the photojournalist is to record a place, a time, and a human condition as fairly and as objectively as possible.

M. Romanos and C. Auffrey (eds.), Managing Intermediate Size Cities, 321—326.

3. UNDERSTANDING THE LOCAL CONTEXT

I prepared by first gaining a basic understanding of the country's history, and becoming aware of the social expectations, habits, and traditions of the various classes within the society. Then I attempted to gain an understanding of the attitudes and laws concerning foreigners and the taking of photographs (i.e. laws concerning the photographing of military sites and personnel, etc.). Understanding the working conditions (heat, humidity, rain, air-conditioning, refrigeration, etc.) that affect camera gear and batteries was important. So, I became aware of the customs regulations, health care and preventive preparation, and finding out how to get out of a country (and/or get film out), if necessary, as quickly as possible.

I also had to get a sense of the city, identify obstacles to doing the fieldwork (availability of transportation and translators, government regulations, laws and customs, etc.), and determine the level of acceptance to the team by Chiang Mai residents (from bureaucrats to slum dwellers). On a personal level, I needed to know where to get food, how to get water on the streets, and where camera equipment can be repaired if necessary.

4. OVERCOMING THE WOW! FACTOR

Regardless of how well prepared I think that I might be, there is often a WOW! factor (physically, emotionally, psychologically, philosophically, culturally) when entering a new, unique, foreign environment. The senses of sight, sound, and smell are overwhelmed with new symbols and letters, unique colors, a new climate, natural and manufactured noises, and a different language. Aromas and odors define the place. What may seem unusual to the visitor is often routine to the native, and one must guard against applying their cultural, social, and political standards and expectations to another culture. So, what is first seen (WOW!) is the difference between the two, not what the reality is to the people.

In Chiang Mai, it would have been very easy to record the unusual or the extreme as a westerner might see it and to incorrectly assume that this is how the Thai culture also perceives it. The result would be a cliché at best *and, more likely, derogatory*. To avoid this dangerous pitfall the photojournalist must be aware of, and respectful of, the dignity of the people.

5. ESTABLISHING OBJECTIVES

For me, the Chiang Mai project consisted of four weeks of in-country shooting with short and long-term objectives. It was essential to immediately talk with the researchers in order to get a basic understanding of the specific issues that each individual was investigating and how they were investigating (interviews, surveys, observations, meetings, etc.) these issues. The researchers began thinking how the issues could be visually represented (a new way for some researchers). The processing of film, selection of images, printing, and the preparation of lecture and display images continued for more than a year after the shoot.

CONTRIBUTORS

Christopher Auffrey is an associate professor in the School of Planning at the University of Cincinnati. He holds a BA in environmental conservation from the University of Colorado, an MA in development economics from San Diego State University, and a Ph.D. in urban, technological, and environmental planning from the University of Michigan. His research interests include sustainable urban development, environmental planning, and community health planning. He has professional experience with commercial development projects in Los Angeles and an environmental consulting firm in Washington, D.C. He also served as a Peace Corps volunteer in Paraguay. He has used geographic information systems (GIS) to analyze the spatial dimension of environmental hazards and their impacts on the health of communities. Professor Auffrey has also used GIS to measure the variation in infant mortality in Cincinnati census tracts, and the variation in reported air quality at southwest Ohio air quality monitors.
e-mail: chris.auffrey@uc.edu

Roger J. Barry is a professor in the School of Planning at the University of Cincinnati, and holds a doctorate of philosophy in economic geography from Indiana State University. He has participated in postgraduate studies at the University of Illinois, Harvard University Institute for Educational Management, the Presidential Academy of the American Association of Community Colleges, and the W. K. Kellogg Executive Leadership Institute sponsored by the League for Innovation. Professor Barry's teaching and research interests include professional values and ethical issues in professional planning, economic development, strategic planning, and project management. He served as a manager of the 1998 University of Cincinnati Summer Faculty Development Field Project in Sustainable Development, Chiang Mai, Thailand. He has served as Executive Director of a Regional Planning and Development Commission, and Dean of Instruction of a Community College in Illinois. He served as Dean and Professor of Geography of the University of Cincinnati Clermont College and acting Dean of Raymond Walters College. He also was Chair of the Clermont Chamber of Commerce Economic Development Council, and currently serves on the board of the Greater Cincinnati Foundation, Clermont Community Fund.
e-mail: roger.barry@uc.edu

Carla Chifos is an assistant professor in the School of Planning at the University of Cincinnati. She holds a BS in Environmental Health Engineering from Northwestern University, a Masters of Community Planning with emphasis in economic development and environmental planning from the University of Cincinnati, and a Ph.D. in urban and regional development from Cornell University. She has lived and worked in Southeast Asia, mainly Indonesia and Thailand, for over four years, focusing on the study of the linkages between environmental issues and poverty in the neighborhoods of large cities through work with the United Nations, World Resources Institute, USAID, and the Asia Foundation. Her doctoral dissertation, "Urban neighborhoods and the natural environment: the case of Jakarta, Indonesia",

was also completed during her stay overseas. She is co-editor as well as author of several chapters in *Southeast Asian Urban Environments: Structure and Spontaneity*, which explores issues of infrastructure provision in the cities of Southeast Asia.
e-mail: carla.chifos@uc.edu

Jon Hughes is professor of English and Journalism in the Department of English at the University of Cincinnati. He is author/photographer of several books and radio plays, and contributor to several journals. He is co-founder of the Writing Program, Radio Repertory Company, Inc., and the Cincinnati Writers Project; executive producer of the public radio series *Dimension Radio Theatre*; National Endowment for the Arts Grant (1990); named one of the eight top photographers by the Association of Alternative Newsweeklies (1999); project director and documentary photographer for *A Year in the Life of Cincinnati Project* (1994-1999); involved with international documentary projects in many countries including Cuba, Thailand, China, Bosnia, Croatia; has photographs widely published ranging from *The Sun* magazine to the *Los Angeles Times*. He has put together eleven one-person exhibits; and was named the first winner of both the Post-Corbett Literary Artist Award (1990) and the Post-Corbett Visual Artist Award (1999).
e-mail: jon.hughes@uc.edu

Krista Kahle is a planner/project manager for sustainable development issues in the Chicago area at Bethel New Life, a nationally-renowned community development corporation. She holds a Bachelor's degree in sociology from Aquinas College and a Masters of Community Planning from the University of Cincinnati. She has a broad and diverse knowledge of planning issues, having worked at the City of Cincinnati, Department of Economic Development and the Center for Public Management & Regional Affairs at Miami University. Her research interests are citizen participation, economic development, and affordable housing. Her Master's thesis won the Jenkins Interdisciplinary Scholarship Award, and she has presented at the American Collegiate Schools of Planning conference.
e-mail: kahlekj@hotmail.com

Lin Liu is an associate professor in the Department of Geography at the University of Cincinnati. He obtained a Ph.D. in geography with a specialization in geographic information systems (GIS) from the Ohio State University. He has been working in the field of GIS and remote sensing since 1984. He participated in the design and development of multiple GIS software packages. His research interests include scientific visualization of spatio-temporal data, spatial data mining, and applications of GIS to urban/economic and environmental problems. He has publications in various GIS related journals, book chapters, and conference proceedings. Dr. Liu has taught many courses in GIS, computer cartography, location analysis, and regional geography. As an ESRI authorized instructor, he has offered various GIS training courses for GIS professionals. Currently he is a fellow in the Institute of Data Sciences and the co-director of the Joint Center for GIS and Spatial Analysis.
e-mail: lin.liu@uc.edu

Johanna W. Looye is an associate professor in the School of Planning at the University of Cincinnati. She holds a Ph.D. and a Master of Regional Planning from Cornell University and a B.A. in Spanish and Latin American Studies from Grinnell College. Looye specializes in international development planning, with a focus on Latin America. From 1979 to 1981, as a Thomas J. Watson Fellow, she examined handicrafts and rural development in Chile and Ecuador. Since 1984, she has worked in Brazil on urban and regional planning. From 1986 to 1984, she conducted research on the metalworking sector in the Northeastern State of Ceará, with emphasis on microenterprise development. In 1992 and 1994, and again in 1999, as a Senior Fulbright Scholar, she lived in Rio de Janeiro and conducted research on urban environmental issues.
e-mail: johanna.looye@uc.edu

Lou Matz is an associate professor and chair of the Department of Philosophy at the University of the Pacific in Stockton, California, and is a certified philosophical counselor. He completed a Ph.D. in philosophy at the University of California, San Diego, and has taught at the University of Redlands, UC San Diego, and Xavier University. He teaches courses in the History of Modern Philosophy, the Philosophy of Religion, Ancient and Medieval Philosophy, and Moral Problems. His research interests focus on ethics, political philosophy, and the philosophy of religion. His publications include "Hegel's Missing Moral Virtues?" and "The Utility of Religious Belief: A Criticism of J.S. Mill's Religion of Humanity." He also completed a course in traditional Thai massage during his stay in Chiang Mai.
e-mail: lmatz@uop.edu

Dwita Hadi Rahmi is a lecturer at the Department of Architecture, Gadjah Mada University, Yogyakarta, Indonesia where she teaches eco-urban design and environmental technology. She has an undergraduate degree in architecture and a master's degree in Geography from the University of Waterloo in Ontario, Canada. Since 1993, Ms. Rahmi has been part of the Women's' Study Center at Gadjah Mada University, and is involved in studies related to gender issues. In 1997, Ms. Rahmi completed advanced training on local environmental management at Karlstad, Sweden. She was also a visiting scholar at the Institute of Asian Studies at the University of British Columbia in Vancouver. Since then, she has been developing her interest in urban sustainability issues. Her current research focuses on urban green space in Yogyakarta.
e-mail: ssbe@indosat.net.id

Hugo Rincón is an assistant professor in the School of Architecture and Design, Universidad del Zulia, Maracaibo, Venezuela. He is an architect (Universidad del Zulia, 1991) and is currently a Ph.D. candidate in the Community and Regional Planning program of the University of Texas at Austin. From 1996 to 1998, he was a Fulbright Scholar at the University of Cincinnati, School of Planning where he earned a Masters of Community Planning degree in 1998. His graduate research evaluated conservation and redevelopment interventions in historic centers and urban complexes. His doctoral research focuses on the relationship between state

and civil society as preconditions for effective urban redevelopment in Latin American cities. He participated in the Thailand field project in the summer of 1998 under a grant from the University of Cincinnati.
e-mail: rinconhr@mail.utexas.edu

Michael Romanos is professor of Economic Development in the School of Planning at the University of Cincinnati. He holds a Diploma in Architectural Engineering, a Master of Science in Regional Planning, and a Ph.D. in Regional Science from Cornell University. He has served in the faculties of Cornell University, the University of Illinois at Urbana-Champaign, and the School of Planning at the University of Cincinnati, where he has been since 1984. In the early to mid-90s, he was Senior Fulbright Scholar to Indonesia, Malaysia, Thailand and Singapore, and Asia Foundation Fellow to Thailand. Since 1997, he has been director of the Sustainable Development Project funded by the U.S. Information Agency, and coordinator of the University of Cincinnati Sustainable Development Interest Group. Romanos has served as a Trustee of the Greek University of the Aegean (1990-1998) by Prime-Ministerial decree, Senior Advisor to the Government of Indonesia on Regional Development, and Advisor to the Office of the Prime Minister of Malaysia.
e-mail: michael.romanos@uc.edu

Brenda Case Scheer AIA, AICP, is associate professor of Urban Design and Planning at the University of Cincinnati, where she has taught since 1990. Her research specializations are urban morphology and urban design policy, especially design review. She has published many articles on design review, architecture, housing, and suburban form. Professor Scheer has a long record of professional practice, including as a Principle of Scheer & Scheer, Inc., where her projects include master planning, urban design and design guidelines as well as urban architectural projects. Before coming to Cincinnati, she was a Loeb Fellow at the Harvard Graduate School of Design and director of urban design, City of Boston.
e-mail: brenda.scheer@uc.edu

David R. Scheer is an architect who combines an active design practice with research and teaching in architecture. He received his M. Arch. from Yale University in 1984 and studied at the Institute for Architecture and Urban Studies in New York. He is a principal of Scheer & Scheer, Inc., a Cincinnati-based architecture and urban design firm. Before founding this office, he worked with several prominent design practices in New York, Paris, and Los Angeles. Mr. Scheer's recent buildings include the Corryville Recreation Center in Cincinnati and the Center City project in Terre Haute, Indiana, a mixed use retail and residential building. His research interests include comparative practice, urban morphology, and site-based design. He currently teaches architecture at the University of Cincinnati and Miami University of Ohio.
e-mail: dscheer@scheerandscheer.com

Judith Meyer Schultz is a professor of Biology and Environmental Science in the Department of Biology, College of Raymond Walters at the University of

Cincinnati. She holds a B.S. in Zoology, an M.S. in Biological Sciences, and a Ph.D. in Environmental Studies from the University of Cincinnati. At the University of Cincinnati, Dr. Schultz initiated and directed the Environmental Control/Protection Technology Program from 1970-1992, and chaired the Environmental Studies Committee of The McMicken College of Arts and Sciences initiating the Environmental Studies Program in 1993. She has been appointed and served on the United States National Advisory Council on Environmental Education, and has received the Distinguished Alumna Award from the University of Cincinnati Alumni Association and Friends of Women's Studies for contributions in environmental education and service. She has served as President of the North American Association for Environmental Education.
e-mail: judith.schultz@uc.edu

Bobi Setiawan is an urban planner and lecturer in the Department of Architecture, Gadjah Mada University in Yogyakarta, Indonesia. He teaches urban planning and housing in the graduate program in Urban and Regional Planning. He is also a researcher at the Center for Environmental Studies, Gadjah Mada University. In 1992, Dr. Setiawan received his master's degree from the School of Urban and Regional Planning, University of Waterloo, Canada, and in 1998 received his Ph.D. from the School of Community and Regional Planning, the University of British Columbia, Canada. Dr. Setiawan has extensive professional planning experience throughout Indonesia, including Java, Bali, Sumatra, and Kalimantan. He is also involved in several community development projects, particularly in the Yogyakarta area. Presently, he works with several NGOs in Yogyakarta working on urban poverty issues.
e-mail: flamboyn@idola.net.id

Howard A. Stafford is a professor and chair of the Department of Geography at the University of Cincinnati. Since earning his doctorate at the University of Iowa he has also served on the faculties of Southern Illinois University and the University of Wisconsin, and been a visiting scholar at the Universities of Wales, Exeter, and Sheffield in the United Kingdom, and at Gothenburg University in Sweden. Dr. Stafford specializes in economic and urban geography, with emphases on the location of commercial facilities and on urban spatial restructuring.
e-mail: howard.stafford@uc.edu

Suwattana Thadaniti is deputy director for Research Affairs at the Institute for Social Research and an assistant professor of urban and regional planning in the faculty of architecture at Chulalongkorn University, Bangkok, Thailand. She received an MA in geography from the University of Melbourne, a Masters of Urban Planning from Chulalongkorn University, a Certificate in Urban Land Policy from the Land Training Institute in Taiwan, and a Ph.D. in town and country planning from the Technical University of Cracow in Poland. Professor Thadaniti's research interests include land use and sustainable development approaches.
e-mail: tsuwattana@chula.ac.th

Thomas E. Wagner is University Professor Emeritus of Planning and Urban Studies at the University of Cincinnati. Professor Wagner holds graduate degrees in urban education and American history from the University of Cincinnati and Miami University. He has undertaken postgraduate studies at the University of Cincinnati Law School, Yale School of Organization and Management, Harvard University Institute for Educational Management, and University of Wales. Professor Wagner has primary instructional responsibility for undergraduate and graduate courses related to professional values and ethical issues, conflict resolution and mediation, organization theory, citizen involvement, and social planning. His primary research activity is a collaborative study of Appalachian migrant groups and organizations in several mid-western cities. His other research interests include citizen involvement in planning the clean up of Department of Energy nuclear facilities and the emerging concept of citizen participation in Thailand.
e-mail: thomas.wagner@uc.edu

Xinhao Wang is an associate professor in the School of Planning at the University of Cincinnati. He holds a Ph.D. in City and Regional Planning from the University of Pennsylvania, a Master of Community Planning from the University of Rhode Island, a Master of Science in Geo-Sciences, and a Bachelor of Science in Geography from Peking University, Beijing, China. He teaches courses in GIS, environmental planning, statistics, and planning research methods. His research interests are in the areas of environmental planning and GIS application in planning. Examples are: analysis of the spatial pattern of water quality indicators; the relationship between land use and water quality; GIS-based water quality modelling; and, spatial distribution of human health and environmental indicators. Currently, Dr. Wang is a fellow in the Institute of Data Sciences and the Co-Director of the Joint Center of Geographic Information Systems and Spatial Analysis.
e-mail: wangxo@email.uc.edu

INDEX

The GeoJournal Library

---

1. B. Currey and G. Hugo (eds.): *Famine as Geographical Phenomenon.* 1984
ISBN 90-277-1762-1
2. S.H.U. Bowie, F.R.S. and I. Thornton (eds.): *Environmental Geochemistry and Health.* Report of the Royal Society's British National Committee for Problems of the Environment. 1985 ISBN 90-277-1879-2
3. L.A. Kosiński and K.M. Elahi (eds.): *Population Redistribution and Development in South Asia.* 1985 ISBN 90-277-1938-1
4. Y. Gradus (ed.): *Desert Development.* Man and Technology in Sparselands. 1985
ISBN 90-277-2043-6
5. F.J. Calzonetti and B.D. Solomon (eds.): *Geographical Dimensions of Energy.* 1985
ISBN 90-277-2061-4
6. J. Lundqvist, U. Lohm and M. Falkenmark (eds.): *Strategies for River Basin Management.* Environmental Integration of Land and Water in River Basin. 1985
ISBN 90-277-2111-4
7. A. Rogers and F.J. Willekens (eds.): *Migration and Settlement.* A Multiregional Comparative Study. 1986 ISBN 90-277-2119-X
8. R. Laulajainen: *Spatial Strategies in Retailing.* 1987 ISBN 90-277-2595-0
9. T.H. Lee, H.R. Linden, D.A. Dreyfus and T. Vasko (eds.): *The Methane Age.* 1988
ISBN 90-277-2745-7
10. H.J. Walker (ed.): *Artificial Structures and Shorelines.* 1988 ISBN 90-277-2746-5
11. A. Kellerman: *Time, Space, and Society.* Geographical Societal Perspectives. 1989
ISBN 0-7923-0123-4
12. P. Fabbri (ed.): *Recreational Uses of Coastal Areas.* A Research Project of the Commission on the Coastal Environment, International Geographical Union. 1990
ISBN 0-7923-0279-6
13. L.M. Brush, M.G. Wolman and Huang Bing-Wei (eds.): *Taming the Yellow River: Silt and Floods.* Proceedings of a Bilateral Seminar on Problems in the Lower Reaches of the Yellow River, China. 1989 ISBN 0-7923-0416-0
14. J. Stillwell and H.J. Scholten (eds.): *Contemporary Research in Population Geography.* A Comparison of the United Kingdom and the Netherlands. 1990
ISBN 0-7923-0431-4
15. M.S. Kenzer (ed.): *Applied Geography.* Issues, Questions, and Concerns. 1989
ISBN 0-7923-0438-1
16. D. Nir: *Region as a Socio-environmental System.* An Introduction to a Systemic Regional Geography. 1990 ISBN 0-7923-0516-7
17. H.J. Scholten and J.C.H. Stillwell (eds.): *Geographical Information Systems for Urban and Regional Planning.* 1990 ISBN 0-7923-0793-3
18. F.M. Brouwer, A.J. Thomas and M.J. Chadwick (eds.): *Land Use Changes in Europe.* Processes of Change, Environmental Transformations and Future Patterns. 1991
ISBN 0-7923-1099-3

The GeoJournal Library

19. C.J. Campbell: *The Golden Century of Oil 1950–2050.* The Depletion of a Resource. 1991 ISBN 0-7923-1442-5
20. F.M. Dieleman and S. Musterd (eds.): *The Randstad: A Research and Policy Laboratory.* 1992 ISBN 0-7923-1649-5
21. V.I. Ilyichev and V.V. Anikiev (eds.): *Oceanic and Anthropogenic Controls of Life in the Pacific Ocean.* 1992 ISBN 0-7923-1854-4
22. A.K. Dutt and F.J. Costa (eds.): *Perspectives on Planning and Urban Development in Belgium.* 1992 ISBN 0-7923-1885-4
23. J. Portugali: *Implicate Relations.* Society and Space in the Israeli-Palestinian Conflict. 1993 ISBN 0-7923-1886-2
24. M.J.C. de Lepper, H.J. Scholten and R.M. Stern (eds.): *The Added Value of Geographical Information Systems in Public and Environmental Health.* 1995 ISBN 0-7923-1887-0
25. J.P. Dorian, P.A. Minakir and V.T. Borisovich (eds.): *CIS Energy and Minerals Development.* Prospects, Problems and Opportunities for International Cooperation. 1993 ISBN 0-7923-2323-8
26. P.P. Wong (ed.): *Tourism vs Environment: The Case for Coastal Areas.* 1993 ISBN 0-7923-2404-8
27. G.B. Benko and U. Strohmayer (eds.): *Geography, History and Social Sciences.* 1995 ISBN 0-7923-2543-5
28. A. Faludi and A. der Valk: *Rule and Order. Dutch Planning Doctrine in the Twentieth Century.* 1994 ISBN 0-7923-2619-9
29. B.C. Hewitson and R.G. Crane (eds.): *Neural Nets: Applications in Geography.* 1994 ISBN 0-7923-2746-2
30. A.K. Dutt, F.J. Costa, S. Aggarwal and A.G. Noble (eds.): *The Asian City: Processes of Development, Characteristics and Planning.* 1994 ISBN 0-7923-3135-4
31. R. Laulajainen and H.A. Stafford: *Corporate Geography.* Business Location Principles and Cases. 1995 ISBN 0-7923-3326-8
32. J. Portugali (ed.): *The Construction of Cognitive Maps.* 1996 ISBN 0-7923-3949-5
33. E. Biagini: *Northern Ireland and Beyond.* Social and Geographical Issues. 1996 ISBN 0-7923-4046-9
34. A.K. Dutt (ed.): *Southeast Asia: A Ten Nation Region.* 1996 ISBN 0-7923-4171-6
35. J. Settele, C. Margules, P. Poschlod and K. Henle (eds.): *Species Survival in Fragmented Landscapes.* 1996 ISBN 0-7923-4239-9
36. M. Yoshino, M. Domrös, A. Douguédroit, J. Paszynski and L.D. Nkemdirim (eds.): *Climates and Societies – A Climatological Perspective.* A Contribution on Global Change and Related Problems Prepared by the Commission on Climatology of the International Geographical Union. 1997 ISBN 0-7923-4324-7
37. D. Borri, A. Khakee and C. Lacirignola (eds.): *Evaluating Theory-Practice and Urban-Rural Interplay in Planning.* 1997 ISBN 0-7923-4326-3

The GeoJournal Library

38. J.A.A. Jones, C. Liu, M-K. Woo and H-T. Kung (eds.): *Regional Hydrological Response to Climate Change.* 1996 ISBN 0-7923-4329-8
39. R. Lloyd: *Spatial Cognition.* Geographic Environments. 1997 ISBN 0-7923-4375-1
40. I. Lyons Murphy: *The Danube: A River Basin in Transition.* 1997 ISBN 0-7923-4558-4
41. H.J. Bruins and H. Lithwick (eds.): *The Arid Frontier.* Interactive Management of Environment and Development. 1998 ISBN 0-7923-4227-5
42. G. Lipshitz: *Country on the Move: Migration to and within Israel, 1948–1995.* 1998 ISBN 0-7923-4850-8
43. S. Musterd, W. Ostendorf and M. Breebaart: *Multi-Ethnic Metropolis: Patterns and Policies.* 1998 ISBN 0-7923-4854-0
44. B.K. Maloney (ed.): *Human Activities and the Tropical Rainforest.* Past, Present and Possible Future. 1998 ISBN 0-7923-4858-3
45. H. van der Wusten (ed.): *The Urban University and its Identity.* Roots, Location, Roles. 1998 ISBN 0-7923-4870-2
46. J. Kalvoda and C.L. Rosenfeld (eds.): *Geomorphological Hazards in High Mountain Areas.* 1998 ISBN 0-7923-4961-X
47. N. Lichfield, A. Barbanente, D. Borri, A. Khakee and A. Prat (eds.): *Evaluation in Planning.* Facing the Challenge of Complexity. 1998 ISBN 0-7923-4870-2
48. A. Buttimer and L. Wallin (eds.): *Nature and Identity in Cross-Cultural Perspective.* 1999 ISBN 0-7923-5651-9
49. A. Vallega: *Fundamentals of Integrated Coastal Management.* 1999 ISBN 0-7923-5875-9
50. D. Rumley: *The Geopolitics of Australia's Regional Relations.* 1999 ISBN 0-7923-5916-X
51. H. Stevens: *The Institutional Position of Seaports.* An International Comparison. 1999 ISBN 0-7923-5979-8
52. H. Lithwick and Y. Gradus (eds.): *Developing Frontier Cities.* Global Perspectives – Regional Contexts. 2000 ISBN 0-7923-6061-3
53. H. Knippenberg and J. Markusse (eds.): *Nationalising and Denationalising European Border Regions, 1800–2000.* Views from Geography and History. 2000 ISBN 0-7923-6066-4
54. R. Gerber and G.K. Chuan (eds.): *Fieldwork in Geography: Reflections, Perspectives and Actions.* 2000 ISBN 0-7923-6329-9
55. M. Dobry (ed.): *Democratic and Capitalist Transitions in Eastern Europe.* Lessons for the Social Sciences. 2000 ISBN 0-7923-6331-0
56. Y. Murayama: *Japanese Urban System.* 2000 ISBN 0-7923-6600-X
57. D. Zheng, Q. Zhang and S. Wu (eds.): *Mountain Geoecology and Sustainable Development of the Tibetan Plateau.* 2000 ISBN 0-7923-6688-3

58. A.J. Conacher (ed.): *Land Degradation.* Papers selected from Contributions to the Sixth Meeting of the International Geographical Union's Commission on Land Degradation and Desertification, Perth, Western Australia, 20–28 September 1999. 2001 ISBN 0-7923-6770-7
59. S. Conti and P. Giaccaria: *Local Development and Competitiveness.* 2001 ISBN 0-7923-6829-0
60. P. Miao (ed.): *Public Places in Asia Pacific Cities.* Current Issues and Strategies. 2001 ISBN 0-7923-7083-X
61. N. Maiellaro (ed.): *Towards Sustainable Buiding.* 2001 ISBN 1-4020-0012-X
62. G.S. Dunbar (ed.): *Geography: Discipline, Profession and Subject since 1870.* An International Survey. 2001 ISBN 1-4020-0019-7
63. J. Stillwell and H.J. Scholten (eds.): *Land Use Simulation for Europe.* 2001 ISBN 1-4020-0213-0
64. P. Doyle and M.R. Bennett (eds.): *Fields of Battle.* Terrain in Military History. 2002 ISBN 1-4020-0433-8
65. C.M. Hall and A.M. Williams (eds.): *Tourism and Migration.* New Relationships between Production and Consumption. 2002 ISBN 1-4020-0454-0
66. I.R. Bowler, C.R. Bryant and C. Cocklin (eds.): *The Sustainability of Rural Systems.* Geographical Interpretations. 2002 ISBN 1-4020-0513-X
67. O. Yiftachel, J. Little, D. Hedgcock and I. Alexander (eds.): *The Power of Planning.* Spaces of Control and Transformation. 2001 ISBN Hb; 1-4020-0533-4 ISBN Pb; 1-4020-0534-2
68. K. Hewitt, M.-L. Byrne, M. English and G. Young (eds.): *Landscapes of Transition.* Landform Assemblages and Transformations in Cold Regions. 2002 ISBN 1-4020-0663-2
69. M. Romanos and C. Auffrey (eds.): *Managing Intermediate Size Cities.* Sustainable Development in a Growth Region of Thailand. 2002 ISBN 1-4020-0818-X

KLUWER ACADEMIC PUBLISHERS – DORDRECHT / BOSTON / LONDON

MIX
Papier aus verantwortungsvollen Quellen
Paper from responsible sources
FSC® C105338

If you have any concerns about our products,
you can contact us on
ProductSafety@springernature.com

In case Publisher is established outside the EU,
the EU authorized representative is:
Springer Nature Customer Service Center GmbH
Europaplatz 3, 69115 Heidelberg, Germany

Printed by Libri Plureos GmbH
in Hamburg, Germany